Lectures on Mathematical Logic
Volume III : The Logic of Arithmetic

Lectures on Mathematical Logic
Volume III : The Logic of Arithmetic

Walter Felscher

University of Tuebingen
Germany

CRC Press
Taylor & Francis Group
Boca Raton London New York

CRC Press is an imprint of the
Taylor & Francis Group, an **informa** business

CRC Press
Taylor & Francis Group
6000 Broken Sound Parkway NW, Suite 300
Boca Raton, FL 33487-2742

First issued in paperback 2019

ISBN-13: 978-90-5699-268-2 (hbk)
ISBN-13: 978-0-367-39857-6 (pbk)

British Library Cataloguing in Publication Data

Felscher, Walter
Lectures on mathematical logic
Vol. III: The logic of arithmetic
1. Logic, Symbolic and mathematical
I. Title
511.3

Visit the Taylor & Francis Web site at
http://www.taylorandfrancis.com

and the CRC Press Web site at
http://www.crcpress.com

Contents

Dependences

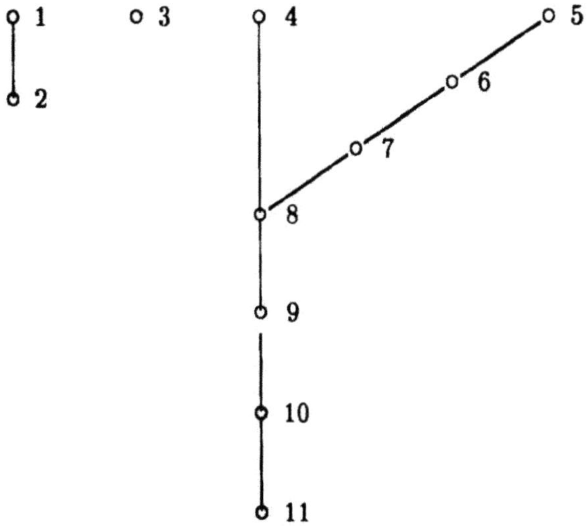

Preface

This is the third of three books of Lectures on Mathematical Logic, destined for students of mathematics or computer science, in their third or fourth year at the university, as well as for their instructors. It is written as the traditional combination of textbook and monograph: while some titles of chapters and sections will sound familiar to those moderately acquainted with logic, their content and its presentation is likely to be new also to the more experienced reader. In particular, concepts have been set up such that the proofs do not act as mousetraps, from which the reader cannot escape without acknowledging the desired result, but make it obvious *why* they entail their result with an inner necessity. In so far, the book expects from the reader a certain maturity; while accessible to students, it also will be helpful to lecturers who look for motivations of procedures which they may only have come to know in the form of dry prescriptions.

The book is mainly about logical properties of arithmetic, namely the decidability and consistency of some of its weaker fragments, and the undecidability and the unprovability of consistency for stronger ones. For the arithmetic of successor and order in Chapter 1, and for Presburger arithmetic in Chapter 2, decidability is obtained from quantifier elimination, and this, being established syntactically, will entail consistency proofs. In Chapter 3 the Liar paradox is discussed and is used to analyze, in a non-technical manner, the undefinability of truth and the incompleteness of provability; this Chapter requires no mathematical knowledge and may be accessible to the educated layman. General incompleteness theorems are set up in Chapter 4 ; here encodings are assumed to be given and certain encoding functions are assumed to be representable. In Chapters 5 and 6 basic facts about recursive functions and relations are developed. In Chapter 8 undecidability of provability from axioms in an arithmetized language is connected with the representability of recursive relations; this is applied in order to find a uniformization of recursive functions, and also in order obtain the SMN-property, needed to conclude that this uniformization is isomorphic to other uniformizations known from computability theory. Chapter 9 discusses Robinson's arithmetic and Church's result on the undecidability of pure logic.

In Chapter 10 it is shown that Peano arithmetic PA (actually, already Heyting arithmetic) can be extended conservatively by function symbols for primitive recursive functions. There follows a discussion of primitive recursive arithmetic in which it is emphasized that recursion equations and their consequences now become provable as free variable formulas. In Chapter 11 the unprovability of the consistency of PA is shown by verifying the

Bernays–Löb provability conditions. In order to see that Σ_1-formulas imply their provability, the process of *internalization* is isolated and studied. It is hoped that the presentation of the basic results of these last two chapters, both in their strength and their limitations, will make them accessible to a wider readership.

More about the book's content will be said in its introduction and in the introductory sections of its chapters.

This book hardly depends on the earlier books of these Lectures.

It is divided into Chapters 1 to 11, and each chapter is divided into numbered sections. Theorems, Corollaries and Lemmas are numbered successively within each chapter. A reference such as Theorem j.k is to Theorem k in Chapter j. A reference to Chapter i.j and Theorem i.j.k with $i<2$ is to Chapter j or to Theorem j.k of Book i.

I am indebted to Professor Dov Gabbay and to the publishers of Gordon & Breach whose efforts made the appearance of these Lectures possible.

Introduction

In this Book 3, various logical properties of the arithmetic of natural numbers will be studied: decidability and undecidability, completeness and incompleteness, consistency and the limitations to establish it.

1. The Problem of Consistency

In naive language, a collection of statements - say in a legal contract - is called consistent if these statements fit together, in the sense that it is not the case that one of them says things contradicted by another one. In logic, therefore, a set of sentences C is called *consistent* if it is not the case that there is a formula v such that both v and its negation ¬v are provable from the sentences in C.

Provability in that case refers to formal deducibility, i.e. within one of the many (equivalent) logical calculi; it does *not* mean deducibility as a semantical consequence. Because semantical consequence concerns truth, be it defined with propositional truth values or be it truth in structures (or models), and just as 0 is distinct from 1, the relations to be found on semantical structures are supposed to be such that, given elements from that structure (considered as to whether they satisfy a formula), they either *do* or *do not* stand in that relation. In simpler words: the semantical world is, from the outset, assumed not to admit contradictory observations and, thereby, not to realize contradictory statements. Provided that a formal calculus be *correct* (in the sense that what is deducible from true hypotheses must be true as well), what is true (or has a model) must also be consistent.

Still, to show the consistency of a set C of statements by presenting a model of C is a perfectly legitimate undertaking, provided such model is built with constructions which, for good reason, may be held as *not more* complex as those expressed in the statements C themselves. For instance, the axioms of non-Euclidean geometry first were shown consistent by constructing models for them within the framework of Euclidean geometry. On the other hand, it is quite a different situation if a model for arithmetic or for the real numbers is constructed from the axioms of set theory: there the reduction is one to notions of a complexity likely to be higher than that of the objects to be explained.

The attentive reader might object that, after all, we *know* that truth coincides with provability, i.e. that a formula v holds in all models of C if, and only if, v is deducible from C in (one of) the (various equivalent) calculi for quantifier logic. In particular, if v is of the form w ∧ ¬w then it will hold in no structure; hence such v can hold in all models of C only if there are not any. But now this is the case if, and only if, v is deducible from C, i.e. if C is *inconsistent* (not consistent). Arguing by classical contraposition, it follows that C *does* have a model if, and only if, it is *not* inconsistent, i.e. is consistent.

And the attentive reader, of course, is right: we know all that – but by which means? That provability implies truth, i.e. that the calculi are *correct,* follows immediately from definitions, once we are prepared to speak about satisfiability in semantical structures at all. But that truth implies provability, that the calculi are *complete*: that requires heavier tools indeed. Let me forget about uncountable languages for which I would need the Boolean Prime Ideal axiom; in most practical cases countable languages will do. Let me further forget about finding a 0–1–valuation separating two elements of a Boolean algebra: for the countable case, the proofs both of the separation Theorem A in **1.3** and the extension theorem **1.10**.1c use recursive constructions. But do not forget that all proofs of completeness are indirect: in order to show for a given calculus, a set of formulas C and a formula v that

if (a_0) every model of C is a model of v
then (a_1) there exists a deduction of v from C

they do not exhibit such proof, but rather show that

if (b_0) for every deduction from C: it does not produce v
then (b_1) there exists a model A of C which is not a model of v .

Again, forget about the use of set theoretical notions used (i) in order to form A as a canonical structure from the algebra of terms, and (ii) in order to find a valuation, not satisfying v, by applying a separation theorem to the Boolean algebra of formulas modulo semantical equivalence. But let me emphazise the indirect argument which proves the implication (a) from the implication (b); it leads from the negation of (b_1) to that of (b_0). The former clearly is equivalent to (a_0). The latter is

it is not the case that for every deduction from C: it does not produce v.

Angels and demons (fallen angels) could search through all deductions from C and, having unlimited time, they would have to come upon a deduction from C for which it is not the case that it does not produce v, i.e. which produces v. Human beings have only a negligeable chance that their search may terminate before humanity comes to an end. And clearly, arguments of this type – to be referred to as *inconstructive* in future – will be viewed as being of rather limited value where foundational matters such as consistency or decidability are under discussion.

[It may be useful to notice that, in the algebraic treatment given in Book 1, in place of (a_1), (b_0) I used statements saying that v is a member, or is not a member, of the *set* of *all* formulas obtained by deductions from C. It is the use of this completed totality of *all* deductions which makes two such statements trivial negations of each other, and it hides the inconstructiveness which becomes apparent when I formulate the statements as speaking about the deductions themselves.]

The above observations make it clear that every consistency proof should be accompanied by stating the tools employed to carry it out.

2. Completeness and Decidability

A domain of knowledge is described completely if every question relating to it can be answered - either by Yes or by No. In that case the description, e.g. by a system of axioms, is *complete*. But completeness expresses a *state* of information, not the *means* by which this information comes to be ascertained - a proof of completeness, for instance, may be indirect. Yet there are questions for which the answer Yes or No can be obtained by an algorithmic decision method, and then these questions will be called *decidable*.

Consider a language for quantifier logic. A set of formulas is called *semantically complete* if, for every sentence v, one of $v \in ct(C)$ or $\neg v \in ct(C)$. A set C of formulas is semantically complete if, and only if, the set $(\forall)C$ of universal closures is so ; hence I may restrict myself to semantically complete sets of *sentences*. Examples abound of sets which are *not* semantically complete - for instance the axioms of groups which do neither imply the commutative law nor its negation. Still, in Chapter 1.16 a model-theoretical, inconstructive argument produced the example of a semantically complete set C_1 of formulas, for which the structure $N = \langle \omega, 0, s, <, id^{\omega} \rangle$ was a model. That semantical completeness has drastic consequences is illustrated by the following two observations.

If a structure A is a model of a set C then $(\forall)ct(C) \subseteq Ths(A)$; if C is semantically complete then $Ths(A) = (\forall)ct(C)$, because for a sentence v true in A the sentence $\neg v$ cannot be in ct(C) whence v is in ct(C). Thus any two models of a semantically complete set are elementarily equivalent.

For every structure A the set Ths(A) is semantically complete, but being defined from A by semantical notions, it will, by itself, not be accessible to an algorithmic description. However, if A happens to be a model of a semantically complete set of axioms C (e.g. N of C_1 above) then any algorithmic description of the operator ct - e.g. by some modus ponens calculus - will produce an algorithmic description of Ths(A) from C.

Let ccqt be one of the (modus ponens) calculi for classical logic. I then define that a set C be *complete* if, for every sentence v, one of $v \in ccqt(C)$ or $\neg v \in ccqt(C)$; this only formally depends on the particular chosen ccqt, since any two of these calculi derive the same formulas. This time it will follow from the rules of ccqt that, again, C is complete if, and only if, the set $(\forall)$C is so. The sets Ths(A) are complete since ccqt(Th(A)) = Th(A). Now the inclusion ccqt(C) $\subseteq$ ct(C) involves not more than the notion of structure and satisfaction, present already in the definition of ct; the inclusion ct(C) $\subseteq$ ccqt(C), however, depends on the inconstructive argument discussed in Section 1. It follows from the definitions that complete sets are semantically complete, and that semantically incomplete sets are incomplete, while the reverse implications only hold inconstructively.

The notion of *algorithm* has appeared repeatedly in this work when referring to a method which produces certain objects or data by a schematic formal computation – e.g. parsing algorithms for strings of characters, algorithms producing normal forms of Boolean terms, algorithms generating subalgebras and congruence relations, and algorithms generating derivations and deductions in calculi of equations, sequents and formulas. I have used this notion in a purely descriptive manner, not as a defined mathematical terminus, and this I shall continue as long as I shall not have to make general statements about algorithms – just as it suffices to point at an oak, a maple or a pine when saying that these are trees. Matters will become different when, in later Chapters, I want to state that for certain problems there a *no* algorithms to solve them, because then a precise notion of algorithmic presentability will be required.

Yet one distinction can be made already informally here: that between algorithms working *with parameters* and algorithms working *absolutely*. Of the latter type are the algorithms for normal forms; an algorithm, however, generating the consequences of a set C, of equations or formulas, will necessarily refer to the various elements of C themselves – e.g. when saying that if $c \in C$ then c be deducible from C – and so it will be parametric in C. Only if C itself is algorithmically presented – i.e. if C is finite or if there is an algorithmic function f on ω producing C as its image – then the algorithm producing C, inserted into the algorithm with parameters, makes that into an absolute one.

Let S be the set of all sentences of a language; every subset M of S determines its characteristic function χ_M on S with $\chi_M(s) = 1$ if $s \in M$ and $\chi_M(s) = 0$ if not $s \in M$. An algorithm is said to *decide membership* in M if it computes the function χ_M. A set C of formulas is called (algorithmically) *decidable* if there is an algorithm deciding membership for the sentences in ccqt(C); such algorithm then is said to be a *decision method* for C. The set C is called *semantically* decidable (or decidable with respect to *validity*) if there is an algorithm deciding membership for the sentences in ct(C); this implies the above decidability with respect to provability, but the reverse implication would require the inconstructively established inclusion ct(C) $\subseteq$ ccqt(C).

If a set C of formulas is inconsistent then all sentences are in ccqt(C). If C is consistent and is complete then the algorithm, generating ccqt(C) = M from C, produces, for every sentence s, either $s \epsilon M$ or $\neg s \epsilon M$, i.e. $\chi_M(s) = 1$ or $\chi_M(\neg s) = 1$, and consistency leads from $\chi_M(\neg s) = 1$ to $\chi_M(s) = 0$. This means that the algorithm produces both the sentences in ccqt(C) and the sentences *not* in ccqt(C), and it then is a matter of routine to produce from it an algorithm computing χ_M (for a precise proof cf. Theorem 1 of Chapter 8). In general, the algorithm depends on the members of C as parameters, but these can be eliminated if C is algorithmically presented. Thus if C is algorithmically presented and is complete then C is decidable.

In particular, for the semantically complete set C_1, mentioned above, and assuming inconstructively that $ct(C_1) = ccqt(C_1)$, the set C_1 is complete and, being finite, also is decidable. As C_1 is complete, the set of sentences in $ct(C_1)$ is $Ths(N) = (\forall)ct(C_1)$ for $N = \langle \omega, 0, s, <, id^{\omega} \rangle$, and is complete as well; under the assumption $ct(C_1) = ccqt(C_1)$ therefore, also the set Ths(N) becomes algorithmically presented and decidable. In view of the inconstructive arguments employed here, no decision algorithm can actually be exhibited; however, in Chapter 1 both completeness and decidability of C_1 will be shown constructively. In general, of course, a set Ths(A) for a structure A is anything but algorithmically describable.

3. The Historical Setting

The Greek mathematicians of antiquity discovered very early that in geometry there arose magnitudes (e.g. a side and a diagonal of a quadrangle) whose relationsships to each other could not be expressed by ratios between integers. Their solution to this problem of *incommensurabilities* we find in Book V of Euclid's Elements, namely a theory of proportions between magnitudes which, together with the axioms of the fourth proportional and the Archimedian axiom, permitted a logically impeccable method to compare the ratios between straight distances and the ratios between areas, say those under parabolas (cf. KRULL 60 for a characterization of this theory by modern algebraic concepts). This *Method of the Ancients*, or the proceeding *more geometrico*, was still observed and paid lip service to until the 17th century, but it did not suffice to express the results of the powerful usage of limit processes brought forward by James Gregory, Wallis, Newton and Leibniz. Logical rigour was lacking in the treatment of the new numbers defined by such processes (cf. the commentaries of Bishop BERKELEY 34), and it remained so until, by the mid-19th century, the theories of Dedekind and Weierstraß gave a new foundation to real numbers. The key instrument making this possible was to postulate the results of mental experiments (i.e. objects described by limiting processes, such as sequences and functions) as *well defined* through the infinite processes by which they arise, and to apply to them, as the *completed* and *finished* results of such processes, the laws of classical logic. Dedekind's characterization of real numbers as *cuts* amounted to their description by infinite sets (of rationals), and in this way quite generally the completed results of infinite processes could be reduced to *sets* obtained by comprehension with respect to their defining properties. Yet while inconstructive arguments about infinite sequences were used and accepted in analysis, an inconstructive argument about infinite sets, in the proof of HILBERT's 90 theorem on the existence of a finite basis for a set of algebraic forms, appeared novel enough to draw Paul Gordan's famous comment that *this* was not mathematics but theology.

In this situation, it was most disturbing when B.RUSSELL 03 discovered an error in a system of higher order logic presented in FREGE 03, because stated in the related terminology of sets (and known then as Russell's paradox) it offered an example where unrestricted set formation would lead to a contradiction. Consistency of mathematical concepts appeared in doubt, and it now was Hilbert who set out to begin a rescue operation. Motivated possibly by conversations with L.E.J.Brouwer 1909 (cf. BROUWER 28), Hilbert proposed (i) to translate, for this purpose, mathematical argumentation into the formal language of logic and (ii) to attempt, for the axiom systems of various mathematical theories, consistency proofs by a combinatorial study of finite proof figures (trees), hopefully resulting in the impossibility to have proofs both for a sentence and for its negation. Hilbert's program was pursued during the 1920ies and 1930ies by his collaborators Bernays, Ackermann, his students Gentzen and Schütte, and by Jaques Herbrand. (For Hilbert's program cf. HILBERT 22, 23, 26, BERNAYS 22, 35, KREISEL 58.)

Hilbert's program asked for consistency proofs of mathematical axiom systems, and for some weaker fragments of arithmetic such proofs were found. In the meantime, however, GÖDEL 31 discovered that consistency of full arithmetic, i.e. of addition and multiplication together with induction axioms, could not be proved with the tools of arithmetic itself, but would require the employment of stronger principles; Gödel's result (sometimes called his 2nd incompleteness theorem) will be presented in Chapter 11 of this book. Consequently, a consistency proof for full arithmetic as found by GENTZEN 36/38 does require more than that, namely a principle of induction up to the transfinite 'ordinal' ε_0; for an account of Gentzen's proof and a discussion of ordinals from a finitist standpoint cf. TAKEUTI 75/87, and for further developments in this direction cf. SCHÜTTE 60/77.

The question of completeness arises most naturally when an axiomatization is intended to describe a particular situation completely; the question of decidability supplements that of completeness by asking for a uniform method to decide provabilities. Beginning with LÖWENHEIM 15 , considerable effort was spent upon (a) establishing cases of decidability of validity (in arbitrary structures) for classes of formulas which have a special prenex normal form, and (b) methods *reducing* decidability of validity from the case of general formulas to that of those having a particular prenex normal form (cf. §§46–47 of CHURCH 56 for a first introduction); the results of the type (a) can be found in ACKERMANN 54 and those of the type (b) in SURANYI 59 (cf. also BOERGER-GRAEDEL-GUREVICH 97). However, the particular prenex normal forms treated under (b) remained a much larger class than that of the special ones treated successfully under (a), and that this gap could not be bridged became clear only when CHURCH 36 succeeded to show that provability in pure logic was undecidable (in the present book, this will be Theorem 7 of Chapter 9).

As for systems of mathematical axioms which are decidable, already LÖWENHEIM 15 had established semantical decidability of pure equality logic. SKOLEM 19 invented the use of quantifier elimination to prove the decidability of a particular class of Boolean algebras; LANGFORD 27 used it to show the decidability of the axioms of various kinds of linear orders. Tarski in 1927 established decidability for certain systems of plane geometry, developed the general principle underlying quantifier elimination, and Tarski's student M.PRESBURGER 30 in 1928 proved the decidability of the addition of integers, since then called *Presburger arithmetic*. SKOLEM 31 obtained the same decidability by a different approach and then proved the decidability of the multiplication of integers (cf. the presentation in SMORINSKY 91). While most of these investigations were carried out for semantical decidability, Bernays, noticing that their results remained in effect also for decidability of provability, employed them in HILBERT-BERNAYS 34 for consistency proofs of respective fragments of arithmetic; the presentation of these results will be the content of Chapters 1 and 2.

Just as Gödel's theorem about the improvability of consistency had shown the limits of consistency proofs, another theorem of GÖDEL 31 (often called his 1st incompleteness theorem) pointed out limits for the type of axiom systems which might be proved to be decidable: full arithmetic, it taught, is not decidable, and neither is any theory in which full arithmetic may be defined. The statements about the consistency of certain systems, claiming the unavailability of proofs of contradictions, had required an analysis of the notion of *proof* (as here developed in the second book of these Lectures); statements about undecidabilities now, claiming the unavailability of algorithms which decide, will require an analysis of the notion of algorithm and algorithmic computation, and it will be in Chapters 5 and 6 that the necessary prerequites about algorithmic computations will be presented in a study of *recursive functions* and *recursively enumerable* sets. Undecidability of arithmetic is closely related to its incompleteness, and the basic ideas underlying the *impossibility* to find certain algorithms can as well be used to formulate the impossibility of completeness. Moreover, in order to make these ideas work, only a few key properties of algorithmetic-arithmetical descriptions are needed. For this reason, it will be possible to present, in Chapters 3 and 4, the methods leading to the theorems about incompleteness (and to related ones about undefinability) already in an abstract form, depending only on a few properties of *representability*; hopefully, this presentation will permit to understand those basic ideas without the burden of technical details about recursive functions. Once these details have been secured in Chapters 5 and 6, the subjects of undecidability and incompleteness will be taken up, and concluded, in Chapters 7 to 9.

Returning to the question about the decidability of mathematical axiom systems, it should at least be mentioned that, surprising after Gödel's results, it could be anwered in the positive both for Euclidian and non-Euclidian geometries as well as for the theories of real closed fields (the algebraic substratum of the real numbers) and of algebraically closed fields. Here the principal source is TARSKI 39, 48 , and the geometrical work is described comprehensively in SCHWABHÄUSER-SZMILIEW-TARSKI 83. The efficiency of Tarski's decision method for real closed fields, outlined e.g. in KREISEL-KRIVINE 66, has been considerably improved by G.E.COLLINS , cf. COLLINS 82 and ARNON-COLLINS-McCALLUM 84 .

References

W. Ackermann: Solvable Cases of the Decision Problem. Amsterdam 1954

D.S. Arnon, G.E.Collins, S.McCallum: Cylindrical Algebraic Decomposition I: The Basic Algorithm. SIAM J.Comput. 13 (1984) 866–877 : II. An Adjacency Algorithm for the Plane. id. 878–889

G. Berkeley: The Analyst, or a discourse addressed to an Infidel Mathematician. London 1734 . Reprint in: The works of George Berkeley, vol. 4 . London 1951

P. Bernays: Über Hilberts Gedanken zur Grundlegung der Arithmetik. Jahresber.D.M.V 31 (1922) 10–19

P. Bernays: Hilberts Untersuchungen über die Grundlagen der Arithmetik. pp.196–216 in: D. Hilbert: Gesammelte Abhandlungen. Band III . Berlin 1935

E. Boerger, E.Grädel, Y.Gurevich: The Classical Decision Problem. Berlin 1997

L.E.J. Brouwer: Intuitionistische Betrachtungen über den Formalismus. Koninkl.Nederl.Akad. Wetenschap.Amsterdam, Proc. 31 (1928) 374–379

A. Church: A note on the Entscheidungsproblem. J.Symb.Logic 1 (1936) 40–41 and 101–102

A. Church: Introduction to Mathematical Logic I . Princeton 1956

G.E. Collins: Quantifier Elimination for Real Closed Fields: A Guide to the Literature. pp. 80–81 in: B. Buchberger, G.E. Collins, R. Loos: Computer Algebra: Symbolic and Algebraic Computation. Berlin 1982

G.Frege: Grundgesetze der Arithmetik. Begriffsschriftlich abgeleitet. Vol. II . Jena 1893

G.Gentzen: Untersuchungen über das logische Schliessen I, II. Math.Z. 39 (1934/35) 176–210 and 405–431

G.Gentzen: Die Widerspruchsfreiheit der reinen Zahlentheorie. Math.Ann. 122 (1936) 493–565

G.Gentzen: Neue Fassung des Widerspruchsfreiheitsbeweises für die reine Zahlentheorie. Forsch.z.Logik Grundl.ex.Wiss. NF 4 (1938) 19–44

K. Gödel: Über formal unentscheidbare Sätze der Principia Mathematica und verwandter Systeme I . Monatshefte Math.Phys. **38** (1931) 173–198

D. Hilbert: Über die Theorie der algebraischen Formen. Math.Ann. 36 (1890) 473–534

D. Hilbert: Neubegründung der Mathematik. Erste Mitteilung. Abh. aus dem Math.Semin.der Hamburg Univ. **1** (1922) 157–177

D. Hilbert: Die logischen Grundlagen der Mathematik. Math.Ann. **88** (1923) 151–165

D. Hilbert: Über das Unendliche. Math.Ann. **95** (1926) 161–190

D. Hilbert, P. Bernays: Grundlagen der Mathematik I . Berlin 1934

G. Kreisel: Hilbert's Programme. Dialectica **12** (1958) 346–372

G. Kreisel, J.-L.Krivine: Eléments de logique mathématique. Paris 1966 . English translation: Elements of Mathematical Logic. Amsterdam 1971 . Deutsche Übersetzung: Modelltheorie. Berlin 1972

W. Krull: Über die Endomorphismen von total geordneten Archimedischen Abelschen Gruppen. Math.Z. **74** (1960) 81–90 .

C.H. Langford: Some theorems on deducibility. Ann.of Math.II **28** (1927) 16–40 , 459–471

L. Löwenheim: Über Möglichkeiten im Relativkalkul. Math.Ann. **76** (1915) 447–470

M. Presburger: Über die Vollständigkeit eines gewissen Systems der Arithmetik ganzer Zahlen, in welchen die Addition als einzige Operation hervortritt. Compt. Rend. I. Congr.des Mathémath.des Pays Slaves. Warszawa 1930 , 92–101 and supplement p. 395

B. Russell: The Principles of Mathematics. Cambridge 1903

K. Schütte: 60/77 Beweistheorie. Berlin 1960 . 2nd. ed : Proof Theory. Berlin 1977

W. Schwabhäuser, W.Szmiliew, A.Tarski: Metamathematische Methoden in der Geometrie. Berlin 1983

Th. Skolem: Untersuchungen über die Axiome des Klassenkalkuls und über Produktations- und Summationsprobleme, welche gewisse Klassen von Aussagen betreffen. Skrifter Videnskab. Kristiana **3** (1919) 1–37

Th. Skolem: Über einige Satzfunktionen in der Arithmetik. Skrifter Videnskap. Oslo, I, no. 7 (1931) 1–28

C. Smorinsky: Logical Number Theory I . Berlin 1991

J.Surányi: Reduktionstheorie des Entscheidungsproblems im Prädikatenkalkul der ersten Stufe. Budapest 1959

G. Takeuti: 75 Proof Theory. Amsterdam 1975 . 2nd ed. Amsterdam 1987

A. Tarski: The completeness of elementary algebra and geometry . Manuscript 1939 of a book which should have appeared in 1940. Reprinted 1967 Paris

A. Tarski: A decision method for elementary algebra and geometry. 1948 Santa Monica, Rand Corp., 1951 Univ. of California

Chapter 1. Consistency, Decidability, Completeness for the Arithmetic of Order with Successor

1. Examples of simple Consistency Proofs

I shall employ the calculi ccqt_i presented in Chapter 2.9, but as I consider sets C of *sentences*, I may as well use the calculi ccqs_i (cf. Lemma 2.9.2) or a classical sequent calculus and the derivabiliy of $C \Rightarrow v$ (cf. Theorem 2.10.1). Working with ccqt_i, together with v also the sentence $(\forall)v$ is provable; together with $\neg v$ also $(\exists)\neg v$ is provable (already in ccqs), hence also $\neg(\forall)v$. Consequently, consistency of C will be secured if for no *sentence* v both v and $\neg v$ are provable.

My first example of a consistent set of sentences shall be the empty one. As the formulas provable from it are those provable by logic alone, consistency of the empty set amounts to the consistency of pure logic itself. Observe first that a formula v is provable in pure logic if the sequent $\blacktriangle \Rightarrow v$ is derivable in the sequent calculus . Assume now that both v and $\neg v$ were provable; then the derivations of the corresponding sequences could be extended with help of $(I\neg)$, and then by a cut with $\neg v$, to a derivation of the empty sequent, i.e. the sequent with empty antecedent and empty succedent:

$$\dfrac{\dfrac{\blacktriangle \Rightarrow v}{\neg\, v \Rightarrow \blacktriangle} \qquad\qquad \blacktriangle \Rightarrow \neg\, v}{\blacktriangle \Rightarrow \blacktriangle}\ .$$

Eliminating this cut, I would arrive at a cut free derivation of $\blacktriangle \Rightarrow \blacktriangle$, and it is evident from the logical rules of the calculus MK, say, that they cannot produce this empty sequent. Consequently, classical logic (and a fortiori the weaker logics) is consistent. – As with every consistency proof, I should ask which price I did pay when performing it: the possibility to parse formulas and sequents, elementary constructions with trees of derivations, and the constructions required for the proof of cut elimination. The latter requires a double induction, i.e. an induction on the lexicographically ordered pairs of $\omega \times \omega$.

Clearly, this very simple proof requires that Gentzen's sequent calculus be available. Without that, consistency of pure logic in a language L without function symbols can be established as follows. I enrich L to to a language L_k with constants $c_0, \dots, c_{k-1}$ for a fixed positive integer k (k = 1 suffices). Let S be an L_k-deduction of a formula $v(\xi)$ in which eigenvariables have been made distinct; then S becomes an L_k-deduction of $v(c_0, c_0, \dots)$ if the variables in ξ are replaced by c_0 throughout S. I define a map τ for sentences; τ shall be homomorphic for propositional connectives, and for sentences starting with quantifiers I set

$$\tau(\forall x\, v(x,\alpha)) = \bigwedge <v(c_i,\alpha) \mid i<k> \ , \quad \tau(\exists x\, v(x,\alpha)) = \bigvee <v(c_i,\alpha) \mid i<k> .$$

I define $\tau^*(v)$ for sentences by iterating τ until all quantifiers have been removed; clearly, $\tau^*(\neg v) = \neg\, \tau^*(v)$ and

$$\tau^*(\forall x\, v(x,\alpha)) = \bigwedge <\tau^*(v(c_i,\alpha)) \mid i<k> \quad ,$$
$$\tau^*(\exists x\, v(x,\alpha)) = \bigvee <\tau^*(v(c_i,\alpha)) \mid i<k> .$$

I now shall show that an L_k-deduction S of an L_k-sentence v, again with distinct eigenvariables, can be transformed into a propositional deduction S^* of $\tau^*(v)$ containing only sentences without quantifiers. As this then holds in particular for L-deductions of L-sentences v, the consistency of quantifier logic will have been reduced to that of propositional logic. The proof is by induction on the lenght of S. If S has lenght 1 and consists of a propositional axiom then $\tau^*(v)$ is a propositional axiom of the same form. If v is an axiom $\forall x\, w(x,\alpha) \rightarrow w(c_j,\alpha)$ then $\tau^*(v)$ is

$$\bigwedge <\tau^*(w(c_i,\alpha)) \mid i<k> \ \rightarrow \ \tau^*(w(c_j,\alpha))$$

and so a propositional tautology; the case of $w(c_i,\alpha) \rightarrow \exists x w(x,\alpha)$ is analogous. If the last rule of S is an instance $u(\xi)$, $u(\xi) \rightarrow v \vdash v$ of modus ponens then I replace in the subdeductions of $u(\xi)$ and $u(\xi) \rightarrow v$ all variables in $u(\xi)$ by c_0. Thus I have deductions of $u(c_0,\ldots)$ and $u(c_0,\ldots) \rightarrow v$ from which the inductive hypothesis leads to propositional deductions both of $\tau^*(u(c_0,\ldots))$ and of $\tau^*(u(c_0,\ldots)) \rightarrow \tau^*(v)$. If the last rule is an instance of the quantifier rule $u \rightarrow w(x,\alpha) \vdash u \rightarrow \forall x\, w(x,\alpha)$, where u must be a sentence, then from the subdeduction of $u \rightarrow w(x,\alpha)$ I find deductions of the sentences $u \rightarrow w(c_i,\alpha)$, hence by inductive hypothesis also propositional deductions of $\tau^*(u) \rightarrow \tau^*(w(c_i,\alpha))$ for every i with $i<k$, and from them I obtain a propositional deduction of $\tau^*(u) \rightarrow \bigwedge <\tau^*(w(c_i,\alpha)) \mid i<k>$ which is $\tau^*(v)$ for $v = u \rightarrow \forall x\, w(x,\alpha)$.

It now is an immediate consequence that also equality logic is consistent, since by Theorem 2.10.2 it is a conservative extension of pure logic.

As a further example, I discuss the consistency of an axiom system whose models are infinite sets, viewed as those which permit an injective function mapping them onto a proper subset. So let L be an equality L language with a 1-ary function symbol s and a constant c_0; I write $s^0(x)$ for x and $s^{n+1}(x)$ for $s(s^n(x))$. Working with a modus ponens calculus ccqs, my axiom system S_1 shall consist of (the universal closures of)

b0. $\quad s(x) \equiv s(y) \ \rightarrow \ x \equiv y$

b1. $\quad \neg\, s(x) \equiv c_0$.

That S_1 is consistent was shown already in HILBERT 22. Again, I prefer the convenience of a sequent calculus which I shall construct from the calculus XE_1 for equality (cf. Chapter 2.10.3). The inital sequents of XE_1 form a set ET, obtained from the logical axiom sequents $r_0 \equiv r_1 \Longrightarrow r_0 \equiv r_1$ and the sequents

e0 $\blacktriangle \Longrightarrow r \equiv r$ e2 $r_0 \equiv r_1, r_1 \equiv r_2 \Longrightarrow r_0 \equiv r_2$
e1 $r_0 \equiv r_1 \Longrightarrow r_1 \equiv r_0$ e3 $r_0 \equiv r_1 \Longrightarrow s(r_0) \equiv s(r_1)$.

by closing up under cuts and structural rules. My new calculus XS_1 shall have initial sequents from the set ET_s, obtained from the sequents used for ET and the new sequents

sb0 $s(r_0) \equiv s(r_1) \Longrightarrow r_0 \equiv r_1$ sb1 $s(r_0) \equiv c_0 \Longrightarrow \blacktriangle$

by closing up under cuts and structural rules. Initial sequents being closed under cuts, the cut elimination algorithm carries over to XS_1 without changes. In order to show that XS_1 be consistent, it will again suffice to see that the empty sequent $\blacktriangle \Longrightarrow \blacktriangle$ cannot be derived.

As such derivation could be found cut free from initial sequents, it will suffice to show that $\blacktriangle \Longrightarrow \blacktriangle$ does not occur as an initial sequent in ET_s. If it would occur, it could only appear as produced by a cut, hence as cut of two sequents $\blacktriangle \Longrightarrow t \equiv r$ and $t \equiv r \Longrightarrow \blacktriangle$.

Now the terms t of my language are of the forms $s^n(x)$ or $s^n(c_0)$, and I may call n here the *rank* of t. An equation $t \equiv r$ I shall call *even* if t and r have the same rank; otherwise I shall call it *odd*. I now show:

(1) if a sequent $E \Longrightarrow t \equiv r$ is in ET_s and if $t \equiv r$ is odd
then E contains at least one odd equation .

This is evident for e0, e1, e3, sb0, and it holds for e2 since if both $r_0 \equiv r_1$, $r_1 \equiv r_2$ are even then so is $r_0 \equiv r_2$. So I can use induction on the number of cuts required to obtain $E \Longrightarrow t \equiv r$ in ET_s. If in

$$\frac{E_0 \Longrightarrow t_0 \equiv r_0 \qquad t_0 \equiv r_0, E_1 \Longrightarrow t_1 \equiv r_1}{E_0, E_1 \Longrightarrow t_1 \equiv r_1}$$

$t_1 \equiv r_1$ is odd then, by inductive hypothesis, either E_1 contains an odd equation or $t_0 \equiv r_0$ is odd. In the latter case, again by inductive hypothesis, E_0 will contain an odd equation, and so in any case E_0, E_1 contains an odd equation. This completes the proof of (1). – In the same manner I show:

(2) if a sequent $E \Longrightarrow \blacktriangle$ is in ET_s
then E contains at least one odd equation .

This is evident for sb1, and again I can use induction on the number of cuts required to obtain $E \Longrightarrow \blacktriangle$ in ET_s. If

$$\frac{E_0 \Longrightarrow t_0 \equiv r_0 \qquad t_0 \equiv r_0, E_1 \Longrightarrow \blacktriangle}{E_0, E_1 \Longrightarrow \blacktriangle}$$

then, by inductive hypothesis, either E_1 contains an odd equation or $t_0 \equiv r_0$

is odd. In the latter case, E_0 will contain an odd equation by (1), and so in any case E_0, E_1 contains an odd equation. This completes the proof of (2).

It follows from (1) that $\blacktriangle \Longrightarrow t \equiv r$ can only be in ET_s if $t \equiv r$ is even, and it follows from (2) that $t \equiv r \Longrightarrow \blacktriangle$ can only be in ET_s if $t \equiv r$ is odd. Hence the empty sequent cannot be in ET_s. Thus XS_1 is consistent.

As for the connection between S_1 and XS_1, I show that $v \in ccqs(S_1 \cup EA_u)$ is equivalent to the derivability of $\blacktriangle \Longrightarrow v$ in XS_1. First, by Theorem 2.10.1 $v \in ccqs(S_1 \cup EA_u)$ is equivalent to the derivability of $S_1 \Longrightarrow v$ in XE_1. Clearly,

$$\blacktriangle \Longrightarrow s(x) \equiv s(y) \rightarrow x \equiv y \quad , \quad \blacktriangle \Longrightarrow \neg\, s(x) \equiv c_0 \;,$$
$$\blacktriangle \Longrightarrow \forall x\, \forall y\, (s(x) \equiv s(y) \rightarrow x \equiv y) \quad , \quad \blacktriangle \Longrightarrow \forall x\, \neg\, s(x) \equiv c_0$$

are derivable in XS_1; hence together with $S_1 \Longrightarrow v$ also $\blacktriangle \Longrightarrow v$ is derivable in XS_1C and so in XS_1. Conversely, sb0 and sb1 are derivable in XE_1 from the axioms S_1. First

$$s(t_0) \equiv s(t_1) \rightarrow t_0 \equiv t_1,\, s(t_0) \equiv s(t_1) \Longrightarrow t_0 \equiv t_1$$
$$\forall x\, \forall y\, (s(x) \equiv s(y) \rightarrow x \equiv y),\, s(t_0) \equiv s(t_1) \Longrightarrow t_0 \equiv t_1$$

shows this for sb0. Since also

$$s(t_0) \equiv c_0 \Longrightarrow s(t_0) \equiv c_0$$
$$\neg\, s(t_0) \equiv c_0 \,,\, s(t_0) \equiv c_0 \Longrightarrow \blacktriangle \;,$$

a cut with $\neg\, s(t_0) \equiv c_0$ derives

$$S_1 \,,\, \neg\, s(t_0) \equiv c_0 \,,\, s(t_0) \equiv c_0 \Longrightarrow \blacktriangle$$
$$S_1 \,,\, \forall x\, \neg\, s(x) \equiv c_0 \,,\, s(t_0) \equiv c_0 \Longrightarrow \blacktriangle$$
$$S_1 \,,\, s(t_0) \equiv c_0 \Longrightarrow \blacktriangle \;,$$

and so also the sb1 are derivable from the axioms S_1. Hence it follows that for every sequent $M \Longrightarrow v$ derivable in XS_1 also $S_1, M \Longrightarrow v$ is derivable in XE_1.

Consequently, also the axiom system S_1 is consistent.

The consistency of S_1 and XS_1 means that it is possible to *speak consistently* about certain properties held by infinite totalities only.

The consistency proofs to be discussed from now on do not appear to profit from the use of sequent calculi. Instead, they will depend on a technique, the *elimination of quantifiers*, which also provides access to decision methods.

In the model theoretic developments of Chapter 1.14, I reduced certain semantical questions about formulas containing quantifiers to related questions about open formulas; for open formulas then these questions could be easily answered. While there a formula was semantically related to open formulas, I now wish to mention a technique which, for suitable sets C of formulas (axioms), permits to reduce both decidability and consistency to the case of open sentences (formulas).

Let C be a set of formulas (axioms); a *quantifier elimination* map $^\#$ *for* C shall be a map defined for all formulas v such that $v^\#$ is open, $(\neg v)^\# = \neg v^\#$ and $v \leftrightarrow v^\#$ is provable from C; I further require $v^\# = v$ for open formulas v and also that $fr(v^\#) \subseteq fr(v)$. Thus if v is a sentence then so is $v^\#$.

Consequently, if C admits a quantifier elimination map then, v being equivalent to $v^\#$ and $\neg v$ being equivalent to $\neg v^\#$, the question whether sentences v or/and $\neg v$ are provable from C reduces to that question asked for the open sentences $v^\#$ and $\neg v^\#$.

It depends very much on the special content of the formulas in C whether it admits a quantifier elimination map.

As a particular case, let A be a structure and let C be the set Th(A) of all formulas true in A. Then $v \leftrightarrow v^\#$ is provable from C if (and only if) it is true in A, and in order to establish a map $^\#$ as quantifier elimination map, one may argue about truth instead of provability.

2. The Arithmetic C_1 of Order with Successor

Let L be a language with the constant c_0, the 1-ary operation symbol s and predicate symbols $\equiv$ and $<$, and with the abbreviation $x \lesssim y$ for $x<y \vee x \equiv y$. Let C_{01} be the set of axioms

c1. $\neg\, x<x$ (irreflexivity)
c2. $x<y \wedge y<z \rightarrow x<z$ (transitivity)
c3. $x<y \vee y<x \vee x \equiv y$ (trichotomy)
c4. $c_0<x \vee c_0 \equiv x$
c5. $x<s(x)$
c6. $x<y \rightarrow s(x)<y \vee s(x) \equiv y$ (successor)

C_{01} is semantically incomplete, hence also incomplete, because for

c7. $c_0<y \rightarrow \exists y_1\, s(y_1) \equiv y$ (predecessor).

the universal closure is true in the standard model on the set ω of natural numbers, but does not hold in the model on $2\times\omega$, ordered lexicographically and with the successor function s defined as $s[x,y] = [x,y+1]$.

Let C_1 be the axiom system obtained from C_{01} by adding the axiom c7. The following three sections are devoted to the proof that C_1 is complete, decidable and consistent.

Provability, for the moment, shall mean provability *from* C_{01} in a calculus $ccqt_=$; $\vdash v$ shall denote that a formula v is provable. For terms t and natural numbers n, I shall write t+0 instead of t, and t+(n+1) instead of $s(t+n)$. Thus all my terms are of the form c_0+n or x+n where x is a variable. The constant terms c_0+n I abbreviate as c_n. Observe that (t+n)+m =

t+(n+m) since (t+n)+0 = t+n and (t+n)+(m+1) = s((t+n)+m) = s(t+(n+m)) = t+(n+m)+1 = t+n+(m+1). During the following proofs I shall use without explicit reference that the rule (Rσ): u $\vdash$ sub(η|u) is admissible (cf. Chapter 2.9). Further, I shall use that $\neg a \wedge (a \vee b) \rightarrow b$ holds intuitionistically as it follows from $\neg a \wedge a \rightarrow b$, $\neg a \wedge b \rightarrow b$.

0. $\vdash x \equiv y \rightarrow \neg x<y$. Apply c1 and $\vdash \neg x<x \wedge x \equiv y \rightarrow \neg x<y$.

1. $\vdash x<y \rightarrow \neg x \equiv y$. Propositionally (m0) .

2. $\vdash x<y \rightarrow \neg y<x$.

$\vdash x<y \rightarrow (y<x \rightarrow x<x)$ by c2
$\vdash (y<x \rightarrow x<x) \rightarrow (\neg x<x \rightarrow \neg y<x)$ propositionally (m2)
$\vdash x<y \rightarrow (\neg x<x \rightarrow \neg y<x)$
$\vdash \neg x<x \rightarrow (x<y \rightarrow \neg y<x)$ Now apply c1 .

3. $\vdash x<y+1 \rightarrow x \lesssim y$.

$\vdash \neg x \lesssim y \rightarrow y<x$ from c3 intuitionistically
31 $\vdash \neg y<x \rightarrow x \lesssim y$
$\vdash y<x \rightarrow y+1 \lesssim x$ by c6
$\vdash x<y+1 \rightarrow (y+1 \lesssim x \rightarrow y+1<y+1)$
$\vdash x<y+1 \rightarrow (y<x \rightarrow y+1<y+1)$
$\vdash x<y+1 \rightarrow (\neg y+1<y+1 \rightarrow \neg y<x)$
$\vdash \neg y+1<y+1 \rightarrow (x<y+1 \rightarrow \neg y<x)$
$\vdash x<y+1 \rightarrow \neg y<x$
$\vdash x<y+1 \rightarrow x \lesssim y$ by 31

4. $\vdash x \equiv y \leftrightarrow x<y+1 \wedge y<x+1$.

$\vdash x<y+1 \wedge y<x+1 \rightarrow (x \equiv y \vee x<y) \wedge (x \equiv y \vee y<x)$
$\vdash x<y+1 \wedge y<x+1$
$\rightarrow x \equiv y \vee (x \equiv y \wedge y<x) \wedge (x<y \vee x \equiv y) \vee (x<y \wedge y<x)$
$\vdash \neg (x \equiv y \wedge y<x)$ by c1
$\vdash \neg (x \equiv y \wedge x<y)$ by c1
$\vdash \neg (x<y \wedge y<x)$ by c1
$\vdash x<y+1 \wedge y<x+1 \rightarrow x \equiv y$.

5. $\vdash x<y \rightarrow x+1<y+1$. Apply c6, c5 and c2 .

6. $\vdash x+1<y+1 \rightarrow x<y$. Apply 3, c5 and c2 .

7. $\vdash x<y \leftrightarrow x+n<y+n$. By induction .

8. $\vdash x \equiv y \leftrightarrow x+n \equiv y+n$.

$\vdash x+1 \equiv y+1 \rightarrow x<y+1 \wedge y<x+1$
$\vdash x+1 \equiv y+1 \rightarrow (x \equiv y \vee x<y) \wedge (y \equiv x \vee y<x)$
$\vdash x+1 \equiv y+1 \rightarrow x \equiv y \vee (x \equiv y \wedge y<x) \vee (x \equiv y \vee x<y)$
$\wedge (x<y \wedge y<x)$
$\vdash \neg (x \equiv y \wedge y<x)$ by c1
$\vdash \neg (x \equiv y \wedge x<y)$ by c1

$\vdash \neg (x<y \wedge y<x)$ by c1
$x+1 \equiv y+1 \rightarrow x \equiv y$ Now induction on n.

9. $\vdash \neg\, x+n \equiv x+k$ for $n \neq k$.

91 $\vdash x+n \equiv x+(m+n) \rightarrow x \equiv x+m$ by 8
92 $\vdash x<x+m$ for $m>0$ by c5, c2 and induction
$\vdash \neg\, x \equiv x+m$ for $m>0$ by 1
$\vdash \neg\, x+n \equiv x+(n+m)$ for $m>0$ by 91

10. $\vdash x+n<x+k$ for $n<k$. Apply 92 with $m = k-n$.

11. $\vdash \neg\, x+n<x+k$ for $k \leq n$.

$\vdash \neg\, x+n<x+n$ by c1.
$\vdash x+k<x+n$ by 10 for $k<n$
$\vdash \neg\, x+n<x+k$ by 2.

The axiom system was discussed already from a model theoretical point of view in Chapter 1.16. Provability now shall mean provability from C_1. I then can show the possibility of limited subtraction:

12. $\vdash c_n<y \rightarrow \exists x\ x+n+1 \equiv y$.

For $n = 0$ this is c7; assume that 12 holds for n. Then

$\vdash c_{n+1}<y \rightarrow c_0<y$
$\vdash c_{n+1}<y \rightarrow \exists y_1\ y_1+1 \equiv y$ by c7
$\vdash c_{n+1}<y \wedge y \equiv y_1+1 \rightarrow c_n<y_1$ by 6
$\vdash c_n<y_1 \rightarrow \exists x\ x+n+1 \equiv y_1$ inductive hypothesis
$\vdash c_{n+1}<y \rightarrow \exists y_1 \exists x\ y_1+1 \equiv y \wedge x+n+1 \equiv y_1$
$\vdash c_{n+1}<y \rightarrow \exists x\ x+n+2 \equiv y$.

LEMMA 1 C_{01} is decidable for atomic sentences.

The only constant terms being the c_a, the only atomic sentences are of the form $c_n \equiv c_m$ and $c_n<c_m$. If $n = m$ then $c_n \equiv c_m$ is provable by 8, if $n \neq m$ then $\neg\, c_n \equiv c_m$ is provable by 9. If $n<m$ then $c_n<c_m$ is provable by 10, if $n \geq m$ then $\neg\, c_n<c_m$ is provable by 11.

3. Reductions and Quantifier Elimination for C_1

The consistency of C_1 will be established will help of a model for which the provable open formulas can be shown to be true. This model will not be formed with the powerful tools of set theory, but with conceptually much simpler constructions. Moreover, truth for open sentences with respect to this model can be defined without explicitly referring to the model, but will use only syntactical properties of the constants c_m. For this reason, I shall define my new notion of truth® and of sentences which are true®. The reader may notice that truth® amounts to classical truth in a standard structure of natural numbers, but he should also notice that the set theoretical notions of model theory are not employed here. Details about the model, to which truth® then can be read as referring, will be discussed after Theorem 1 in section **4**.

I shall classify the open sentences into those which are *true®* and those which are not. For the atomic sentences $c_n \equiv c_m$ and $c_n < c_m$ of L I define

$c_n \equiv c_m$ is true® if $m = n$, and not true® otherwise ,

$c_n < c_m$ is true® if $n < m$, and not true® otherwise .

Atomic sentences, therefore, are *either* true® *or* not true®. Once the notion of *truth®* is available for atomic sentences, I extend it to *open* sentences mirroring the definition of classical truth: if v and w are sentences then

$v \wedge w$ is true® if both v and w are so, and otherwise it is not true®,

$v \vee w$ is true® if one of v or w is so, and otherwise it is not true®,

$\neg v$ is true® if v is not true®, and otherwise it is not true®,

$v \to w$ is true® if v is not true® or if w is true®, and otherwise it is not true®.

It follows that also open sentences are either true® or not true®, and so a composite open sentence is true® *only* if its components are true® or not true® in the indicated way. For an open sentence, therefore, truth® can be determined in the usual manner of truth values, and as I know already that atomic sentences are either true® or not true®, also every open sentence is *either* true® *or* not true®.

LEMMA 2 If an open sentence v is true® then I can construct a proof of v from C_1.

Truth® of v reduces to truth® or non-truth® of its atomic subsentences. Taking the conjunctive normal form NC(v) of v, all its disjunctions d must be true®, and I find a propositional proof of NC(v) and v from these d. It is easily checked which of the atomic subsentences or their negations are true® in such a d, and choosing a first true® one among them, I implant a proof from C_1 of it determined as in Lemma 1. In this way, a proof of v from C_1 has been constructed. - I wish to add the observation that the propositional proof of v from the disjunctions d in NC(v) can always be set up, but if I do not know whether v is true® then it is not clear whether every d contains a member for which there is a proof from C_1. Also, an open sentence which is not true® cannot be provable in this manner.

I now choose a fixed variable x and assign open formulas $R_x(v)$ to all open formulas v such that

A(v) $\vdash \exists x v \leftrightarrow R_x(v)$ and $\mathrm{var}(R_x(v)) = \mathrm{var}(v) - \{x\}$.

If x is the only variable in v, then I further wish that both

B(v) if $R_x(v)$ is true® then I can find a number $\gamma(v)$ such that rep(x, $c_{\gamma(v)}$ | v) is true® ,

C(v) if, for some constant c_k, rep(x, c_k | v) is true® then so is $R_x(v)$

hold. Finally, I require in any case that

D(v) if $y \neq x$ and if t^* is a term not containing x then $R_x(\mathrm{rep}(y, t^* \mid v)) = \mathrm{rep}(y, t^* \mid R_x(v))$,

E(v) If η is a map with $\eta(x) = x$ which maps $\mathrm{var}(v) - \{x\}$ into constants then $R_x(\mathrm{rep}(\eta \mid v)) = \mathrm{rep}(\eta \mid R_x(v))$.

If the properties A(v) − E(v) hold then I shall call $R_x(v)$ a *reduction* of v.

If x does not occur in v then I set $R_x(v) = v$. This is a reduction because $\vdash \exists x v \rightarrow v$ as x is not free in v; B(v), C(v) are empty and D(v), E(v) hold trivially. It remains to consider formulas which do contain x, and I begin with the case of atomic formulas:

(aa) $v = x+n \equiv x+m$: $R_x(v) = \mathrm{rep}(x, c_0 \mid v)$,

(ab) $v = x+n < x+m$: $R_x(v) = \mathrm{rep}(x, c_0 \mid v)$,

(ac) $v = x+n \equiv t$ and x not in t : $R_x(v) = c_n \leq t$,

(ad) $v = x+n < t$ and x not in t : $R_x(v) = \mathrm{rep}(x, c_0 \mid v)$,

(ae) $v = t < x+n$ and x not in t : $R_x(v) = \mathrm{rep}(x, t+1 \mid v)$.

I shall now prove that this are reductions. Clearly, the second part of A(v) holds in any case. Observe that in the cases in which $R_x(v)$ has the form rep(x,t | v), the proof of A(v) requires only to show $\vdash \exists x v \rightarrow R_x(v)$.

Case **(aa)**. If n = m then $c_n \equiv c_m$ is provable by 8, hence $\vdash \exists x v \rightarrow c_n \equiv c_m$ by $\vdash p \rightarrow (q \rightarrow p)$ propositionally. If $n \neq m$ then $\neg$ v is provable by 9, thus also $\forall x \neg v$ and $\neg \exists x v$. Hence $\vdash \exists x v \rightarrow c_n \equiv c_m$ by $\vdash \neg q \rightarrow (q \rightarrow p)$ propositionally. This proves A(v). $R_x(v)$ is true® if, and only if, n = m , and this is the case if, and only if, some, and then any, rep(x, c_k | v) is true®. Thus $\gamma(v)$ may be chosen arbitrarily, and C(v) holds as well. D(v), E(v) hold trivially.

Case **(ab)**. If n < m then $c_n < c_m$ is provable by 10, if $m \leq n$ then $\neg$ v is provable by 11; hence A(v) follows as above. $R_x(v)$ is true® if, and only if, n < m , and this is the case if, and only if, some, and then any, rep(x, c_k | v) is true®. Thus B(v) and C(v) hold as above. D(v), E(v) hold trivially.

Case **(ac)**. First, $\vdash x+n \equiv t \to c_n \leq t$ by c4, 7 and 8. Thus $\vdash \exists x v \to R_x(v)$ as x is not in t. Also $\vdash t+0 \equiv t \to \exists x x+0 \equiv t$, and if $n>0$ then $\vdash c_n \leq t \to c_{n-1}<t$, hence by 12 also $\vdash c_n \leq t \to \exists x\ x+n \equiv t$. This proves A(v). If v contains no variables besides x then t is a constant c_m, and $R_x(v)$ is true$^{\circledR}$ if, and only if, $n \leq m$. Also, rep(x, c_k | v) is $c_k+n \equiv c_m$, and this is true$^{\circledR}$ if, and only if, $k+n = m$. Thus $\gamma(v)$ may be chosen as $m-n$, and C(v) holds since $k+n = m$ implies $k \leq m$. If $\zeta = \zeta(y, t^*)$ then $R_x(\text{rep}(y, t^* \mid v))$ is $c_n \leq h_\zeta(t)$, and this is also $\text{rep}(y, t^* \mid R_x(v))$. This proves D(v), and E(v) holds analogously.

Case **(ad)**. First $\vdash x+n<t \to c_0+n<t$ as in **(ac)**, hence $\vdash \exists x v \to R_x(v)$. This proves A(v). If v contains no variables besides x then t is a constant c_m, and $R_x(v)$ is true$^{\circledR}$ if, and only if, $n<m$. Further, rep(x, c_k | v) is $c_k+n<c_m$, and this is true$^{\circledR}$ if, and only if $k+n<m$. Thus $\gamma(v)$ may be chosen with $\gamma(v)<m-n$, and C(v) holds since $k+n<m$ implies $n<m$. If $\zeta = \zeta(y, t^*)$ then $R_x(\text{rep}(y, t^* \mid v))$ is $c_n<h_\zeta(t)$, and this is $\text{rep}(y, t^* \mid R_x(v))$. This proves D(v), and E(v) holds analogously.

Case **(ae)**. First, $\vdash t<t+1+n$, $\vdash R_x(v)$ and so $\vdash \exists x v \to R_x(v)$. If v contains no variables besides x then t is a constant c_m, and $R_x(v)$ is true$^{\circledR}$; hence C(v) holds trivially. Also, rep(x, c_k | v) is $c_m<c_k+n$, and this is true$^{\circledR}$ if, and only if $m<k+n$. Thus $\gamma(v)$ may be chosen with $m<\gamma(v)+n$. If $\zeta = \zeta(y, t^*)$ then $R_x(\text{rep}(y, t^* \mid v))$ is $h_\zeta(t)<h_\zeta(t)+1+n$, i.e. $\text{rep}(y, t^* \mid R_x(v))$ since $h_\zeta(t+1) = h_\zeta(t)+1$. This proves D(v), and E(v) holds analogously.

Next, I define $R_x(v)$ for the case of conjunctions v of two atomic formulas $t_0<r_0$ and $r_1<t_1$ which are opposite in the sense that x is in r_0 and r_1, but not in t_0 or t_1. Thus

(af) $v = t_0<x+n \wedge x+m<t_1$,

and if r is the larger of n, m and $t_2 = t_0+(r-n)$, $t_3 = t_1+(r-m)$ then v is equivalent to $v' \equiv t_2<x+r \wedge x+r<t_3$ by 7, and I set

$$R_x(v) = c_r<t_3 \wedge t_2+1<t_3 .$$

Clearly, the second part of A(v) holds. Now

$\vdash v \to v'$
$\vdash v' \to c_r<t_3$
$\vdash v' \to t_2+1 \leq x+r$ by c6
$\vdash v' \to t_2+1<t_3$
$\vdash v' \to R_x(v)$
$\vdash v \to R_x(v)$
$\vdash \exists x v \to R_x(v)$

and

$\vdash t_2+1 \leq c_r \wedge c_r<t_3 \to t_2<c_r \wedge c_r<t_3$
$\vdash t_2+1 \leq c_r \wedge R_x(v) \to t_2<c_r \wedge c_r<t_3$
$\vdash t_2+1 \leq c_r \wedge R_x(v) \to \exists x\ (t_2<x+r \wedge x+r<t_3)$

$$\vdash c_r<t_2+1 \rightarrow \exists x\ x+r+1\equiv t_2+1 \qquad \text{by } 12$$
$$\vdash x+r+1\equiv t_2+1 \rightarrow \exists x\ x+r\equiv t_2+1$$
$$\vdash c_r<t_2+1 \rightarrow \exists x\ x+r\equiv t_2+1$$
$$\vdash c_r<t_2+1 \wedge t_2+1<t_3 \rightarrow \exists x\ (x+r\equiv t_2+1 \wedge x+r<t_3)$$
$$\vdash c_r<t_2+1 \wedge t_2+1<t_3 \rightarrow \exists x\ (t_2<x+r \wedge x+r<t_3)$$
$$\vdash c_r<t_2+1 \wedge R_x(v) \rightarrow \exists x\ (t_2<x+r \wedge x+r<t_3)$$

$$\vdash t_2+1\lesssim c_r \vee c_r<t_2+1$$
$$\vdash R_x(v) \rightarrow \exists x\ (t_2<x+r \wedge x+r<t_3)$$
$$\vdash R_x(v) \rightarrow \exists x\ v'$$
$$\vdash R_x(v) \rightarrow \exists x\ v\ .$$

This proves A(v). If v contains no variables besides x then t_0, t_1, t_2, t_3 are constants c_p, c_q, c_u, c_w with $u = p+(r-n)$, $w = q+(r-m)$, and $R_x(v)$ is true$^{\textcircled{n}}$ if, and only if, $r<w$ and $u+1<w$. Also, rep(x, $c_k \mid v$) is $c_p<c_k+n \wedge c_k+m<c_q$, and this is true$^{\textcircled{n}}$ if, and only if $p<k+n$, $k+m<q$, i.e. $u<k+r$, $k+r<w$. This implies $u+1<w$ and $r<w$, and so C(v) holds. Conversely, if $u<r$ then B(v) holds with $\gamma(v) = 0$, and if $r\lesssim u$ then B(v) holds with $\gamma(v) = (u+1)-r$ since then $\gamma(v)+r = u+1$. If $\zeta = \zeta(y, t^*)$ then $R_x(\text{rep}(y, t^* \mid v))$ is $c_r<h_\zeta(t_3) \wedge h(t_2)+1<h(t_3)$, i.e. rep(y, $t^* \mid R_x(v)$). This proves D(v), and E(v) holds analogously. [The reader will immediately recognize the reduction **(af)** as the reason to require denseness for an analogous decision method in the case of a linear order *without* successor as treated by LANGFORD 27.]

Next, I wish to define $R_x(v)$ for the case of conjunctions v of atomic formulas which contain a formula $r\equiv t$ for which x is in r, but not in t. As I may assume this to be the first member of my conjunction, v will have the form $x+n\equiv t \wedge g$.

An atomic formula b occurring in g has the form $r_0\equiv t_0$ or $r_0<t_0$; define $b^\dagger$ as $r_0+n\equiv t_0+n$ and $r_0+n<t_0+n$; then $\vdash b \leftrightarrow b^\dagger$ by 7 and 8. For $g = \bigwedge\langle b_i \mid i<h\rangle$ define $g^\dagger = \bigwedge\langle b_i^\dagger \mid i<h\rangle$; hence $\vdash g \leftrightarrow g^\dagger$. In the formulas $b_i^\dagger$ the variable x occurs only in terms x+k with $n\lesssim k$; let z be a new variable not occurring in v, and let $b_i^\ddagger$ and $g^\ddagger$ be the unique formulas obtained from $b_i^\dagger$, $g^\dagger$ replacing x+k by z+(k−n); hence rep(z, $x+n \mid b_i^\ddagger$) = $b_i^\dagger$ and rep(z, $x+n \mid g^\ddagger$) = $g^\dagger$. Now I set

(ag) $\quad v = x+n\equiv t \wedge g \quad : \quad R_x(v) \;=\;\equiv\; c_{n-1}<t \wedge \text{rep}(z, t \mid g^\ddagger)$.

Again, the second part of A(v) is clear, and the first part follows from

$$\vdash g \leftrightarrow g^\dagger \rightarrow (\exists x\ (x+n\equiv t \wedge g) \leftrightarrow \exists x\ (x+n\equiv t \wedge g^\dagger))$$

and

$$\vdash x+n\equiv t \rightarrow (\text{rep}(z, x+n \mid g^\ddagger) \rightarrow \text{rep}(z, t \mid g^\ddagger))$$
$$\vdash x+n\equiv t \rightarrow (g^\dagger \rightarrow \text{rep}(z, t \mid g^\ddagger))$$
$$\vdash x+n\equiv t \rightarrow (g^\dagger \rightarrow x+n\equiv t \wedge \text{rep}(z, t \mid g^\ddagger))$$
$$\vdash x+n\equiv t \rightarrow (g^\dagger \rightarrow c_{n-1}<t \wedge \text{rep}(z, t \mid g^\ddagger))$$
$$\vdash x+n\equiv t \wedge g^\dagger \rightarrow c_{n-1}<t \wedge \text{rep}(z, t \mid g^\ddagger)$$
$$\vdash \exists x\ (x+n\equiv t \wedge g^\dagger) \rightarrow c_{n-1}<t \wedge \text{rep}(z, t \mid g^\ddagger)$$

and

$$\vdash c_{n-1}<t \wedge rep(z, t \mid g^{\ddagger}) \rightarrow \exists x\, x+n\equiv t \wedge rep(z, t \mid g^{\ddagger}) \qquad \text{by 12}$$
$$\vdash c_{n-1}<t \wedge rep(z, t \mid g^{\ddagger}) \rightarrow \exists x\, (x+n\equiv t \wedge rep(z, t \mid g^{\ddagger}))$$

but

$$\vdash x+n\equiv t \rightarrow (rep(z, t \mid g^{\ddagger}) \rightarrow rep(z, x+n \mid g^{\ddagger}))$$

hence

$$\vdash c_{n-1}<t \wedge rep(z, t \mid g^{\ddagger}) \rightarrow \exists x\, (x+n\equiv t \wedge g^{\dagger}) \ .$$

Assume now that v contains no variables besides x; then t is a constant c_a. Let b be an atomic formula occurring in g, say $x+p<r_0$; then also r_0 is a constant c_e. Then $b^{\dagger}$ is $x+n+p<c_e+n$, $b^{\ddagger}$ is $y+p<c_e+n$ and

$rep(x, c_k \mid b)$	is true® if, and only if,	$k+p<e$
$rep(x, c_k \mid b^{\dagger})$	is true® if, and only if,	$k+n+p<e+n$
$rep(z, c_{k+n} \mid b^{\ddagger})$	is true® if, and only if,	$k+n+p<e+n$.

It follows that these sentences are simultaneosly true® or not, and this relationship remains in effect if b is replaced by g. Thus

$rep(x, c_k \mid v)$ is true® if, and only if, $k+n = a$ and $rep(x, c_k \mid g)$ is true®.

$R_x(v)$ is $c_{n-1}<c_a \wedge rep(z, c_a \mid g^{\ddagger})$ and is true® if, and only if, $n \leq a$ and $rep(x, c_{a-n} \mid g)$ is true®. Thus $\gamma(v)$ may be chosen as $a-n$, and C(v) holds as well.

The formula $g^{\ddagger}$ depends on the chosen variable z and may be written as ${g^{\ddagger}}_z$; yet if another variable a is chosen outside var(v) then $rep(z, t \mid {g^{\ddagger}}_z) = rep(a, t \mid {g^{\ddagger}}_a)$. Now if $\zeta = \zeta(y, t^*)$ then $rep(y, t^* \mid v)$ is $x+n\equiv h_\zeta(t) \wedge rep(y, t^* \mid g)$ and

$$R_x(rep(y, t^* \mid v)) = c_{n-1}<h_\zeta(t) \wedge rep(a, h_\zeta(t) \mid {rep(y, t^* \mid g)^{\ddagger}}_a)$$

where a is not in $rep(y, t^* \mid v)$, and I may also chose a not to be in v (i.e. $a \neq y$). Since x is not in t^*, I first find ${rep(y, t^* \mid g)^{\ddagger}}_a = rep(y, t^* \mid {g^{\ddagger}}_a)$, thus

$$rep(a, h_\zeta(t) \mid {rep(y, t^* \mid g)^{\ddagger}}_a) = rep(a, h_\zeta(t) \mid rep(y, t^* \mid {g^{\ddagger}}_a)) \ .$$

But Lemma 1.13.8 gives

$$rep(y, t^* \mid rep(a, t \mid {g^{\ddagger}}_a)) = rep(a, h_\zeta(t) \mid rep(y, t^* \mid {g^{\ddagger}}_a))$$

since a is not in t^* (and $\zeta(a, t)$ trivially is free for the open formula ${g^{\ddagger}}_a$). As $rep(a, t \mid {g^{\ddagger}}_a) = rep(z, t \mid {g^{\ddagger}}_z)$, there follows

$$R_x(rep(y, t^* \mid v)) = c_{n-1}<h_\zeta(t) \wedge rep(y, t^* \mid rep(z, t \mid {g^{\ddagger}}_z)) \ ,$$

and this is $rep(y, t^* \mid R_x(v))$. This proves D(v), and the proof of E(v) is analogous.

I continue the assigment of reductions $R_x(v)$ to conjunctions $v = b \wedge g$ where b is an atomic formula in which x occurs on both sides ; I here assume that $R_x(g)$ has been defined already and is a reduction :

(ba$_0$) $b = x{+}n \equiv x{+}m \ : \ R_x(v) = \mathrm{rep}(x, c_g \mid b) \wedge R_x(g)$,

(ba$_1$) $b = x{+}n < x{+}m \ : \ R_x(v) = \mathrm{rep}(x, c_g \mid b) \wedge R_x(g)$,

Induction shows immediately that the second part of A(v) holds.

Case **(ba$_0$)** If $n = m$ then $\vdash c_g{+}n \equiv c_g{+}m$ hence $\vdash \exists xg \to c_g{+}n \equiv c_g{+}m \wedge \exists xg$, $\vdash \exists xg \to c_g{+}n \equiv c_g{+}m \wedge R_x(g)$ by A(g), and so $\vdash \exists x\,(b \wedge g) \to R_x(v)$. But also $\vdash b$, hence $\vdash g \to b \wedge g$, $\vdash g \to \exists xv$, $\vdash \exists xg \to \exists xv$, $\vdash R_x(g) \to \exists xv$ by A(g) and $\vdash R_x(v) \to \exists xv$. If $m \neq n$ then both $\vdash \neg b$, $\vdash \neg(b \wedge g)$, $\vdash \neg \exists xv$, hence $\vdash \exists xv \to R_x(v)$ propositionally. Conversely, $\vdash \neg\, c_g{+}n \equiv c_g{+}m$, $\vdash \neg\,(c_g{+}n \equiv c_g{+}m \wedge R_x(g))$ whence $\vdash c_g{+}n \equiv c_g{+}m \wedge R_x(g) \to \exists xv$. This proves A(v).

Assume now that v contains no variables besides x. Then $R_x(v)$ is true® if, and only if, both $c_g{+}n \equiv c_g{+}m$ and $R_x(g)$ are true®, i.e. if $n = m$ and $R_x(g)$ is true®; $\mathrm{rep}(x, c_k \mid v)$ is true® if, and only if, $c_k{+}n \equiv c_k{+}m$ is true®, i.e. $n = m$, and $\mathrm{rep}(x, c_k \mid g)$ is true®, and by C(g) this implies that $R_x(g)$ is true®. This proves C(v), and choosing $\gamma(v)$ as $\gamma(g)$, B(v) will follow from B(g).

Case **(ba$_1$)** If $n<m$ then $\vdash c_g{+}n<c_g{+}m$, if $m \leq n$ then $\vdash \neg\, c_g{+}n<c_g{+}m$, and so A(v) follows as above. The proofs of C(v) and B(v) are analogous. The statements D(v) and E(v) follow immediately from D(g) and E(g).

At this stage, I have defined reductions $R_x(v)$ for all conjunctions v consisting of at most one inequality (cases **ab**, **ad**, **ae**), or of two opposite inequalities (case **af**), together with arbitrarily many equations (cases **aa**, **ac**, **ag**, **ba$_0$**) or inequalities containing x on both sides (cases **ab**, **ba$_1$**). This means that I have reductions for *all* conjunctions v of atomic formulas, except those which contain at least two inequalities, not opposite but with x appearing only on one side.

So I shall complete the definition of $R_x(v)$ for *all* conjunctions v of atomic formulas by recursion of the number $\nu(v)$ of atomic formulas in v: for all conjunctions g with $\nu(g) < \nu(v)$ the reduction $R_x(g)$ has been defined. The case $\nu(g) = 1$ is covered by the cases **aa**–**ae** .

Assume now that v contains two of those not opposite inequalities. As I may assume them to be the first two members of v, it will have one of the forms

(bb$_0$) $v = r_0<t_0 \wedge r_1<t_1 \wedge e$ x in r_0, r_1, but not in t_0, t_1 ,

(bb$_1$) $v = t_0<r_0 \wedge t_1<r_1 \wedge e$ x in r_0, r_1, but not in t_0, t_1 ,

where e is either empty or a conjunction of atomic formulas.

Case **(bb$_0$)**

$$r_0 = x{+}n,\ r_1 = x{+}m\ ,\ m \leq n\ :\ g = r_0<t_0 \wedge e\ ,\ h = r_0<t_1{+}(n{-}m) \wedge e\ ,$$

$$R_x(v) = t_0<t_1{+}(n{-}m) \wedge R_x(g) \ \vee\ \neg t_0<t_1{+}(n{-}m) \wedge R_x(h) \ ;$$

$$r_0 = x+n,\ r_1 = x+m,\ n \leq m\ :\ g = r_0<t_0+(m-n) \wedge e,\ h = r_0<t_1 \wedge e,$$

$$R_x(v) = t_0<t_1+(m-n) \wedge R_x(g) \vee \neg t_0<t_1+(m-n) \wedge R_x(h) .$$

Again, the second part of A(v) follows by induction. Consider now the case $m \leq n$. Then $r_1+(n-m) = r_0$, and v is provably equivalent to

$$v_1 = r_0<t_0 \wedge r_0<t_1+(n-m) \wedge e .$$

By propositional transformations, this is equivalent to

$$(t_0<t_1+(n-m) \vee \neg t_0<t_1+(n-m)) \wedge r_0<t_0 \wedge r_0<t_1+(n-m) \wedge e$$
$$(t_0<t_1+(n-m) \wedge r_0<t_0 \wedge e)$$
$$\vee\ (\neg t_0<t_1+(n-m)) \wedge r_0<t_1+(n-m) \wedge e),$$

$$v_2 = (t_0<t_1+(n-m) \wedge g) \vee (\neg t_0<t_1+(n-m)) \wedge h) .$$

Thus

$$\vdash \exists x v \leftrightarrow \exists x\, (t_0<t_1+(n-m) \wedge g) \vee \exists x\, (\neg t_0<t_1+(n-m)) \wedge h)$$
$$\vdash \exists x v \leftrightarrow (t_0<t_1+(n-m) \wedge \exists x g) \vee (\neg t_0<t_1+(n-m)) \wedge \exists x h)$$
$$\vdash \exists x v \leftrightarrow (t_0<t_1+(n-m) \wedge R_x(g)) \vee (\neg t_0<t_1+(n-m)) \wedge R_x(h))$$
$$\vdash \exists x v \leftrightarrow R_x(v)$$

making use of the fact that x is not in t_0, t_1 and that $\nu(g) = \nu(v)-1$, $\nu(h) = \nu(v)-1$ whence A(g), A(h) hold by inductive hypothesis.

This proves A(v). Assume now that v contains no variables besides x. The disjunction $R_x(v)$ is true$^{\circledR}$ if, and only if, one of its members is so, and this means that

(v_r) $\quad t_0<t_1+(n-m)$ and $R_x(g)$ are true$^{\circledR}$, or
$\quad \neg\, t_0<t_1+(n-m))$ and $R_x(h)$ are true$^{\circledR}$.

The formula $x+m<t_1$ is true$^{\circledR}$ if, and only if, $r_0<t_1+(n-m)$ is so; hence v is true$^{\circledR}$ if, and only if v_1 is so. In particular, rep(x, $c_k | v$) is true$^{\circledR}$ if, and only if, rep(x, $c_k | v_1$) is so. Together with v_1 and v_2, also rep(x, $c_k | v_1$), rep(x, $c_k | v_2$) arise from each other under propositional transformations, and these preserve truth$^{\circledR}$ from atomic formulas. Thus rep(x, $c_k | v$) is true$^{\circledR}$ if, and only if, rep(x, $c_k | v_2$) is so. This is the formula

$$\text{rep}(x, c_k \mid (t_0<t_1+(n-m) \wedge g) \vee (\neg t_0<t_1+(n-m)) \wedge h))$$
$$= (t_0<t_1+(n-m) \wedge \text{rep}(x, c_k | g)) \vee (\neg t_0<t_1+(n-m)) \wedge \text{rep}(x, c_k(h))$$

and so rep(x, $c_k | v$) is true$^{\circledR}$ if, and only if

(v_k) $\quad t_0<t_1+(n-m)$ and rep(x, $c_k | g$) are true$^{\circledR}$, or
$\quad \neg\, t_0<t_1+(n-m)) \wedge$ rep(x, $c_k | h$) are true$^{\circledR}$.

If $R_x(v)$ is true$^{\circledR}$ by the first part of (v_r) then B(g) implies the first part of (v_k) for $c_k = c_{\gamma(g)}$, and so I choose $\gamma(v) = \gamma(g)$. If $R_x(v)$ is true$^{\circledR}$ by the second part of (v_r) then I choose $\gamma(v) = \gamma(h)$. This proves B(v). If rep(x, $c_k | v$) is true$^{\circledR}$ by the first part of (v_k) then C(g) implies the first part of (v_r); if it is true$^{\circledR}$ by the second part of (v_k) then C(h) implies the second part of (v_r). This proves C(v).

The case $n \lesssim m$ is perfectly analogous. The statements D(v) and E(v) follow immediately from D(g), D(h) and E(g), E(h).

The case (bb_1) is perfectly analogous.

Having defined reductions for all conjunctions consisting of atomic formulas, I proceed to conjunctions v consisting of atomic formulas or negated atomic formulas. Let $\nu(v)$ now be the number of negated atomic formulas in v , and assume that for all conjunctions g with $\nu(g) < \nu(v)$ the reduction $R_x(g)$ has been defined.

Assume now that v contains a negated atomic formula. As I may assume it to be the first member of v, there will occur the possibilities

(bc_0) $v = \neg r \equiv t \wedge e : g = r<t \wedge e,\ h = t<r \wedge e,\ R_x(v) = R_x(g) \vee R_x(h)$,
(bc_1) $v = \neg r<t \wedge e : g = r \equiv t \wedge e,\ h = t<r \wedge e,\ R_x(v) = R_x(g) \vee R_x(h)$.

That these are reductions is seen as in the preceding discussion. Because in both cases v is propositionally equivalent to $g \vee h$ whence $\vdash \exists x v \leftrightarrow \exists x g \vee \exists x h$, hence $\vdash \exists x v \leftrightarrow R_x(g) \vee R_x(h)$ by A(g), A(h), and this proves A(v). $R_x(v)$ is true® if $R_x(g)$ or $R_x(h)$ is so. Being propositionally equivalent to rep(x, c_k | g) ∨ rep(x, c_k | h), the formula rep(x, c_k | v) is true® if, and only if, one of rep(x, c_k | g), rep(x, c_k | h) is so. Hence if rep(x, c_k | v) is true® then C(g) or C(h) implies that $R_x(v)$ is true®; thus C(v) holds. If $R_x(v)$ is true® then I set $\gamma(v) = \gamma(g)$ if $R_x(g)$ is true®, and $\gamma(v) = \gamma(h)$ in the other case. Thus B(g) or B(h) implies B(v). The statements D(v) and E(v) follow immediately from D(g), D(h) and E(g), E(h).

Finally, I now can define reductions for arbitray open formulas v as follows. Consider first the case that v is a disjunction of conjunctions of atomic formulas and negated atomic formulas:

$$v = \bigvee \langle d_i \mid i<r \rangle \quad : \quad R_x(v) = \bigvee \langle R_x(d_i) \mid i<r \rangle \ .$$

This is a reduction. Because $A(d_i)$ for $i<r$ implies A(v). $R_x(v)$ is true® if, and only if, at least one $R_x(d_i)$ is so. Also, rep(x, c_k | v) is propositionally equivalent to $\bigvee\langle$rep(x, c_k | d_i) | $i<r\rangle$, hence true® if, and only if, one of rep(x, c_k | d_i) is so. Hence if rep(x, c_k | d_i) is true® then $C(d_i)$ implies that $R_x(v)$ is true®; thus C(v) holds. If $R_x(v)$ is true® then I set $\gamma(v) = \gamma(d_i)$ for the first i such that $R_x(d_i)$ is true®. Thus $B(d_i)$ implies B(v). The statements D(v) and E(v) follow immediately from the $D(d_i)$ and $E(d_i)$.

In order to handle the general case, I first recall a fact from propositional logic (cf. Chapter 1.10). Let u be a propositional formula with the propositional variables $q_0, \ldots, q_{n-1}$. If there are sequences φ in 2^n such that u is satisfied by the map sending q_i to the truth value $\varphi(i)$ for $i<n$, then let I(u) be the set of these φ. Otherwise let I(u) consist of one object δ, and in that case define

$$d_\delta = q_0 \wedge \neg q_0 \wedge \bigwedge \langle q_i \mid 0<i<n \rangle \ .$$

The canonical distributive normal form ND(u) of u is

$$\mathrm{ND}(u) = \mathbb{W}\langle d_\varphi \mid \varphi \varepsilon \mathrm{I}(u)\rangle\ ,\quad d_\varphi = \mathbb{M}\langle b_{\varphi(i)} \mid i<n\rangle$$

where for $\varphi \neq \delta$ then $b_{\varphi(i)} = q_i$ if $i=1$ and $b_{\varphi(i)} = \neg q_i$ if $i=0$. Then ND(u) is propositionally equivalent to u and is uniquely determined by u.

Every open formula v determines a propositional formula u, together with an assigment of its distinct atomic formulas $a_0, \dots, a_{n-1}$ to the propositional variables $q_0, \dots, q_{n-1}$ of u, such that v arises from u under this assignment. The formula u is uniquely determined by v if I choose a fixed enumeration of terms and a fixed enumeration of predicate symbols and then establish the enumeration $a_0, \dots, a_{n-1}$ by the lexicograpic order between the atomic formulas $p(\lambda)$. If ND(u) is defined as above, then v is provably equivalent to the formula ND(v) which arises from ND(u) under the assigment of the a_i to the q_i. For ND(v), the reduction has already been defined, and so I set

$$R_x(v) = R_x(\mathrm{ND}(v)) = \mathbb{W}\langle R_x(d_\varphi) \mid \varphi \varepsilon \mathrm{I}(u)\rangle\ ,\quad d_\varphi = \mathbb{M}\langle b_{\varphi(i)} \mid i<n\rangle$$

where for $\varphi \neq \delta$ then $b_{\varphi(i)} = a_i$ if $i=1$ and $b_{\varphi(i)} = \neg a_i$ if $i=0$. Also this is a reduction. Because $\vdash v \leftrightarrow \mathrm{ND}(v)$ and A(ND(v)) imply A(v). Being equivalent under propositional transformations, v is true$^{\circledR}$ if, and only if, ND(v) is true$^{\circledR}$. But $\mathrm{rep}(x,c_k \mid v)$ is propositionally equivalent to $\mathrm{rep}(x,c_k \mid \mathrm{ND}(v))$, and so C(v) and B(v), as well as D(v) and E(v), follow as above.

This concludes the definition of reductions.

LEMMA 3 C_1 admits a quantifier elimination map $^{\#}$ satisfying

(i) $\vdash v \leftrightarrow v^{\#}$,

(ii) $\mathrm{fr}(v) = \mathrm{fr}(v^{\#})$.

(iii) $\mathrm{rep}(y, t \mid w)^{\#} = \mathrm{rep}(y, t \mid w^{\#})$ with $\zeta(y,t)$ free for w,

(iv) $\mathrm{rep}(\eta \mid w)^{\#} = \mathrm{rep}(\eta \mid w^{\#})$ where η is a map with $\eta(x) = x$ which maps $\mathrm{fr}(v) - \{x\}$ into constants .

For atomic formulas, $^{\#}$ shall be the identity. Provable equivalence being a congruence relation for propositional connectives, $^{\#}$ shall be homomorphic for these; hence in particular $(\neg v)^{\#} = \neg v^{\#}$. It follows that $v = v^{\#}$ if v is open. The case of universal quantifiers I reduce to that of existential ones :

$$(\forall x w)^{\#} = \neg(\exists x \neg w)^{\#}\ ;$$

in order to read this as a recursion on the complexity $|v|$ of formulas v, I have to change the usual definition by $|\forall x w| = |w| + 3$. The case of existential quantifiers I define by recursion and with help of reductions with respect to the quantified variable x:

$$(\exists x w)^{\#} = R_x(w^{\#}) \text{ if } w^{\#} \text{ is open}\ ,\quad (\exists x w)^{\#} = \exists x w \text{ otherwise.}$$

Induction immediately shows that the second case cannot occur and that $v^{\#}$ is always open. – It should be noticed that this recursive definition will, for *every* quantified subformula $\exists x w$ of v, make it necessary to perform a reduction $R_x(w^{\#})$ including, in particular, the setting up of a disjunctive form.

Statement (i) follows by induction. Because v = v26if v is open. If v is $\forall x w$ then $\vdash v \leftrightarrow \neg \exists x \neg w$. Now $\vdash \exists x \neg w \leftrightarrow (\exists x \neg w)^{\#}$ by inductive hypothesis, hence $\vdash \neg \exists x \neg w \leftrightarrow \neg (\exists x \neg w)^{\#}$. But $(\neg \exists x \neg w)^{\#} = \neg (\exists x \neg w)^{\#}$, and so also $\vdash v \leftrightarrow v^{\#}$. If v is $\exists x w$ then $\vdash w \leftrightarrow w^{\#}$ by inductive hypothesis. Thus $\vdash \exists x v \leftrightarrow \exists x v^{\#}$, but $\vdash \exists x w^{\#} \leftrightarrow R_x(w^{\#})$ by $A(w^{\#})$ for the open formula $w^{\#}$. Statement (ii) follows by induction. Because if $fr(w) = fr(w^{\#})$ then $fr(\exists x w^{\#}) = fr(\exists x w)$, but $fr(R_x(w^{\#})) = fr(\exists x w^{\#})$ by the second part of $A(w^{\#})$, hence $fr((\exists x w)^{\#}) = fr(R_x(w^{\#})) = fr(\exists x w)$. Statement (iii) follows by induction. First, if w is open then $rep(x, t \mid w)^{\#} = rep(x, t \mid w^{\#}) = rep(x, t \mid w)$. Also, $rep(x, t \mid \exists x w)^{\#} = (\exists x w)^{\#} = rep(x, t \mid (\exists x w)^{\#})$ since x is not free in $(\exists x w)^{\#} = R_x(w^{\#})$. If $x \neq y$ and $\zeta(y,t)$ is free for $\exists x w$ then

$$
\begin{aligned}
rep(y, t \mid \exists x w)^{\#} &= (\exists x\, rep(y, t \mid w))^{\#} && \text{and x not in t} \\
&= R_x(rep(y, t \mid w)^{\#}) \\
&= R_x(rep(y, t \mid w^{\#})) && \text{inductive hypothesis} \\
&= rep(y, t \mid R_x(w^{\#})) && D(w^{\#}) \text{ for the open formula } w^{\#} \\
&= rep(y, t \mid (\exists x w)^{\#}) .
\end{aligned}
$$

Finally, $rep(x, t \mid \forall x w)^{\#} = (\forall x w)^{\#} = rep(x, t \mid (\forall x w)^{\#})$ and

$$
\begin{aligned}
rep(y, t \mid \forall x w)^{\#} &= (\forall x\, rep(y, t \mid w))^{\#} \\
&= \neg (\exists x \neg rep(y, t \mid w))^{\#} && \text{by definition of } {}^{\#} \\
&= \neg (\exists x\, rep(y, t \mid \neg w))^{\#} \\
&= \neg (rep(y, t \mid (\exists x \neg w)^{\#}) && \text{by what was proved above} \\
&= rep(y, t \mid \neg (\exists x \neg w)^{\#}) \\
&= rep(y, t \mid (\forall x w)^{\#}) .
\end{aligned}
$$

The proof of statement (iv) is analogous.

The presence of a quantifier elimination map reduces the decidabilty of C_1 to its special case concerning open sentences; it also reduces the consistency of C_1 to that special case. Open sentences, however, may well need proofs in $ccqt_{0=}$ which require quantifiers. For instance, already the proof of a simple instantiation from an open axiom, e.g. of $v = c_7 < c_8$ from c_5, requires to reason

$$
\begin{aligned}
&\vdash c_5 \rightarrow ((v \rightarrow v) \rightarrow c_5) && \text{propositional tautology} \\
&\vdash (v \rightarrow v) \rightarrow c_5 && \text{modus ponenes with } c_5 \\
&\vdash (v \rightarrow v) \rightarrow \forall x c_5 && \text{as x is not free in } v \rightarrow v \\
&\vdash \forall x c_5 && \text{modus ponens with the tautology } v \rightarrow v \\
&\vdash \forall x c_5 \rightarrow v && \text{quantifier axiom} \\
&\vdash v && \text{modus ponens .}
\end{aligned}
$$

I now extend the notion of truth® to open formulas:

> An open formula v containing variables shall be true® if, for every assignment η sending the variables of v into constant terms, the sentence rep(η | v) is true®, and otherwise v is not true®.

Clearly, truth® of a propositionally composed open formula is related to the truth® of its components as it was in the case of open sentences.

LEMMA 4 A provable open formula is true®.

To this end, I need to make a definite choice of my calculus: it shall be the calculus $ccqt_{0=}$ and with the system EA_0 of equality axioms, consisting of the open formulas

e_0 $x \equiv x$,
e_1 $x \equiv y \rightarrow y \equiv x$,
e_2 $x \equiv y \rightarrow (y \equiv z \rightarrow x \equiv z)$,
e_3 $x \equiv y \rightarrow s(x) \equiv s(y)$,
e_4 $x_0 \equiv y_0 \wedge x_1 \equiv y_1 \rightarrow (x_0 < x_1 \rightarrow y_0 < y_1)$.

Consider now a proof S from C_1 in this calculus, ending with an open formula u. S can be presented as a tree with formulas at its nodes and, in particular, with u at its root. I can assume that none of these formulas contains the quantifier $\forall$ since I can replace axioms $\forall x v \rightarrow$ rep(x, t | v) and instances of the rule (R00)

$u \rightarrow w$
$u \rightarrow \forall x w$, x not in fr(u)

by subderivations employing the axioms rep(x, t | v) $\rightarrow \exists x v$ and the rule (R01):

$\neg$rep(x, t | v) $\rightarrow \exists x \neg v$
$\neg \exists x \neg v \rightarrow \neg\neg$rep(x, t | v)
$\neg \exists x \neg v \rightarrow$ rep(x, t | v)

$u \rightarrow w$
$\neg w \rightarrow \neg u$
$\exists x \neg w \rightarrow \neg u$, x not in fr(u)
$\neg\neg u \rightarrow \neg\exists x \neg w$
$u \rightarrow \neg\exists x \neg w$.

Let $S^\#$ arise from S by replacing each of its formulas v by $v^\#$. I shall prove the Lemma showing, by induction on the tree of $S^\#$, that each of its formulas is true®; since $u^\# = u$ then u will be true®.

The formulas at the maximal nodes of $S^\#$ come either from EA_0 or from C_1, or they come from axioms of $ccqt_0$. The formulas e_i in EA_0 are open, hence $e_i^\# = e_i$. So e_0 is true® as $c_n \equiv c_n$ is true® for every n. Also, e_1 is true® since $c_m \equiv c_n$ is true® if $m = n$ and then also $c_n \equiv c_m$ is true®. In order to see that e_3 is true®, I have to conclude from the truth® of $c_n \equiv c_m$ upon that of $c_m \equiv c_p \rightarrow c_n \equiv c_p$. The hypothesis implies $n = m$, and $c_m \equiv c_p \rightarrow c_n \equiv c_p$ will be true® if $m = p$ implies $n = p$, which is the case. Further, e_3 is true®, since the truth® of $c_n \equiv c_m$, i.e. $n = m$, implies that of $c_{n+1} \equiv c_{m+1}$, i.e. $n+1 = m+1$. In order to see that e_4 is true® I have to show that the truth® of $c_n \equiv c_p \wedge c_m \equiv c_q$, i.e. $n = p$ and $m = q$, together with that of $c_n < c_m$, i.e. $n < m$, implies the truth® of $c_p < c_q$, i.e. $p < q$, and this again is the case.

The axioms c1 – c6 from C_1 also are open, and the reader will verify in the same manner that they are true®. The axiom c7, i.e.

$$c_0 < y \rightarrow \exists y_1 \, y_1 + 1 \equiv y$$

is replaced by c7* which by (ac) is

$$c_0 < y \rightarrow c_1 \leq y \ .$$

This will be true® if $c_0 < c_n \rightarrow c_1 \leq c_n$ is true® for every c_n. This is the case if $c_0 < c_n$ is non–true®, and if $c_0 < c_n$ is true®, i.e. $0 < n$, then $1 \leq n$ whence $c_1 \leq c_n$ is true®.

Of the axioms of $ccqt_0$, the instances v of propositional axioms are true®. For example, if v is $u \rightarrow (w \rightarrow u)$ and is open then every assignment of truth® values to its components u, w will make v true®. If v is not open then v^* is an instance of the same axiom, e.g. $u^* \rightarrow (w^* \rightarrow u^*)$, and I can reason as before. If v is a quantifier axiom $rep(x, t \mid w) \rightarrow \exists x w$ then $\zeta(x,t)$ is free for w and so $rep(x, t \mid w)^* = rep(x, t \mid w^*)$ by Lemma 3(iii); thus v^* is

$$rep(x, t \mid w^*) \rightarrow R_x(w^*) \ .$$

If η maps the variables from w^* into constants and $\eta(x) = x$, then $h_\eta(t)$ is a constant c_k and

$$rep(\eta \mid rep(x, t \mid w^*)) = rep(x, h_\eta(t) \mid rep(\eta \mid w^*)) = rep(x, c_k \mid rep(\eta \mid w^*))$$

by Lemma 1.13.8. Also $rep(\eta \mid R_x(w^*)) = R_x(rep(\eta \mid w^*))$ by $E(v^*)$. Thus $rep(\eta \mid v^*)$ is

$$rep(x, c_k \mid rep(\eta \mid w^*)) \rightarrow R_x(rep(\eta \mid w^*))$$

and this is true® by $C(rep(\eta \mid w^*))$.

Thus I know that the formulas at the maximal nodes of S^* are true®. The formulas at the non–maximal nodes arise from those at their upper neighbours by the two rules of $ccqt_{0=}$, and so it will suffice to show that they are true® if the formulas at their upper neighbour are so. The first of these rules is modus ponens, and clearly together with $v^* \rightarrow u^*$ and v^* also u^* is true®. The other rule is (R01)

$$\begin{array}{l} w \rightarrow u \\ \exists x w \rightarrow u \quad , \text{x not in fr(u)} \end{array}$$

whose instances in S^* are replaced by the open formulas

$$\begin{array}{l} w^* \rightarrow u^* \\ R_x(w^*) \rightarrow u^* \ . \end{array}$$

Abbreviate as v this last formula. If η sends the variables of v into constant terms then rep(η | v) is

$$\mathrm{rep}(\eta \mid R_x(w^{\#})) \to \mathrm{rep}(\eta \mid u^{\#}) \;=\; R_x(\mathrm{rep}(\eta \mid w^{\#})) \to \mathrm{rep}(\eta \mid u^{\#})$$

by E($w^{\#}$). Now the sentence $R_x(\mathrm{rep}(\eta \mid w^{\#}))$ *is* or *is not* true®, and in the latter case clearly rep(η | v) is true®. If $R_x(\mathrm{rep}(\eta \mid w^{\#}))$ *is* true® then B(rep(η | $w^{\#}$)) implies that rep(x, c_k | rep(η | $w^{\#}$)) = rep(η | rep(x, c_k | $w^{\#}$)) is true® for a suitable k. By inductive hypothesis, the open formula $w^{\#} \to u^{\#}$ is true®, hence

$$\mathrm{rep}(\eta \mid \mathrm{rep}(x, c_k \mid w^{\#})) \to \mathrm{rep}(\eta \mid \mathrm{rep}(x, c_k \mid u^{\#}))$$

is true®, and thus also rep(η | rep(x, c_k | $u^{\#}$)) is so. As x is not in fr(u), hence not in fr($u^{\#}$), the last formula is rep(η | $u^{\#}$), and so also rep(η | v) is true®.

If an (arbitrary) formula v is provable, then so is $v^{\#}$ by Lemma 3(i), and $v^{\#}$ being open, it then is true® by Lemma 4. In particular, if v_0 and v_1 are provably equivalent then $v_0{}^{\#}$ and $v_1{}^{\#}$ are simultenously true® or not true®. Observe now

(CE_0) Let x be free in v and let v_0 be $\exists x\ v(x)$. If $v_0{}^{\#}$ is true® then I can construct a constant c_m such that $v(c_m)^{\#}$ is true®.

Because if $v_0{}^{\#} = R_x(v(x)^{\#})$ is true® then by B($v(x)^{\#}$) I find a constant c_m such that rep(x, c_m | $v(x)^{\#}$) is true®. Making use of Lemma 3(iii), this is $(\mathrm{rep}(x, c_m \mid v(x))^{\#} = v(c_m)^{\#}$.

(CE_1) Let $x_0, x_1, \ldots, x_{n-1}, x_n$ be free in v and let v_1 be the formula $\exists x_0\ \exists x_1 \ldots \exists x_{n-1}\ \exists x_n\ v(x_0, x_1, \ldots, x_{n-1}, x_n)$. If $v_1{}^{\#}$ is true® then I can construct constants $j_0, j_1, \ldots j_{n-1}, j_n$ such that the open sentence $v(j_0, j_1, \ldots, j_{n-1}, j_n)^{\#}$ is true®.

Applying (CE_0) to $w_0(x_0) = \exists x_1 \ldots \exists x_{n-1}\ \exists x_n\ v(x_0, x_1, \ldots, x_{n-1}, x_n)$, I find j_0 such that $w_0(j_0)^{\#} = (\exists x_1 \ldots \exists x_{n-1}\ \exists x_n\ v(j_0, x_1, \ldots, x_{n-1}, x_n))^{\#}$ is true®. Assume now that I have found constants $j_0, j_1, \ldots, j_{n-2}$ such that for $w_{n-1}(x_{n-1}) = \exists x_n\ v(j_0, j_1, \ldots, j_{n-2}, x_{n-1}, x_n)$ the formula $w_{n-1}(x_{n-1})^{\#}$ is true®. Applying (CE_0) to it, I find I find j_{n-1} such that $w_{n-1}(j_{n-1})^{\#} = v(j_0, j_1, \ldots, j_{n-1}, x_n)^{\#}$ is true®. Applying (CE_0) to $w_n(x_n) = v(j_0, j_1, \ldots, j_{n-1}, x_n)$, I find j_n such that $w_n(j_n)^{\#} = v(j_0, j_1, \ldots, j_{n-1}, j_n)^{\#}$ is true®.

(CA_0) Let x be free in v and let v_0 be $\forall x\ v(x)$. If $v_0{}^{\#}$ is true® then for every constant c_m also $v(c_m)^{\#}$ is true®.

If $v_0{}^{\#} = \neg\,(\exists x\ \neg v(x))^{\#} = \neg\ R_x((\neg v(x))^{\#})$ is true® then $R_x((\neg v(x))^{\#})$ is not true®. Then by C($(\neg v(x))^{\#}$) for every c_m the formula rep(x, c_m | $(\neg v(x))^{\#}$) is not true®. But this is $(\mathrm{rep}(x, c_m \mid \neg v(x))^{\#} = (\neg v(c_m))^{\#} = \neg\ v(c_m)^{\#}$, and as it is not true®, $v(c_m)^{\#}$ is true®. - Finally, in analogy to (CE_1) there holds

(CA_1) Let $x_0, x_1, \ldots, x_{n-1}, x_n$ be free in v and let v_1 be the formula $\forall x_0\ \forall x_1 \ldots \forall x_{n-1}\ \forall x_n\ v(x_0, x_1, \ldots, x_{n-1}, x_n)$. If $v_1{}^{\#}$ is true® then for all constants $j_0, j_1, \ldots j_{n-1}, j_n$ also $v(j_0, j_1, \ldots, j_{n-1}, j_n)^{\#}$ is true®.

I shall now supplement Lemma 4 with a *completeness* result :

LEMMA 5 A formula v is provable from C_1 if, and only if, $v^{\#}$ is true®. If $v^{\#}$ is true® then a deduction S of v from C_1 can be exhibited.

I observed already above that the provability of v entails the truth® of $v^{\#}$. by Lemma 4. Assume now, conversely, that $v^{\#}$ is true®. In the special case that v is a sentence, $v^{\#}$ is an open sentence, and this being true®, I can find a proof of $v^{\#}$ by Lemma 2; hence together with $v^{\#}$ also v has a proof by Lemma 3(i). In the general case that v contains free variables, I suffices to find a proof of $v_1 = (\forall)v$ which then leads to a proof of v. As v_1 is equivalent to $\neg v_2$ with $v_2 = (\exists)\neg v$, it will suffice to find a proof of $\neg v_2$ or, in view of Lemma 3(i), a deduction S_1 of $(\neg v_2)^{\#} = \neg v_2^{\#}$. Together with v_2 also $\neg v_2^{\#}$ is a sentence. Thus $\neg v_2^{\#}$ is an open sentence, and repeating the argument from Lemma 2, I find a propositional deduction S_0 of $\neg v_2^{\#}$ from the disjunctions d_k, $k<m$, in the conjunctive normal form $NC(\neg v_2^{\#})$; here m depends on the complexity of $v_2^{\#}$ and, therefore, the complexity of v. As a first case, assume that, for every k the disjunction d_k contains a member which is true®. I then take, for every k, the first such one, b_{kj}, and as this true® b_{kj} is an atomic sentence or a negated atomic sentence, the argument of Lemma 1 provides a deduction S_{ok} of b_{kj} from C_1; implanting the S_{ok} above the d_k in S_0, I obtain the deduction S_1. If the first case does not prevail, then I define S_1 to be trivial deduction from C_1 of its first axiom.

It follows that S_1 is a deduction from C_1, explicitly defined in a finite number of steps. It remains to see that S_1 deduces $\neg v_2^{\#}$, i.e. that every d_k contains a member which is true®. I shall refute the contrary assumption by an explicit construction based on (CE_1). Because if some d_k would not contain a true® member then d_k would not be true®, thus $\neg v_2^{\#}$ would not be true® and $v_2^{\#}$ would be true®. Employing (CE_1), I would construct, for every variable x_i free in $\neg v$, a constant c_i such that $\mathrm{rep}(\eta \mid \neg v)^{\#} = \neg\, \mathrm{rep}(\eta \mid v)^{\#} = \neg\, \mathrm{rep}(\eta \mid v^{\#})$ would be true® for the map η with $\eta(x_i) = c_i$. Hence $\mathrm{rep}(\eta \mid v^{\#})$ would not be true®, and this contradicts the truth® of $v^{\#}$.

In the following, I shall use the phrase that a deduction can be *exhibited* to refer to the situation of the proof above.

LEMMA 6 Let w be a formula and let x be free in w. If $\vdash \mathrm{rep}(x, \dot{c}_n \mid w)$ for every c_n then a deduction of $\forall x w$ can be exhibited.

If $\mathrm{rep}(x,c_n \mid w)$ is provable then $\mathrm{rep}(x,c_n \mid w)^{\#} = \mathrm{rep}(x,c_n \mid w^{\#})$ is true®. If this holds for every c_n then also $w^{\#}$ is true®. By Lemma 5 then a deduction of w can be exhibited and, therefore, also one of $\forall x w$.

4. The Main Theorems of the Arithmetic C_1

It follows from Lemma 4 that an open sentence v is provable from C_1 if, and only if, it is true®. As an open sentence v is either true® or not true®, it follows that it is either provable or not provable from C_1. But if it is provable, hence true®, then v must have a propositional proof from its atomic sentences or their negations. Hence I have a decision algorithm for the provability of v, testing (1) the truth® of the atomic sentences of v and (2) the truth® of v, respectively the propositional provability of v from those atomic sentences and their negations.

It follows from Lemma 3(i) that an arbitrary sentence v is provable from C_1 if, and only if, the open sentence v^* is so. Hence I have a decision algorithm for the provability of v, forming the sentence v^* and deciding its provability. Thus I have proved the first two properties in

THEOREM 1 The axiom system C_1 is complete, decidable and consistent.

As for the consistency of C_1, if both v and $\neg v$ are provable then Lemma 4 implies that the open sentences v^* and $\neg v^* = (\neg v)^*$ both are true®, and this is a contradiction. (A different argument would be that C_1 cannot prove the sentence $c_0 \equiv c_1$ because it is not true®.)

The consistency of the axiom system S_1 (in section 1) was proved by showing explicitly that no proof tree (in the equivalent calculus XS_1) could end with the empty sequent. In contrast, the consistency proof for C_1 employed the intermediate notion of truth® which, however, also made it possible to prove decidability.

As for the tools employed in the consistency proofs, I have at first those concerning formulas and proof trees. In the case of XS_1 this was the (double) induction on proof lengths required for the cut elimination theorem; in the case of C_1 it were (a) inductions on proof trees in Lemma 4, and (b) recursions on the complexity of formulas for the definition of the reductions R_x and during the proof of Lemma 3. The natural numbers used here to measure proof lengths and complexities are the same *exterior* numbers used to count variables or the constants c_m, and the amount of arithmetic required for them during these syntactical constructions is at least addition, together with schemata for induction and recursion. Addition for the numbers counting the c_m was also employed during the proofs of the properties B(v) and C(v) of the reductions R_x.

But it is possible to deal with the constants c_m without employing exterior numbers; this was observed already in HILBERT 22. To this end, I recall the realizations of terms by finite strings of characters (in Chapter 1.5) and describe the c_m as finite strings generated from two characters *0* and *1* as follows :

> the one-member string *1* is the number® c_0 ,
>
> if the string c_n is a number® then the string, obtained by writing *0* to the right of c_n, is a number®,

and that number® then is called c_{n+1}. So every number® is a string beginning with *1* followed by a *tail* consisting entirely of repetitions of *0*. For numbers® c_n and c_m I define $c_n \prec c_m$ if c_n is a proper initial part (read from the left) of c_m; this is equivalent to c_n being shorter than c_m. I define the string $c_n +^{®} c_m$ as that obtained by writing the tail of c_m to the right of c_n; induction on the generation of c_m then shows that $c_n +^{®} c_m$ also is a number® c_{n+m}. It is easily seen by induction on the generation of numbers® that the usual properties of order and addition, as well as principles of induction, hold for (the indices of) these numbers®. This is what is needed if the c_m are to be used as exterior numbers for syntactical constructions.

Moreover, identity of strings is decidable, and induction on the generation of numbers® shows that the relation $c_n \prec c_m$ is decidable. Making the set C of all constants into a structure® with the relations $=$ and $\prec$, the sentences $c_n \equiv c_m$ and $c_n < c_m$ are true® if, and only if, they are true in the usual sense with respect to the structure® of constants. As far as this structure® is used to reduce truth® to usual truth, I do not need addition, nor have I used an induction principle. While the notion of truth® for open sentences is decidable, truth® for open formulas v with free variables requires knowledge about the behaviour of *all* possible sentences rep($\eta\,|\,$v). However, during the proofs of Lemma 4 and Lemma 5 I have used truth® for open formulas only in cases where it could be explicitly verified; I have *not* used that open formulas with free variables be either true® or not true®.

It follows from (CE_0) that if $\exists$x v(x) is provable then I can find a witness c_m such that $v(c_m)^{\#}$ is true® and, therefore, provable as well. In this sense provability from C_1 can be *numerically realized*. More generally, for a prenex formula

$$\forall x_0\, \exists x_1\, \forall x_2\, \exists x_3 \ldots \forall x_{2n}\, \exists x_{2n+1}\, v(x_0, x_1, \ldots, x_{2n}, x_{2n+1})$$

there holds

> for every j_0: there can be found a j_1:
> for every j_2: there can be found a j_3: ...
> ... for every j_{2n}: there can be found a j_{2n+1}:
> $v(j_0, j_1, j_2, j_3, \ldots, j_{2n}, j_{2n+1})^{\#}$ is true® .

It is a further interesting fact about C_1 that it admits the *induction rule*, stating for every formula v

(IRA) $\quad v(c_0),\ v(x) \rightarrow v(x+1) \vdash v(y)$

where v(t) abbreviates rep(x,c_0 | v). The variable x is called the *eigenvariable* of that rule; about y nothing needs to be specified. I shall show that, given deductions of the premisses of (IRA), a deduction of its conclusion can be exhibited.

Because together with a formula w also ∀xw is provable, and since ζ(x, c_n) is free for w also every rep(x, c_n | w) is provable. Thus if v(x) → v(x+1) is provable then so is every v(c_n) → v(s(c_n)) = v(c_n) → v(c_{n+1}). Under the hypotheses of (IRA) therefore, every v(c_n) is provable: this is assumed for v(c_0), and has it been shown for v(c_n) then together with v(c_n) → v(c_{n+1}) also v(c_{n+1}) has a deduction. This being established, it follows from Lemma 6 that a deduction of ∀xv can be exhibited, hence also one of v(y).

More on the induction rule will be said in section 4 of the following Chapter.

Chapter 2. Consistency, Decidability, Completeness for the Arithmetic of Addition and Order

1. The Arithmetic C_2 of Addition and Order

Let L be a language with the constants c_0, c_1, the 2-ary operation symbol + and predicate symbols $\equiv$ and $<$, and with the abbreviation $x \lesssim y$ for $x<y \vee x\equiv y$. Let C_{02} be the set of axioms

c1.	$\neg\, x<x$	
c2.	$x<y \wedge y<z \rightarrow x<z$	
c3.	$x<y \vee y<x \vee x\equiv y$	
c4.	$c_0<x \vee c_0\equiv x$	
$c4_1$.	$c_0<c_1$	
c6.	$x<y \rightarrow x+c_1 \lesssim y$	(successor)
a1.	$x+(y+z)\equiv(x+y)+z$	
a2.	$x+y\equiv y+x$	
a3.	$x+c_0\equiv x$	
a4.	$x+y\equiv x+z \rightarrow y\equiv z$	(cancellation)
b1.	$y<z \rightarrow x+y<x+z$	(monotonicity)
b2.	$y<z \rightarrow \exists x\, y+x\equiv z$.	

The following three sections are devoted to the proof that an axiom system C_2, obtained from C_{02} by adding axioms for division with remainder, is complete, decidable and consistent. On the approach to this result, a certain extension C_2^* of C_{02} will be introduced which is formulated in an extension L^* of L by certain defined predicate symbols.

C_{02} is semantically incomplete, hence also incomplete: neither the sentence $v = \forall x \exists y\, (x\equiv y+y \vee x+1= y+y)$ nor its negation is a consequence of C_{02}. Because let N be the standard model of C_{02} with the ordered semigroup of natural numbers; let A be the semigroup N×N on which the order is defined lexicographically and in which c_1 is interpreted by the pair [0,1]. Then A is a model of C_{02}, but v does not hold because $[1,a] = [b,c]+[b,c]$ cannot be satisfied with any b, hence neither can $[1,a]+[0,1] = [b,c]+[b,c]$.

Provability, for the moment, shall mean provability *from* C_{02} in a calculus $ccqt_=$; $\vdash v$ shall denote that a formula v is provable.

In Chapter 1, section 2, the proofs of the statements 0 to 2 require only the axioms c1 to c3; hence 1.2.0–2 are available here. If I define $s(x)$ as $x+c_1$, then the axioms c6 of C_{01} and C_{02} coincide, and c5 is provable from C_{02} by $c4_1$, b1 and a3 ; hence 1.2.3–6 are available as well. Also, in view of associativity, it will be convenient to abbreviate terms as, e.g., $x+y+z$

when $x+(y+z)$ or $(x+y)+z$ is meant. There are various formulas provable from C_{02} with familiar arguments:

0a. $c_0+x \equiv x$ by a3, a2 .
0b. $y<z \rightarrow y+x<z+x$ by b1, a2 .

1. $\vdash y<z \leftrightarrow \exists x\,(y+x \equiv z \wedge \neg\, x \equiv c_0)$.

First, $\vdash y+x \equiv z \wedge x \equiv c_0 \rightarrow y \equiv z$ by a3, $\vdash y+x \equiv z \wedge \neg\, y \equiv z \rightarrow \neg\, x \equiv c_0$, but $\vdash y<z \rightarrow \neg\, y \equiv z$ whence $\vdash y<z \rightarrow \exists x\,(y+x \equiv z \wedge \neg\, y \equiv z)$ by b2. Thus I can derive the implication from left to right. On the other hand, $\vdash \neg\, x \equiv c_0 \rightarrow c_0<x$ by c4 and $\vdash c_0<x \rightarrow c_0+y<x+y$ by b1, hence also $\vdash y+x \equiv z \wedge \neg\, x \equiv c_0 \rightarrow y<z$.

2. $\vdash \exists x\,(y+x \equiv z \vee z+x \equiv y)$.

Because $\vdash y+c_0 \equiv z \vee \exists x\, y+x \equiv z \vee \exists x\, z+x \equiv y$ by c3 and 1, and there holds $\vdash (\exists x\, v \vee \exists x\, w) \rightarrow \exists x\,(v \vee w)$.

3. $\vdash x+y < x+z \rightarrow y<z$.

Because $\vdash x+y+r \equiv x+z \rightarrow y+r \equiv z$, $\vdash x+y+r \equiv x+z \rightarrow \exists r\; y+r \equiv z$ whence also $\vdash \exists r\; x+y+r \equiv x+z \rightarrow \exists r\; y+r \equiv z$.

I define $c_m = c_{m-1}+c_1$ for integers m with $m>1$; clearly $\vdash c_0<c_m$ for $m>0$. As in section 1.2, I define $t+0 = t$ and $t+(m+1) = s(t+m)$; then the proofs of the formulas 1.2.7 to 1.2.11 remain in effect.

4. $\vdash x+m \equiv x+c_m$.

This is clear for $m=0$ and holds for $m+1$ since $\vdash x+(m+1) \equiv (x+m) + c_1$ and $\vdash x+c_{m+1} = x+(c_m+c_1) \equiv (x+c_m)+c_1$ by associativity.

5. $\vdash c_{m+n} \equiv c_m+c_n$.

Because $\vdash c_{m+n} \equiv c_0+(m+n) \equiv (c_0+m)+n \equiv c_m+n \equiv c_m+c_n$ by 4.

6. $\vdash \neg\, c_m \equiv c_{m+n}$ and $\vdash c_m<c_{m+n}$ for $n>0$.

This follows from 4 and 1.2.9, 1.2.10 .

7. $\vdash c_m<c_n$ and $\vdash \neg\, c_n<c_m$ and $\vdash \neg\, c_m \equiv c_n$ for $m<n$.

8. $\vdash c_0<y<c_n \rightarrow \mathbb{W}\langle y \equiv c_i \mid 0<i<n\rangle$.

This holds for $n=1$ since $\vdash c_0<y \rightarrow c_1 \leq y$ by c6. But $\vdash y<c_{n+1} \rightarrow (y<c_n \vee y \equiv c_{n+1})$ by 1.2.3. So I may proceed by induction on n.

I define $0 \cdot x = c_0$ and $1 \cdot x = x$ and, for $m>0$, $(m+1) \cdot x = m \cdot x + x$. Then

9. $\vdash (m+n) \cdot x \equiv m \cdot x + n \cdot x$.

This follows for every m and $n=0$ from $0 \cdot x = c_0$; for for every m and $n=1$ it holds by definition, and if it holds for m and n then $\vdash (m+(n+1)) \cdot x =$ $= ((m+n)+1) \cdot x \equiv (m+n) \cdot x + x \equiv m \cdot x + n \cdot x + x \equiv m \cdot x + (n+1) \cdot x$.

10a. $\vdash x \equiv y \leftrightarrow m \cdot x \equiv m \cdot y$ for $m>0$ by induction with 0a .
10b. $\vdash x<y \leftrightarrow m \cdot x < m \cdot y$ for $m>0$ by induction with 0b .

11. $\vdash n \cdot (x+y) \equiv n \cdot x + n \cdot y$.

For $n=0$ this is $\vdash c_0 \equiv c_0 + c_0$; if it holds for n then $\vdash (n+1) \cdot (x+y) = n \cdot (x+y) + x+y \equiv (n \cdot x + n \cdot y) + x+y \equiv (n \cdot x + x) + (n \cdot y + y) \equiv (n+1) \cdot x + (n+1) \cdot y$ by a1, a2.

12. $\vdash n \cdot c_1 = c_n$ and $\vdash n \cdot c_0 \equiv c_0$.

Since $0 \cdot x = c_0$ and $(n+1) \cdot c_1 = n \cdot c_1 + c_1 = c_n + c_1 = c_{n+1}$, $\vdash (n+1) \cdot c_0 = n \cdot c_0 + c_0 = c_0 + c_0 \equiv c_0$.

13. $\vdash n \cdot c_k \equiv c_{nk}$ for $k \geq 0$.

For $k=0$, hence $nk=0$, $\vdash n \cdot c_0 \equiv c_0$ by 12. If $\vdash n \cdot c_k \equiv c_{nk}$ then $\vdash c_{n(k+1)} = c_{nk+n} \equiv c_{nk} + c_n \equiv n \cdot c_k + n \cdot c_1 \equiv n \cdot (c_k + c_1) = n \cdot c_{k+1}$ by 5, 12 and 11.

14. If $n>1$ and $k>1$ then $\vdash n \cdot x \equiv c_k \rightarrow \mathbb{V}\langle x \equiv c_i \mid 0<i<k \textit{ and } ni=k\rangle$

and $\vdash n \cdot x \equiv c_k \rightarrow \neg \mathbb{V}\langle x \equiv c_i \mid 0<i \leq k \textit{ and } ni \neq k\rangle$,

and if C_{02} is consistent then the first disjunction is not empty.

First, $\vdash x \equiv c_0 \rightarrow n \cdot x \equiv c_0$ by induction, hence $\vdash c_0 < n \cdot x \rightarrow c_0 < x$, and so $c_0 < c_k$ leads to $\vdash n \cdot x \equiv c_k \rightarrow c_0 < x$ for $k>0$.

Next, the second disjunction is not empty since $nk \neq k$. Also, as $\vdash n \cdot c_i \equiv c_{ni}$ by 13, there follows for $ni \neq k$ that $\vdash \neg n \cdot c_i \equiv c_k$ by 7, hence $\vdash n \cdot x \equiv c_k \rightarrow \neg n \cdot x \equiv n \cdot c_i$ and $\vdash n \cdot x \equiv c_k \rightarrow \neg x \equiv c_i$. Consequently, $\vdash n \cdot x \equiv c_k \rightarrow \mathbb{A}\langle \neg x \equiv c_i \mid 0<i \leq k \textit{ and } ni \neq k\rangle$ and

$$\vdash n \cdot x \equiv c_k \rightarrow \neg \mathbb{V}\langle x \equiv c_i \mid 0<i \leq k \textit{ and } n \cdot i \neq k\rangle .$$

Further, $\vdash c_k < x \rightarrow c_k < n \cdot x$ for $n>0$, hence $\vdash c_k \equiv n \cdot x \rightarrow x \leq c_k$. Thus $\vdash c_k \equiv n \cdot x \rightarrow \mathbb{V}\langle x \equiv c_i \mid 0<i \leq k\rangle$ by 8, whence

$$\vdash n \cdot x \equiv c_k \rightarrow \mathbb{V}\langle x \equiv c_i \mid 0<i \leq k \textit{ and } ni=k\rangle \vee \mathbb{V}\langle x \equiv c_i \mid 0<i \leq k \textit{ and } ni \neq k\rangle .$$

If the first disjunction is empty then C_{02} is inconsistent. If it is not empty, then the tautology $(p \vee q \wedge \neg q) \rightarrow p$ implies

$$\vdash n \cdot x \equiv c_k \rightarrow \mathbb{V}\langle x \equiv c_i \mid 0<i<k \textit{ and } ni=k\rangle .$$

While the only terms not containing variables are the c_m, there holds

15. Let t be a term and let $x_0, \ldots, x_{k-1}$ be the variables occurring in t. Then there are positive numbers $a_0, \ldots, a_{k-1}$, and a number b such that

$$\vdash t \equiv a_0 \cdot x_0 + \ldots + a_{k-1} \cdot x_{k-1} + c_b .$$

The presentation holds for c_0 with $k=b=0$, for c_1 with $k=0$, $b=1$, for x_i with $a_i=1$, $b=0$. It is preserved under the operation $+$.

16. Let t be a term. There is a term t_x not containing x and an integer a such that $\vdash t \equiv t_x + a \cdot x$; x occurs in t if, and only if, a = 0.

It follows from 7 that the atomic sentences $c_m \equiv c_m$ and $c_m < c_n$ with $m < n$ are provable, that $\neg\, c_m \equiv c_n$ is provable for $m \neq n$ and that $\neg\, c_n < c_m$ is provable for not $n < m$. Thus C_{02} is decidable for atomic sentences.

In section 1.2, I extended the incomplete axiom system C_{01} to C_1, adding the axiom c7 which permitted to prove a property of limited *subtraction*. This was essential for the reductions (ac) and (af) in section 1.3 and made it possibile to define a quantifier elimination map. Here now the incomplete axiom system C_{02} shall be similarly extended to an axiom system C_2^* which admits a quantifier elimination map. While in the earlier situation it was, among others, the reductions $R_x(x+n \equiv t)$ which required limited subtraction, the analogous reductions $R_x(n \cdot x \equiv t)$ here will require properties of limited *division*.

Let C_2 be the axiom system obtained by enlarging C_{02} with the axioms

a_{6n}. $\exists z \bigvee \langle x \equiv n \cdot z + c_i \mid i < n \rangle$ for every n with $n > 1$;

this is the principle of division with remainder. C_2 may be called the *Presburger arithmetic* of *natural numbers*, although PRESBURGER 30 studied the simpler case of an axiom system for integers and formulated it without employing the notion of order (but with infinitely many axioms about multiples $m \cdot x$).

Yet matters are more complicated here than in Chapter 1, and I still am not able to find a quantifier elimination map for C_2. It will also be necessary to extend my language L by new predicate symbols in order, e.g., to define reductions $R_x(n \cdot x \equiv t)$. So the discussion of C_2 will delayed until section 3.

I first enlarge my language L to a language $L^\dagger$, introducing, for every n with $n > 1$, a 1-ary predicate symbol $n|$ with the definition

$a_{5n}^\dagger$. $n|y \leftrightarrow \exists z_0\, n \cdot z_0 \equiv y$.

Observe that $n \cdot z_1 \equiv y \rightarrow \exists z_0\, n \cdot z_0 \equiv y$ is an axiom whence $\vdash \exists z_1\, n \cdot z_1 \equiv y \rightarrow \exists z_0\, n \cdot z_0 \equiv y$; thus $n|y$ is also equivalent to any $\exists z_1\, n \cdot z_1 \equiv y$ with $z_1 \neq y$.

I next enlarge the language L to a language L^*, introducing, for every n with $n > 1$, a 2-ary predicate symbol $\approx[n]$ with the definition

a_{5n}^*. $x \approx[n]\, y \leftrightarrow \exists z_0\, (x + n \cdot z_0 \equiv y \vee x \equiv n \cdot z_0 + y)$;

again, z_0 here may be replaced by any z_1 distinct from x and y. Observe also that, by quantifier logic, $x \approx[n]\, y$ is equivalent to $\exists z_0\, (x + n \cdot z_0 \equiv y) \vee \exists z_0\, (x \equiv n \cdot z_0 + y)$. For L^* now $t \approx[n]\, t'$ will be an atomic formula of L^* for any t, t' (and for every such n). In elementary number theory, two numbers are called *congruent mod*(n) if divided by m they have the same residue, i.e. if they satisfy $x \approx[n]\, y$ in the standard model.

While L^* is the language for which I shall find a quantifier elimination map, I shall sometimes formulate statements in a common extension $L^‡$ of both $L^†$ and L^* by adding $\approx[n]$ and $n|$ respectively with their respective definitions. For the time being, provability shall remain provability from C_{02} and the defining axioms of my definitorial extensionms. From these then it follows that

$$\vdash\ x \approx[n]\, y$$
$$\leftrightarrow\ (x \leq y \wedge \exists z_0\, (x+z_0 \equiv y \wedge n|z_0)) \vee (y \leq x \wedge \exists z_0\, (y+z_0 \equiv x \wedge n|z_0))$$

and

$$\vdash\ n|y\ \leftrightarrow\ c_0 \approx[n]\, y\ .$$

I first observe

17a. $\vdash x \approx[n]\, x$.
17b. $\vdash x \approx[n]\, y\ \rightarrow\ y \approx[n]\, x$.

18. $\vdash x \approx[n]\, y\ \wedge\ y \approx[n]\, z\ \rightarrow\ x \approx[n]\, z$.

The proof requires to distinguish four cases.

a.
$$\vdash x+n\cdot z_0 \equiv y\ \wedge\ y+n\cdot z_1 \equiv z\ \rightarrow\ x+n\cdot(z_0+z_1) \equiv z$$
$$\vdash x+n\cdot z_0 \equiv y\ \wedge\ y+n\cdot z_1 \equiv z\ \rightarrow\ \exists z_2\, x+n\cdot z_2 \equiv z$$
$$\vdash x+n\cdot z_0 \equiv y\ \wedge\ \exists z_1\, y+n\cdot z_1 \equiv z\ \rightarrow\ \exists z_2\, x+n\cdot z_2 \equiv z$$
$$\vdash \exists z_0\, (x+n\cdot z_0 \equiv y)\ \wedge\ \exists z_1\, (y+n\cdot z_1 \equiv z)\ \rightarrow\ \exists z_2\, (x+n\cdot z_2 \equiv z)\ ,$$

b.
$$\vdash x \equiv n\cdot z_0+y\ \wedge\ y \equiv n\cdot z_1+z\ \rightarrow\ x \equiv n\cdot(z_0+z_1)+z$$
$$\vdash \exists z_0\, (x \equiv n\cdot z_0+y)\ \wedge\ \exists z_1\, (y \equiv n\cdot z_1+z)\ \rightarrow\ \exists z_2\, (x \equiv n\cdot z_2+z)\ ,$$

c.
$$\vdash z_0+z_2 \equiv z_1\ \wedge\ x+n\cdot z_0 \equiv y\ \wedge\ y \equiv n\cdot z_1+z$$
$$\rightarrow\ x+n\cdot z_0 \equiv n\cdot z_0+n\cdot z_2+z$$
$$\vdash z_0+z_2 \equiv z_1\ \wedge\ x+n\cdot z_0 \equiv y\ \wedge\ y \equiv n\cdot z_1+z\ \rightarrow\ x \equiv n\cdot z_2+z$$
$$\vdash z_0+z_2 \equiv z_1\ \wedge\ x+n\cdot z_0 \equiv y\ \wedge\ y \equiv n\cdot z_1+z\ \rightarrow\ \exists z_2\, x \equiv n\cdot z_2+z$$
$$\vdash z_0 \leq z_1\ \wedge\ x+n\cdot z_0 \equiv y\ \wedge\ y \equiv n\cdot z_1+z\ \rightarrow\ \exists z_2\, x \equiv n\cdot z_2+z$$

$$\vdash z_0 \equiv z_2+z_1\ \wedge\ x+n\cdot z_0 \equiv y\ \wedge\ y \equiv n\cdot z_1+z$$
$$\rightarrow\ x+n\cdot z_2+n\cdot z_1 \equiv n\cdot z_1+z$$
$$\vdash z_0 \equiv z_2+z_1\ \wedge\ x+n\cdot z_0 \equiv y\ \wedge\ y \equiv n\cdot z_1+z\ \rightarrow\ x+n\cdot z_2 \equiv z$$
$$\vdash z_0 \equiv z_2+z_1\ \wedge\ x+n\cdot z_0 \equiv y\ \wedge\ y \equiv n\cdot z_1+z\ \rightarrow\ \exists z_2\, x+n\cdot z_2 \equiv z$$
$$\vdash z_1 \leq z_0\ \wedge\ x+n\cdot z_0 \equiv y\ \wedge\ y \equiv n\cdot z_1+z\ \rightarrow\ \exists z_2\, x+n\cdot z_2 \equiv z$$

$$\vdash x+n\cdot z_0 \equiv y\ \wedge\ y \equiv n\cdot z_1+z$$
$$\rightarrow\ \exists z_2\, (x \equiv n\cdot z_2+z)\ \vee\ \exists z_2\, (x+n\cdot z_2 \equiv z)$$
$$\vdash \exists z_0\, (x+n\cdot z_0 \equiv y)\ \wedge\ \exists z_1\, (y \equiv n\cdot z_1+z)$$
$$\rightarrow\ \exists z_2\, (x \equiv n\cdot z_2+z)\ \vee\ \exists z_2\, (x+n\cdot z_2 \equiv z)\ ,$$

d.
$$\vdash z_0+z_2 \equiv z_1\ \wedge\ x \equiv n\cdot z_0+y\ \wedge\ y+n\cdot z_1 \equiv z$$
$$\rightarrow\ x+n\cdot z_2 \equiv y+n\cdot z_0+n\cdot z_2 \equiv z$$
$$\vdash z_0 \leq z_1\ \wedge\ x \equiv n\cdot z_0+y\ \wedge\ y+n\cdot z_1 \equiv z\ \rightarrow\ \exists z_2\, (x+n\cdot z_2 \equiv z)$$

$$\vdash z_0 \equiv z_2+z_1 \;\wedge\; x \equiv n\cdot z_0+y \;\wedge\; y+n\cdot z_1 \equiv z$$
$$\rightarrow \quad x \equiv y+n\cdot z_1+n\cdot z_2 \equiv n\cdot z_2+z$$
$$\vdash z_1 \leq z_0 \;\wedge\; x \equiv n\cdot z_0+y \;\wedge\; y+n\cdot z_1 \equiv z \quad\rightarrow\quad \exists z_2\,(x \equiv n\cdot z_2+z)$$
$$\vdash x \equiv n\cdot z_0+y \;\wedge\; y+n\cdot z_1 \equiv z$$
$$\rightarrow \quad \exists z_2\,(x+n\cdot z_2 \equiv z) \;\vee\; \exists z_2\,(x \equiv n\cdot z_2+z)$$
$$\vdash \exists z_0\,(x \equiv n\cdot z_0+y) \;\wedge\; \exists z_1\,(y+n\cdot z_1 \equiv z)$$
$$\rightarrow \quad \exists z_2\,(x+n\cdot z_2 \equiv z) \;\vee\; \exists z_2\,(x \equiv n\cdot z_2+z) .$$

19. $\vdash x \approx[n]\, y \;\longleftrightarrow\; x+z \approx[n]\, y+z$.

$$x+n\cdot z_0 \equiv y \;\rightarrow\; x+z+n\cdot z_0 \equiv y+z$$
$$x+n\cdot z_0 \equiv y \;\rightarrow\; \exists z_0\, x+z+n\cdot z_0 \equiv y+z$$
$$\exists z_0\,(x+n\cdot z_0 \equiv y) \;\rightarrow\; \exists z_0\,(x+z+n\cdot z_0 \equiv y+z) ,$$

$$x+z+n\cdot z_0 \equiv y+z \;\rightarrow\; x+n\cdot z_0 \equiv y$$
$$x+z+n\cdot z_0 \equiv y+z \;\rightarrow\; \exists z_0\, x+n\cdot z_0 \equiv y$$
$$\exists z_0\,(x+z+n\cdot z_0 \equiv y+z) \;\rightarrow\; \exists z_0\,(x+n\cdot z_0 \equiv y) ,$$

$$x \equiv \;+n\cdot z_0+y \;\rightarrow\; x+z \equiv \;+n\cdot z_0+y+z$$
$$\exists z_0\,(x \equiv \;+n\cdot z_0+y) \;\rightarrow\; \exists z_0\,(x+z \equiv \;+n\cdot z_0+y+z)$$

$$x+z \equiv \;+n\cdot z_0+y+z \;\rightarrow\; x \equiv \;+n\cdot z_0+y$$
$$\exists z_0\,(x+z \equiv \;+n\cdot z_0+y+z) \;\rightarrow\; \exists z_0\,(x \equiv \;+n\cdot z_0+y) .$$

20. $\vdash x \approx[n]\, x+n\cdot y$.

21. $\vdash x \approx[n]\, y \;\longleftrightarrow\; m\cdot x \approx[mn]\, m\cdot y$ for $m>0$.

Because e.g. $\vdash x+n\cdot z_0 \equiv y \;\longleftrightarrow\; m\cdot x+mn\cdot z_0 \equiv m\cdot y$.

22. For $n>1$ and $k\geq 0$: If n divides k then $\vdash \exists z_0\, n\cdot z_0 \equiv c_k$;
if n does not divide k then $\vdash \neg \exists z_0\, n\cdot z_0 \equiv c_k$.

If $ni=k$ then $\vdash n\cdot c_i \equiv c_{ni} \equiv c_k$ by 13. If there is no i such that $ni=k$ then the formula $\bigvee\langle x \equiv c_i \mid 0<i\leq k$ *and* $n\cdot i \neq k\rangle$ becomes $\bigvee\langle x \equiv c_i \mid 0<i\leq k\rangle$ and, in particular, is not empty. Thus by the second part of 14

$$\vdash n\cdot y \equiv c_k \;\rightarrow\; \neg \bigvee\langle y \equiv c_i \mid 0<i\leq k\rangle .$$

But also $\vdash n\cdot y \equiv c_k \;\rightarrow\; \bigvee\langle y \equiv c_i \mid 0<i\leq k\rangle$ by 8. Hence $\vdash n\cdot y \equiv c_k \rightarrow \neg\, c_0 \equiv c_0$ and $\vdash c_0 \equiv c_0 \;\rightarrow\; \neg\, n\cdot y \equiv c_k$, $\vdash \neg\, n\cdot y \equiv c_k$. Thus also $\vdash \forall y \neg\, n\cdot y \equiv c_k$, $\vdash \neg \exists y \neg\, n\cdot y \equiv c_k$, $\vdash \neg\, n \mid c_k$.

23. For $n>1$ and $p>q\geq 0$: If n divides $p-q$ then $\vdash c_p \approx[n]\, c_q$;
if n does not divide $p-q$ then $\vdash \neg\, c_p \approx[n]\, c_q$.

Setting $k=p-q$, it follows from 22 that $\vdash \exists z_0\, n\cdot z_0 \equiv c_{p-q}$ if n divides k. But $\vdash c_p \equiv c_{p-q}+c_q$ by 5, hence $\vdash \exists z_0\, n\cdot z_0 \equiv c_{p-q} \;\rightarrow\; \exists z_0\, c_p \equiv n\cdot z_0+c_q$ and so $\vdash c_p \approx[n]\, c_q$. If n does not divide k then $\vdash \neg \exists z_0\, n\cdot z_0 \equiv c_{p-q}$, and since $\vdash c_p \equiv n\cdot z_0+c_q \rightarrow n\cdot z_0 \equiv c_{p-q}$, $\vdash \exists z_0\,(c_p \equiv n\cdot z_0+c_q) \;\rightarrow\; \exists z_0\,(n\cdot z_0 \equiv c_{p-q})$, there follows $\vdash \neg \exists z_0\, c_p \equiv n\cdot z_0+c_q$. On the other hand,

$$\vdash c_p + n\cdot z_0 \equiv c_q \;\rightarrow\; (c_p < c_q \vee c_p \equiv c_q) \qquad \text{by } 1$$
$$\vdash \exists z_0\, c_p + n\cdot z_0 \equiv c_q \;\rightarrow\; (c_p < c_q \vee c_p \equiv c_q)$$
$$\vdash \neg\, c_p < c_q \;\wedge\; \neg\, c_p \equiv c_q \qquad \text{by } 7$$
$$\vdash \neg\, \exists z_0\, c_p + n\cdot z_0 \equiv c_q$$

and so $\vdash \neg\, c_p \approx[n]\, c_q$.

24. $\vdash \neg\, x \approx[n]\, y \;\vee\; \neg\, x \approx[n]\, y{+}k$ for $0<k<n$.

Because $\vdash x \approx[n]\, y \wedge x \approx[n]\, y{+}k \;\rightarrow\; y \approx[n]\, y{+}k$ by 17b and 18. Now $y<y{+}k$ implies again

$$\vdash \neg\, \exists z_0\, y \equiv n\cdot z_0 + y{+}k \qquad \text{hence}$$
$$\vdash y \approx[n]\, y{+}k \;\rightarrow\; \exists z_0\, y + n\cdot z_0 \equiv y{+}k$$
$$\vdash y \approx[n]\, y{+}k \;\rightarrow\; \exists z_0\, n\cdot z_0 \equiv c_k\,.$$
$$\vdash \neg\, \exists z_0\, n\cdot z_0 \equiv c_k \;\rightarrow\; \neg\, y \approx[n]\, y{+}k\,.$$

But n does not divide k whence $\vdash \neg\, \exists z_0\, n\cdot z_0 \equiv c_k$ by 22.

25. $\vdash y \lesssim x \wedge x \approx[n]\, y{+}j \;\rightarrow\; y{+}j \lesssim x$ for $j<n$.

If $y \lesssim x$ means $y \equiv x$ then $x \approx[n]\, y{+}j$ becomes $x \approx[n]\, x{+}j$. If $0<j$ then $\vdash \neg\, x \approx[n]\, x{+}j$ by 24, hence $\vdash x \approx[n]\, x{+}j \rightarrow x{+}j \lesssim x$ by *ex absurdo quodlibet*. If $j = 0$ then $\vdash x{+}j \equiv x$. Thus I now may restrict myself to $y<x$.

Clearly, $\vdash x \equiv n\cdot x_1 + y{+}j \;\rightarrow\; y{+}j \lesssim x$; hence it will suffice to show

$$\vdash y<x \;\wedge\; x + n\cdot x_1 \equiv y{+}j \;\rightarrow\; x \equiv y{+}j\,.$$

First, $\vdash x \equiv y{+}z_0 \;\wedge\; x{+}z \equiv y{+}j \;\rightarrow\; z{+}y{+}z_0 \equiv y{+}j$,

$$\vdash x \equiv y{+}z_0 \;\wedge\; x{+}z \equiv y{+}j \;\rightarrow\; z{+}z_0 \equiv c_j\,.$$
$$\vdash x \equiv y{+}z_0 \;\wedge\; x{+}z \equiv y{+}j \;\rightarrow\; z_0 \lesssim c_j\,.$$

Hence by 8

$$\vdash \neg\, z_0 \equiv c_0 \;\wedge\; x \equiv y{+}z_0 \;\wedge\; x{+}z \equiv y{+}j \;\rightarrow\; \mathbb{V}\langle z_0 \equiv c_i \mid 0<i\lesssim j\rangle\,.$$

Since $\vdash z{+}z_0 \equiv c_j \wedge z_0 \equiv c_i \;\rightarrow\; z \equiv c_{j-i}$, there also follows

$$\vdash \neg\, z_0 \equiv c_0 \;\wedge\; x \equiv y{+}z_0 \;\wedge\; x{+}z \equiv y{+}j \;\rightarrow\; \mathbb{V}\langle z \equiv c_{j-i} \mid 0<i\lesssim j\rangle$$
$$\vdash y<x \;\wedge\; x{+}z \equiv y{+}j \;\rightarrow\; \mathbb{V}\langle z \equiv c_i \mid 0\lesssim i<j\rangle$$
$$\vdash y<x \;\wedge\; x{+}z \equiv y{+}j \;\rightarrow\; z \equiv c_0 \vee \mathbb{V}\langle z \equiv c_i \mid 0<i<j\rangle$$
$$\vdash y<x \;\wedge\; x + n\cdot x_1 \equiv y{+}j \;\rightarrow\; n\cdot x_1 \equiv c_0 \vee \mathbb{V}\langle n\cdot x_1 \equiv c_i \mid 0<i<j\rangle$$
$$\vdash y<x \;\wedge\; x + n\cdot x_1 \equiv y{+}j \;\rightarrow\; n\cdot x_1 \equiv c_0 \vee \mathbb{V}\langle \exists x_1\, n\cdot x_1 \equiv c_i \mid 0<i<j\rangle$$
$$\vdash \neg\, n\cdot x_1 \equiv c_0 \;\wedge\; \mathbb{A}\langle \neg\, \exists x_1\, n\cdot x_1 \equiv c_i \mid 0<i<j\rangle \;\rightarrow\; \neg\,(y<x \;\wedge\; x + n\cdot x_1 \equiv y{+}j)\,.$$

It follows from 23 that $\vdash \neg\, c_0 \approx[n]\, c_i$ for $i<n$, hence $\vdash \neg\, \exists x_1\, n\cdot x_1 \equiv c_i$ and $\vdash \mathbb{A}\langle \neg\, \exists x_1\, n\cdot x_1 \equiv c_i \mid 0<i<j\rangle$. Thus

$\vdash \neg\, n\cdot x_1 \equiv c_0 \;\rightarrow\; \neg\,(y<x \wedge x+n\cdot x_1 \equiv y+j)$

$\vdash y<x \wedge x+n\cdot x_1 \equiv y+j \;\rightarrow\; n\cdot x_1 \equiv c_0$

$\vdash y<x \wedge x+n\cdot x_1 \equiv y+j \;\rightarrow\; x \equiv y+j$.

Working in L^*, I shall enlarge C_{02} to C_2^* by adding two series of new axioms, the first of which are the definitions $a_{5n}{}^*$, and then the axioms

$a_{6n}{}^*$. $\quad \vdash x \approx[n]\, y \;\vee\; x \approx[n]\, y+1 \;\vee \ldots \vee\; x \approx[n]\, y+(n-1)$

for every n with $n>0$. It follows from 24 that $a6_n{}^*$ is equivalent to

26. $\vdash \neg\, x \approx[n]\, y \;\longleftrightarrow\; x \approx[n]\, y+1 \vee x \approx[n]\, y+2 \vee \ldots \vee x \approx[n]\, y+(n-1)$.

In L^*, the new predicate symbols $\approx[n]$ give rise to new atomic formulas, and 26 illuminates the technical meaning of the axioms $a6_n{}^*$: they secure that the negations of the new atomic formulas become equivalent to combinations of other atomic formulas.

While C_{02} was finite, C_2^* is infinite, but clearly C_2^* still is algorithmically presented. In the following, $\vdash^*$ shall denote provability from C_2^*.

I observed already above that C_{02} is decidable for the atomic sentences of L. It follows from 23 that C_2^* is decidable for the atomic sentences of L^*, formed with the new predicate symbols $\approx[n]$.

It should be noticed that the observations 14 and 22/23 are the basic tools when setting up proofs from C_2^*. For instance, it follows from 22 that

27. If C_{02} is consistent and $\vdash^* n \mid c_k$ then n divides k .

In KREISEL-KRIVINE 66 it is claimed that

28. If n_0, n_1 are coprime and $\vdash^* n_0 \mid c_k \wedge n_1 \mid c_k$ then $\vdash^* n_0 n_1 \mid c_k$.

This can be proved with help of 27, because then the hypotheses of 28 imply that the coprime numbers n_0, n_1 both divide k whence also their product divides k .

2. Reductions and Quantifier Elimination for C_2^*

The definition of truth®, as given in section 1.3, I extended to the language L^*, adding that

$c_p \approx[n]\, c_q$ is true® if p, q are congruent *mod*(n), and not true® otherwise.

Then the properties of truth®, stated in 1.3, remain in effect for L^*.

I again choose a fixed variable x and assign open formulas $R_x(v)$ to all open formulas v such that the properties A(v)−E(v) from section 1.3 hold; I also require that

F(v) If z is a variable not in v then $R_x(v) = R_z(\mathrm{rep}(x, z \mid v))$.

If the properties A(v)−F(v) hold then I shall call $R_x(v)$ a *reduction* of v. [Actually, F(v) holds also for the reductions in 1.3, but there was no reason to make use of that fact.]

If x does not occur in v then I set again $R_x(v) = v$. So I assume now that v does contain x, and I begin with a few special cases in all of which x shall not be in t_0, t, t_1:

(ca) $v = t_0 < t+x \;:\; R_x(v) = \mathrm{rep}(x, t_0+1 \mid v)$

(cb) $v = t+x < t_0 \;:\; R_x(v) = \mathrm{rep}(x, c_0 \mid v)$

(cc) $v = t_0 < t+x \wedge t+x < t_1 \;:\; R_x(v) = t_0+1 < t_1 \wedge t < t_1$.

F(v) is evident in all these cases since x is not in t_0, t, t_1.

Case (ca) $\vdash \exists x v \to R_x(v)$ follows from $\vdash R_x(v)$ which follows from $\vdash t_0 < t+t_0+1$; this proves A(v). If v contains no variables besides x then t_0, t are constants c_a, c_b; $R_x(v)$ is always true®, and for $\gamma(v) > a$ also $\mathrm{rep}(x, c_{\gamma(v)} \mid v)$ is true®; this proves C(v) and B(v). If $\zeta = \zeta(x, t^*)$ then $R_x(\mathrm{rep}(x, t^* \mid v))$ is $h_\zeta(t_0) < h_\zeta(t) + h_\zeta(t_0)+1$ which is $\mathrm{rep}(x, t^* \mid R_x(v))$. This proves D(v), and E(v) holds analogously.

Case (cb) $\vdash \exists x v \to R_x(v)$ follows from $\vdash t+x < t_0 \to t < t_0$ which follows from c4 . The remainder the proof is analogous to the above.

Case (cc) $\vdash \exists x v \to R_x(v)$ follows from $\vdash (t_0 < t+x \wedge t+x < t_1) \to (t_0+1 < t_1 \wedge t < t_1)$. I prove $\vdash R_x(v) \to \exists x v$ with help of c3 :

$$\vdash t_0 < t \wedge R_x(v) \to t_0 < t+0 \wedge t+0 < t_1 \to \exists x v$$
$$\vdash t_0 \equiv t \wedge R_x(v) \to t_0 < t+1 \wedge t+1 < t_1 \to \exists x v$$

$$\vdash t+x \equiv t_0 \to t+x+1 \equiv t_0+1$$
$$\vdash t+x \equiv t_0 \wedge t_0+1 < t_1 \to t_0 < t+x+1 \wedge t+x+1 < t_1$$
$$\vdash t+x \equiv t_0 \wedge t_0+1 < t_1 \to \exists x v$$
$$\vdash \exists x\, t+x \equiv t_0 \wedge t_0+1 < t_1 \to \exists x v$$
$$\vdash t < t_0 \to \exists x\, t+x \equiv t_0 \qquad \text{by b2}$$
$$\vdash t < t_0 \wedge R_x(v) \to \exists x v .$$

The remainder the proof is analogous to that of the above cases.

(cd_0) $v = t_0 \lesssim t+x \wedge t+x < t_1 \wedge \bigwedge\langle x \approx[n_i]\, e_i \mid i<h\rangle$

where for i with i<h the n_i are natural numbers with $n_i > 1$, the e_i are constants $c_{d(i)}$ with $d(i) < n_i$, and x is not in t_0, t, t_1. Then I choose a fixed common multiple n of the n_i (for instance their product) and set

$$R_x(v) = (t_0 \lesssim t \wedge D_0) \vee (t \lesssim t_0 \wedge D_1)$$

$$D_0 = \bigvee\langle t+j < t_1 \wedge \bigwedge\langle t+j \approx[n_i]\, t+e_i \mid i<h\rangle \mid j<n\rangle$$
$$D_1 = \bigvee\langle t_0+j < t_1 \wedge \bigwedge\langle t_0+j \approx[n_i]\, t+e_i \mid i<h\rangle \mid j<n\rangle .$$

The second part of A(v) is clear as is F(v). Observe that $\vdash^* x \approx[n] y \to x \approx [n_i] y$. I shall first prove $\vdash^* \exists x v \to D_0$ and $\vdash^* \exists x v \to D_1$.

It follows from 25 that for $j<n$ both

$$\vdash^* t{+}x \approx[n]\, t_0{+}j \ \wedge\ t_0 \leq t{+}x \ \wedge\ t{+}x < t_1 \ \to\ t_0{+}j < t_1$$
$$\vdash^* t{+}x \approx[n]\, t{+}j \ \wedge\ t{+}x < t_1 \ \to\ t{+}j < t_1 \,.$$

Thus if t^* is one of t_0 or t then a fortiori

(x) $\quad \vdash^* t{+}x \approx[n]\, t^*{+}j \ \wedge\ t_0 \leq t{+}x \ \wedge\ t{+}x < t_1 \ \to\ t^*{+}j < t_1 \,.$

Now

$$\vdash^* t{+}x \approx[n]\, t^*{+}j \ \wedge\ x \approx[n_i]\, e_i \ \to\ t^*{+}j \approx[n_i]\, t{+}e_i \,,$$
$$\vdash^* t{+}x \approx[n]\, t^*{+}j \wedge \mathbb{A}\langle x \approx[n_i]\, e_i \mid i<h\rangle$$
$$\to\ \mathbb{A}\langle t^*{+}j \approx[n_i]\, t{+}e_i \mid i<h\rangle \,,$$

hence with (x)

$$\vdash^* t{+}x \approx[n]\, t^*{+}j \wedge\ \mathbb{A}\langle x \approx[n_i]\, e_i \mid i<h\rangle \wedge t_0 \leq t{+}x \wedge t{+}x < t_1$$
$$\to\ t^*{+}j < t_1 \ \wedge\ \mathbb{A}\langle t^*{+}j \approx[n_i]\, t{+}e_i \mid i<h\rangle \,.$$

This holds for every $j<n$. Hence

$$\vdash^* t{+}x \approx[n]\, t^*{+}j \wedge \mathbb{A}\langle x \approx[n_i]\, e_i \mid i<h\rangle \wedge t_0 \leq t{+}x \wedge t{+}x < t_1$$
$$\to\ \mathbb{W}\langle t^*{+}j < t_1 \wedge \mathbb{A}\langle t^*{+}j \approx[n_i]\, t{+}e_i \mid i<h\rangle \mid j<n\rangle \,.$$

$$\vdash^* \mathbb{W}\langle t{+}x \approx[n]\, t^*{+}j \mid j<n\rangle \wedge \mathbb{A}\langle x \approx[n_i]\, e_i \mid i<h\rangle$$
$$\wedge\ t_0 \leq t{+}x \wedge t{+}x < t_1$$
$$\to\ \mathbb{W}\langle t^*{+}j < t_1 \wedge \mathbb{A}\langle t^*{+}j \approx[n_i]\, t{+}e_i \mid i<h\rangle \mid j<n\rangle \,.$$

But $\vdash^* \mathbb{W}\langle t{+}x \approx[n]\, t^*{+}j \mid j<n\rangle$ by $a_{6n}{}^*$. Hence

$$\vdash^* \mathbb{A}\langle x \approx[n_i]\, e_i \mid i<h\rangle \wedge t_0 \leq t{+}x \wedge t{+}x < t_1$$
$$\to\ \mathbb{W}\langle t^*{+}j < t_1 \wedge \mathbb{A}\langle t^*{+}j \approx[n_i]\, t{+}e_i \mid i<h\rangle \mid j<n\rangle \,,$$

$$\vdash^* \exists x\, (\mathbb{A}\langle x \approx[n_i]\, e_i \mid i<h\rangle \wedge t_0 \leq t{+}x \wedge t{+}x < t_1)$$
$$\to\ \mathbb{W}\langle t^*{+}j < t_1 \wedge \mathbb{A}\langle t^*{+}j \approx[n_i]\, t{+}e_i \mid i<h\rangle \mid j<n\rangle \,.$$

Taking t^* as t, this is $\exists x v \to D_0$; taking t^* as t_0, it is $\exists x v \to D_1$.

I shall prove next that $\vdash^* t_0 \leq t \wedge D_0 \ \to\ \exists x v$:

$$\vdash^* t{+}j \approx[n_i]\, t{+}e_i \ \to\ c_j \approx[n_i]\, e_i \qquad \text{by 19}$$
$$\vdash^* t_0 \leq t \wedge t{+}j < t_1 \wedge \mathbb{A}\langle t{+}j \approx[n_i]\, t{+}e_i \mid i<h\rangle$$
$$\to\ t_0 \leq t{+}j \wedge t{+}j < t_1 \wedge \mathbb{A}\langle c_j \approx[n_i]\, e_i \mid i<h\rangle$$
$$\vdash^* t_0 \leq t \wedge t{+}j < t_1 \wedge \mathbb{A}\langle t{+}j \approx[n_i]\, t{+}e_i \mid i<h\rangle$$
$$\to\ \exists x\ t_0 \leq t{+}x \wedge t{+}x < t_1 \wedge \mathbb{A}\langle x \approx[n_i]\, e_i \mid i<h\rangle$$
$$\vdash^* t_0 \leq t \wedge \mathbb{W}\langle t{+}j < t_1 \wedge \mathbb{A}\langle t{+}j \approx[n_i]\, t{+}e_i \mid i<h\rangle \mid j<n\rangle$$
$$\to\ \exists x\ t_0 \leq t{+}x \wedge t{+}x < t_1 \wedge \mathbb{A}\langle x \approx[n_i]\, e_i \mid i<h\rangle \,.$$

Finally, I shall prove that $\vdash^* t \leq t_0 \wedge D_1 \ \to\ \exists x v$:

$$\vdash^* t{+}x_1 \equiv t_0 \wedge t_0{+}j{+}n_i{\cdot}x_0 \equiv t{+}e_i \ \to\ t{+}x_1{+}j{+}n_i{\cdot}x_0 \equiv t{+}e_i$$
$$\vdash^* t{+}x_1 \equiv t_0 \wedge t_0{+}j{+}n_i{\cdot}x_0 \equiv t{+}e_i \ \to\ x_1{+}j{+}n_i{\cdot}x_0 \equiv e_i$$

$$\vdash^* t+x_1 \equiv t_0 \wedge t_0+j \equiv n_i \cdot x_0+t+e_i \;\rightarrow\; t+x_1+j \equiv n_i \cdot x_0+t+e_i$$
$$\vdash^* t+x_1 \equiv t_0 \wedge t_0+j \equiv n_i \cdot x_0+t+e_i \;\rightarrow\; x_1+j \equiv n_i \cdot x_0+e_i$$

$$\vdash^* t+x_1 \equiv t_0 \wedge (t_0+j+n_i \cdot x_0 \equiv t+e_i \vee t_0+j \equiv n_i \cdot x_0+t+e_i)$$
$$\rightarrow\; x_1+j+n_i \cdot x_0 \equiv e_i \vee x_1+j \equiv n_i \cdot x_0+e_i$$
$$\vdash^* t+x_1 \equiv t_0 \wedge t_0+j \approx[n_i]\, t+e_i \;\rightarrow\; x_1+j \approx[n_i]\, e_i$$
$$\vdash^* t+x_1 \equiv t_0 \wedge t_0+j<t_1 \wedge t_0+j \approx[n_i]\, t+e_i$$
$$\rightarrow\; t_0+j \equiv t+x_1+j \wedge t_0 \leq t+x_1+j \wedge t+x_1+j<t_1 \wedge x_1+j \approx[n_i]\, e_i$$
$$\vdash^* t+x_1 \equiv t_0 \wedge t_0+j<t_1 \wedge \mathbb{A}<t_0+j \approx[n_i]\, t+e_i \mid i<h>$$
$$\rightarrow\; t_0+j \equiv t+x_1+j \wedge t_0 \leq t+x_1+j \wedge t+x_1+j<t_1$$
$$\wedge\; \mathbb{A}<x_1+j \approx[n_i]\, e_i \mid i<h>$$
$$\vdash^* t \leq t_0 \wedge t_0+j<t_1 \wedge \mathbb{A}<t_0+j \approx[n_i]\, t+e_i \mid i<h>$$
$$\rightarrow\; \exists x\, (t_0 \leq t+x \wedge t+x<t_1 \wedge \mathbb{A}<x \approx[n_i]\, e_i \mid i<h>)$$

whence also

$$\vdash^* t \leq t_0 \wedge \mathbb{V}<t_0+j<t_1 \wedge \mathbb{A}<t_0+j \approx[n_i]\, t+e_i \mid i<h> \mid j<n>$$
$$\rightarrow\; \exists x\, (t_0 \leq t+x \wedge t+x<t_1 \wedge \mathbb{A}<x \approx[n_i]\, e_i \mid i<h>)\,.$$

This completes the proof of A(v). Assume now that v contains no variables besides x. Then t_0, t, t_1 are constants c_m where m is a_0, a, a_1, and e_i is the constant $c_{d(i)}$ with $d(i)<n_i$. Then rep(x, c_k | v) is true$^{\textcircled{N}}$ if, and only if,

$a_0 \leq a+k<a_1$ and, for any $i<h$, k is congruent with d_i *mod*(n_i)

or, equivalently,

(y) $a_0 \leq a+k<a_1$ and, for any $i<h$, $a+k$ is congruent with $a+d_i$ *mod*(n_i).

Assume now that $a_0 \leq a$. If $k=qn+j$ with $j<n$ then $a+j$ is congruent with $a+k$ *mod*(n), and as n is a common multiple of the n_i, this remains congruent *mod*(n_i). Thus (y) implies that $t_0 \leq t \wedge D_0$ is true$^{\textcircled{N}}$ which proves C(v). Conversely, if $t_0 \leq t \wedge D_0$ is true$^{\textcircled{N}}$ then I choose $\gamma(v)$ as the first of the j with $j<n$ for which $t+j<t_1 \wedge \mathbb{A}<t+j \approx[n_i]\, t+e_i \mid i<h>$ is true$^{\textcircled{N}}$; then also rep(x, c_j | v) is true$^{\textcircled{N}}$ which proves B(v).

Assume next that $a \leq a_0$, $a_0 \leq a+k$. If $a+k - a_0 = qn+j$ with $j<n$ then $a+k - a_0$ is congruent j *mod*(n), $a+k$ congruent a_0+j *mod*(n), and so a_0+j will be congruent with $a+d_i$ *mod*(n_i). Then (y) implies that $t \leq t_0 \wedge D_1$ is true$^{\textcircled{N}}$ which proves C(v). Conversely, if $t \leq t_0 \wedge D_1$ is true$^{\textcircled{N}}$ then there is a first $j<n$ such that $t_0+j<t_1 \wedge \mathbb{A}<t_0+j \approx[n_i]\, t+e_i \mid i<h>$ is true$^{\textcircled{N}}$. Choosing $\gamma(v)$ as $(a_0-a)+j$, hence $a_0 \leq a_0+j = a+\gamma(v)<a_1$, there follows (y) for $k=\gamma(v)$ which proves B(v).

If $\zeta = \zeta(y, t^*)$ then rep(y, t^* | v) is

$$h_\zeta(t_0) \leq h_\zeta(t)+x \wedge h_\zeta(t)+x<h_\zeta(t_1) \wedge \mathbb{A}<x \approx[n_i]\, e_i \mid i<h>$$

where $h_\zeta(t_0)$, $h_\zeta(t)$, $h_\zeta(t_1)$ continue not to contain x. Thus R_x(rep(y, t^* | v)) is rep(y, t^* | R_x(v). This proves D(v), and E(v) holds analogously. Again, F(v) is evident.

(cd$_1$) $v = t_0 \lesssim t+x \wedge \bigwedge\!\!\!\bigwedge<x \approx[n_i]\, e_i \mid i<h>$

where for i with $i<h$ the n_i are natural numbers with $n_i>1$, the e_i are constants $c_{d(i)}$ with $d(i)<n_i$, and x is not in t_0, t. Then I choose a fixed common multiple n of the n_i (for instance their product) and set

$$R_x(v) = (t_0 \lesssim t \wedge D_0) \vee (t \lesssim t_0 \wedge D_1)$$

$$D_0 = \bigvee\!\!\!\bigvee<\bigwedge\!\!\!\bigwedge<t+j \approx[n_i]\, t+e_i \mid i<h> \mid j<n>$$
$$D_1 = \bigvee\!\!\!\bigvee<\bigwedge\!\!\!\bigwedge<t_0+j \approx[n_i]\, t+e_i \mid i<h> \mid j<n> .$$

The proof of $A(v)-F(v)$ proceeds by the same arguments as in **(ca$_0$)**, but now the statement (x) is not required and all atomic formulas containing the term t_1 can simply be omitted.

(ce$_0$) $v = \bigwedge\!\!\!\bigwedge<x \approx[n_i]\, e_i \mid i<h>$,

(ce$_1$) $v = t_0<t+x \wedge \bigwedge\!\!\!\bigwedge<x \approx[n_i]\, e_i \mid i<h>$,

(ce$_2$) $v = t+x<t_1 \wedge \bigwedge\!\!\!\bigwedge<x \approx[n_i]\, e_i \mid i<h>$,

(ce$_3$) $v = t_0<t+x \wedge t+x<t_1 \wedge \bigwedge\!\!\!\bigwedge<x \approx[n_i]\, e_i \mid i<h>$.

In **(ce$_0$)** v is equivalent to $c_0<c_1+x \wedge \bigwedge\!\!\!\bigwedge<x \approx[n_i]\, e_i \mid i<h>$ and so is a special case of **(ce$_1$)**. In **(ce$_2$)** v is equivalent to

$$c_0<t+1+x \wedge t+1+x<t_1+1 \wedge \bigwedge\!\!\!\bigwedge<x \approx[n_i]\, e_i \mid 1<i<h>$$

and so is a special case of **(ce$_3$)**. Replacing t_0 by t_0+1 in **(ce$_1$)**, **(ce$_3$)**, these formulas become equivalent to **(cd$_0$)** and **(cd$_1$)**, and so $R_x(v)$ is defined from them.

Consider now the conjunctions v of atomic formulas b_i of the form

(Da) $t_{i0}<t+x$, $t+x<t_{i4}$, $x \approx [n_i]\, e_i$

where t is the same in all b_i of the first two types, the e_i are constants $c_{d(i)}$ with $d(i)<n_i$, and x does not occur in t or in the t_{i0}, t_{i4}. I shall define $R_x(v)$ by recursion on the number of atomic formulas in v.

(da$_0$) $v = t+x<t_{04} \wedge t+x<t_{14}$,

$g = t+x<t_{04} \wedge e$, $h = t+x<t_{14} \wedge e$,

$R_x(v) = t_{04}<t_{14} \wedge R_x(g) \vee t_{14} \lesssim t_{04} \wedge R_x(h)$.

(da$_1$) $v = t_{00}<t+x \wedge t_{10}<t+x \wedge e$,

$g = t_{10}<t+x \wedge e$, $h = t_{00}<t+x \wedge e$,

$R_x(v) = t_{00}<t_{10} \wedge R_x(g) \vee t_{10} \lesssim t_{00} \wedge R_x(h)$.

The proofs that $R_x(v)$ is a reduction are special cases of those of **(bb$_0$)**, **(bb$_1$)** in section 1.3, again together with the fact that F(v) is evident.

It follows that reductions are defined for all conjunctions v of formulas **(Da)**. If they consist of one inequality only then I have one of cases **(ca)**, **(cb)**; if they consist of two inequalities of different type then I have case **(ac)**. If to these situations an arbitrary number of congruences are added, then I have cases **(ce_1)**, **(ce_2)**, **(ce_3)**, and if there are only congruences then this is case **(ce_0)**. In all other situations I have the cases **(da_0)**, **(da_1)**.

Consider next a conjunction v of atomic formulas b_i of the form

(Db) $\quad t_{i0} < t+x \quad , \quad t+x < t_{i4} \quad , \quad x \approx [n_i]\, t_i$

where t is the same in all b_i of the first two types, and x does not occur in t or in the t_{i0}, t_{i4}, t_i. This can be reduced to **(Da)** as follows.

First, $\vdash \mathbb{V}\langle t_i \approx [n_i]\, c_0 + j \mid j < n_i \rangle$ by $a_{6n}{}^*$, hence $\vdash \mathbb{V}\langle t_i \approx [n_i]\, c_j \mid j < n_i \rangle$. Thus

$$\vdash^* x \approx [n_i]\, t_i \;\leftrightarrow\; x \approx [n_i]\, t_i \wedge \mathbb{V}\langle t_i \approx [n_i]\, c_j \mid j < n_i \rangle$$
$$\vdash^* x \approx [n_i]\, t_i \;\leftrightarrow\; \mathbb{V}\langle x \approx [n_i]\, t_i \wedge t_i \approx [n_i]\, c_j \mid j < n_i \rangle$$
$$\vdash^* x \approx [n_i]\, t_i \;\leftrightarrow\; \mathbb{V}\langle x \approx [n_i]\, c_j \wedge t_i \approx [n_i]\, c_j \mid j < n_i \rangle$$

making use of 18 and 17b. Now I apply recursion on the number of formulas occurring in v which are of the third type in **(Db)**, but are not from **(Da)**. So if $v = x \approx [n_i]\, t_i \wedge g$ where g is a conjunction of formulas **(Db)**, then I define

$$R_x(v) = \mathbb{V}\langle R_x(x \approx [n_i]\, c_j \wedge t_i \approx [n_i]\, c_j \wedge g) \mid j < n_i \rangle \; .$$

In order to see that this is a reduction, I abbreviate $b_i = x \approx [n_i]\, t_i$, $b_{ij} = x \approx [n_i]\, c_j \wedge t_i \approx [n_i]\, c_j$. Thus $\vdash b_i \wedge g \;\leftrightarrow\; \mathbb{V}\langle b_{ij} \wedge g \mid j < n_i \rangle$, hence $\vdash \exists x\, b_i \wedge g \;\leftrightarrow\; \mathbb{V}\langle \exists x\, (b_{ij} \wedge g) \mid j < n_i \rangle$. So if $R_x(b_{ij} \wedge g)$ is a reduction then $\vdash \mathbb{V}\langle \exists x\, (b_{ij} \wedge g) \mid j < n_i \rangle \;\leftrightarrow\; \mathbb{V}\langle R_x(b_{ij} \wedge g) \mid j < n_i \rangle$, and by definition this last formula is $R_x(b_i \wedge g)$; this proves A(v). If v contains no variables besides xand if $R_x(b_i \wedge g)$ is true® then $R_x(b_{ij} \wedge g)$ is true® for some j, and if j is minimal then set $\gamma(b_i \wedge g) = \gamma(b_{ij} \wedge g)$. This proves B(v), and the proofs of C(v)–F(v) are just as straightforward.

Consider next a conjunction v of atomic formulas b_i of the form

(Dc) $\quad t_{i0} < t+a\cdot x \quad , \quad t+a\cdot x < t_{i4} \quad , \quad a\cdot x \approx [n_i]\, t_i$.

where $a>0$, t is the same in all b_i of the first two types, and x does not occur in t or in the t_{i0}, t_{i4}, t_i. If $a>1$ then this can be reduced to **(Da)** as follows. Replacing the term $a\cdot x$ by a new variable z, the formula v changes into a formula v_0 with $\mathrm{rep}(z, a\cdot x \mid v_0) = v$; x does not occur in v_0 as it occurs in v only in the form of $a\cdot x$. So $c_0 \approx [a]\, z \wedge v_0$ is of the form **(Db)**, and I define

(dc) $\quad v = \mathrm{rep}(z, a\cdot x \mid v_0) \;:\; R_x(v) = R_z(c_0 \approx [a]\, z \wedge v_0)$.

This definition is independent of the particular choice of z. Because replacing the term $a \cdot x$ by another new variable u, the formula v changes into a formula v_1 with $\text{rep}(u, a \cdot x \mid v_1) = v$. Then $\text{rep}(z, u \mid v_o) = v_1$ and

$$\text{rep}(z, u \mid c_o \approx [a]\, z \;\wedge\; v_o) = c_o \approx [a]\, u \;\wedge\; v_1 \;;$$

thus $F(c_o \approx [a]\, z \;\wedge\; v_o)$ implies also $R_x(v) = R_u(c_o \approx [a]\, u \;\wedge\; v_1)$.

Observe now that by the rules of provability

$$\vdash^* \text{rep}(z, a \cdot x \mid v_o) \;\leftrightarrow\; \exists z\, (z \equiv x \cdot a \wedge v_o)$$

hence

$$\vdash^* \exists x v \;\leftrightarrow\; \exists x\, \exists z\, (z \equiv x \cdot a \;\wedge\; v_o) \;\leftrightarrow\; \exists z\, \exists x\, (z \equiv x \cdot a \;\wedge\; v_o)$$

hence

$$\vdash^* \exists x v \;\leftrightarrow\; \exists z\, ((\exists x\; z \equiv x \cdot a) \;\wedge\; v_o)$$

as x does not occur in v_o. Thus

$$\vdash^* \exists x v \;\leftrightarrow\; \exists z\, (c_o \approx [a]\, z \;\wedge\; v_o)$$

and now A(v) follows from $A(c_o \approx [a]\, z \;\wedge\; v_o)$. Assume next that v contains no variables besides x, hence v_o no variables besides z. As x is not in v_o there holds

$$\text{rep}(x, c_k \mid v) = \text{rep}(x, c_k \mid \text{rep}(z, x \cdot a \mid v_o)) = \text{rep}(z, c_k \cdot a \mid v_o)$$

by Lemma 1.13.9(i). So if $\text{rep}(x, c_k \mid v)$ is true® then so is $\text{rep}(z, a \cdot c_k \mid v_o)$, hence $R_z(c_o \approx [a]\, z \;\wedge\; v_o)$ is true® by $C(c_o \approx [a]\, z \;\wedge\; v_o)$, and this is $R_x(v)$. This proves C(v), and the proof of B(v) is analogous.

As for D(v), I now use that the definition **(dc)** of $R_x(v)$ is independent of the particular z. So I may choose z such that, in addition to being new for v, also $z \neq y$ and z does not occur in t^*; then z is new for $\text{rep}(y, t^* \mid v)$. As x is not in v_o, it will not be in $\text{rep}(y, t^* \mid v_o)$ since it is not in t^*. Also

$$\text{rep}(z, a \cdot x \mid \text{rep}(y, t^* \mid v_o)) = \text{rep}(y, t^* \mid \text{rep}(z, a \cdot x \mid v_o)) = \text{rep}(y, t^* \mid v)$$

by Lemma 1.13.9(i). Thus the definition **(dc)** leads to

$$\begin{aligned} R_x(\text{rep}(y, t^* \mid v)) &= R_x(\text{rep}(z, a \cdot x \mid \text{rep}(y, t^* \mid v_o)) \\ &= R_z(c_o \approx [a]\, z \;\wedge\; \text{rep}(y, t^* \mid v_o)) \end{aligned}$$

by **(dc)**. On the other hand, $D(c_o \approx [a]\, z \;\wedge\; v_o)$ gives

$$\begin{aligned} \text{rep}(y, t^* \mid R_x(v)) &= \text{rep}(y, t^* \mid R_z(c_o \approx [a]\, z \;\wedge\; v_o)\;. \\ &= R_z(\text{rep}(y, t^* \mid c_o \approx [a]\, z \;\wedge\; v_o)) \\ &= R_z(c_o \approx [a]\, z \;\wedge\; \text{rep}(y, t^* \mid v_o))\;. \end{aligned}$$

This proves D(v), and the proof of E(v) is analogous.

This time, also F(v) needs a proof. Recall that x occurs in v only in the terms $a \cdot x$. If p is a new variable then $\text{rep}(x, p \mid v)$ differs from v only by $a \cdot x$ having been changed into $a \cdot p$. If q is a variable new for $\text{rep}(x, p \mid v)$ as well as for v_o, and if $q \neq z$, then replacing $a \cdot p$ by q in $\text{rep}(x, p \mid v)$ gives rise to a formula v_2, and now also $\text{rep}(z, p \mid v_o) = v_2$. Thus

$$R_z(c_o \approx[a]\,z \wedge v_o) = R_p(\text{rep}(z,p\,|\,c_o \approx[a]\,z \wedge v_o))$$
$$= R_p(c_o \approx[a]\,p \wedge \text{rep}(z,p\,|\,v_o))$$
$$= R_p(c_o \approx[a]\,p \wedge v_2).$$

Hence $R_x(v) = R_p(\text{rep}(x, p\,|\,v))$.

It should be noticed that it is this process of homogenization which gives rise to the appearance of new atomic formulas with the predicate symbols $\approx [n]$.

Consider next a conjunction v of atomic formulas of the form

(Dd) $t_{i0} < t_{i1} + a_i \cdot x$, $t_{i3} + a_i \cdot x < t_{i4}$, $a_i \cdot x \approx [n_i]\, t_i$

with $a_i > 0$ where x does not occur in the terms t_{ik} and t_i. This can be reduced to **(Dc)** as follows.

First I replace v by an equivalent formula v_1 in which the a_i are all the same number a. Namely let a be a common multiple of the a_i such that $k_i a_i = a$. Making use of 10, 11, 21, the formulas of the first two types above are equivalent to $k_i \cdot t_{i0} < k_i \cdot t_{i1} + a \cdot x$, $k_i \cdot t_{i4} + a \cdot x < k_i \cdot t_{i3}$ respectively, and those of the third type are equivalent to $a \cdot x \approx [k_i n_i]\, k_i \cdot t_i$.

Next I replace v_1 by a formula v_2 in which the terms t_{i1} and t_{i3} are the same. From 0b and 3 there follow the equivalences

$$\vdash t_{i0} < t_{i1} + a \cdot x \wedge t_{j0} < t_{j1} + a \cdot x$$
$$\leftrightarrow \quad t_{i0} + t_{j1} < t_{i1} + t_{j1} + a \cdot x \wedge t_{j0} + t_{i1} < t_{j1} + t_{i1} + a \cdot x \ ,$$
$$\vdash t_{i0} < t_{i1} + a \cdot x \wedge t_{j3} + a \cdot x < t_{j4}$$
$$\leftrightarrow \quad t_{i0} + t_{j3} < t_{i1} + t_{j3} + a \cdot x \wedge t_{j3} + t_{i1} + a \cdot x < t_{j4} + t_{i1} \ ,$$
$$\vdash t_{i3} + a \cdot x < t_{i4} \wedge t_{j3} + a \cdot x < t_{j4}$$
$$\leftrightarrow \quad t_{i3} + t_{j3} + a \cdot x < t_{i4} + t_{j3} \wedge t_{j3} + t_{i3} + a \cdot x < t_{j4} + t_{i3} \ ,$$

Induction therefore shows that, replacing atomic formulas under these equivalences, the number of inequalities in v_1 with distinct t_{i1} or t_{i3} can be lowered to zero, and the resulting formula then in v_2.

As v_2 is of the form **(Dc)**, I can set $R_x(v) = R_x(v_2)$. Then $\vdash v \leftrightarrow v_2$ and $A(v_2)$ implies $A(v)$. As truth® is preserved under propositional equivalences, B(v), C(v) follow from $B(v_2)$, $C(v_2)$; and D(v)−F(v) coincide with $D(v_2)-F(v_2)$.

Consider next a conjunction v of atomic formulas b_i of the form

(De) $t_{i0} + a_{i0} \cdot x < t_{i1} + a_{i1} \cdot x$, $t_{i0} + a_{i0} \cdot x \approx [n]\, t_{i1} + a_{i1} \cdot x$

with $a_{ik} > 0$ where x does not occur in the terms t_{ik}. This can be reduced to **(Dd)** as follows.

A formula b_i of the first type with $a_{i0} = a_{i1}$ is equivalent to $t_{i0} < t_{i1}$; if

$a_{io}<a_{i1}$ then it is equivalent $t_{io}<t_{i1}+(a_{i1}-a_{io})\cdot x$, and if $a_{io}>a_{i1}$ then it is equivalent $t_{io}+(a_{io}-a_{i1})\cdot x<t_{i1}$.

A formula b_i of the second type with $a_{io}=a_{i1}$ is equivalent to $t_{io} \approx [n]\, t_{i1}$ by 19 ; if $a_{io}<a_{i1}$ then it is equivalent to $t_{i1}+ (a_{i1}-a_{io})\cdot x \approx [n]\, t_{io}$, and if $a_{io}>a_{i1}$ then it is equivalent $t_{io}+ (a_{io}-a_{i1})\cdot x \approx [n]\, t_{i1}$. But a formula $t_i^0+a_i^0\cdot x \approx [n]\, t_i^1$ by 20 is equivalent to $t_i^0+a_i^0\cdot x \approx [n]\, t_i^1+n\cdot t_i^0$ and by 19 to $a_i^0\cdot x \approx [n]\, t_i^1+(n-1)\cdot t_i^0$.

Replacing atomic formulas under these equivalences, v is transformed into an equivalent formula v_2 of the form **(Dd)**, and the same arguments as used there then show that $R_x(v)=R_x(v_2)$ is a reduction.

The case **(De)** covers all conjunctions of atomic formulas formed with the predicate symbols < and $\approx [n]$. A conjunction v of arbitrary atomic formulas may also contain atomic formulas $r\equiv t$; there are equivalent to $r<t+1 \wedge t<r+1$ by 1.2.4. Replacing them under this equivalence, v is transformed into an equivalent formula v_2 of the form **(De)**, and the same arguments as used there then show that $R_x(v)=R_x(v_2)$ is a reduction.

Negated atomic formulas are equivalent to disjunctions of atomic formulas: for $\equiv$ and < this follows from c3, and for $\approx [n]$ it follows from 26. For a conjunction v of atomic formulas and their negations, the reduction $R_x(v)$ then will be defined by supplementing **(bc_0)**, **(bc_1)** of section 1.3 with

$$v = \neg r \approx [n]t \wedge e \;:\; R_x(v) = \mathbb{V}<R_x(r \approx [n]\, t+i \wedge e)\,|\,0<i<n>.$$

The proofs from section 1.3 extend without change.

Finally, reductions for arbitrary open formulas are defined verbally as in section 1.3.

This concludes the definition of reductions.

LEMMA 1 C_2^* admits a quantifier elimination map * satisfying the properties (i) - (iv) in Lemma 3 of Chapter 1.

The definition of the map * and the proof of Lemma 1.3 from section 1.3 remain in effect without change.

LEMMA 2 A provable open formula is true®.

The only change required in the proof of Lemma 4 from section 1.3 concerns the verification that now the axioms of C_2^* are true®. This is evident for the open axioms c1 - b1 and a_{6n}^*. The axiom b2 reduces as follows :

$$y<z \;\rightarrow\; \exists x\; y+x\equiv z$$
$$y<z \;\rightarrow\; \exists x\,(z<x+y+1 \;\wedge\; y+x<z+1)$$

$$y<z \;\rightarrow\; \exists x\,(z+y<x+y+y+1 \;\wedge\; x+y+y+1<z+y+y+1) \qquad \text{by (Dd)}$$
$$y<z \;\rightarrow\; z+1<z+2 \;\wedge\; y+1<z+2 \qquad \text{by (cc)}$$

which is b2$^{\#}$ and is true$^{\circledR}$. The axioms $a_{5n}{}^*$ are equivalent to

$$x \approx[n]\, y \;\longleftrightarrow\; \exists z_0\,(x+n\cdot z_0 \equiv y) \;\vee\; \exists z_0\,(x \equiv n\cdot z_0+y)\,,$$

and $\exists z_0\,(x+n\cdot z_0 \equiv y)$ reduces to

$$\exists z_0\,(y<x+n\cdot z_0+1 \;\wedge\; x+n\cdot z_0<y+1)$$
$$\exists z_0\,(y<n\cdot z_0+x+1 \;\wedge\; n\cdot z_0+x+1<y+2) \qquad \text{by (Dd)}$$
$$\exists u\,(c_0 \approx[n]\, u \;\wedge\; y<u+x+1 \;\wedge\; u+x+1<y+2)$$
$$\exists u\,(c_0 \approx[n]\, u \;\wedge\; y+1 \lesssim u+x+1 \;\wedge\; u+x+1<y+2)$$
$$(y+1 \lesssim x+1 \;\wedge\; \mathbb{W}<x+1+j<y+2 \;\wedge\; x+1+j \approx[n]\, x+1 \mid j<n>)$$
$$\vee\; (x+1 \lesssim y+1 \;\wedge\; \mathbb{W}<y+1+j<y+2 \;\wedge\; y+1+j \approx[n]\, x+1 \mid j<n>)$$

by $(\mathbf{cd_0})$. Write this last formula as $v(x,y) = v_0(x,y) \vee v_1(x,y)$ and its two disjunctions as $d_0(x,y)$, $d_1(x,y)$. If $d_0(c_a, c_b)$ is true$^{\circledR}$ then $j=0$, hence $v_0(c_a, c_b)$ is true$^{\circledR}$ if, and only if, $a=b$. If $d_1(c_a, c_b)$ is true$^{\circledR}$ then $j=0$ and b+1 is congruent a+1 *mod*(n), hence $v_1(c_a, c_b)$ is true$^{\circledR}$ if, and only if, $a \lesssim b$ and a is congruent b *mod*(n).

The second formula, $\exists z_0\,(x \equiv n\cdot z_0+y)$, arises from the previous one by interchanging x and y, hence the same holds for its reduction. Thus for the reduction $w(x,y)$ of the right side of the equivalence $a_{5n}{}^*$ it holds that $w(c_a, c_b)$ is true$^{\circledR}$ if, and only if,

$a=b$
or $a \lesssim b$ and a is congruent b *mod*(n)
$a=b$
or $b \lesssim a$ and b is congruent a *mod*(n)

which is equivalent to a being congruent to b *mod*(n). Consequently, the reduction $x \approx[n]\, y \longleftrightarrow w(x,y)$ of $a_{5n}{}^*$ is true$^{\circledR}$. This completes the proof of Lemma 2.

The observations (CE_0), (CE_1), (CA_0), (CA_1) from section 1.3 remain in effect without change, as do the Lemmata 1.5 and 1.6 which, for $C_2{}^*$, I now may call Lemmata 3 and 4.

3. The Main Theorems of the Arithmetic $C_2{}^*$ and C_2

Without a change in the proof given for Theorem 1.1, there now holds the

THEOREM 1 The axiom system $C_2{}^*$ is complete, decidable and consistent.

As for the tools employed here, those concerning formulas and proof trees remain the same as for Theorem 1.1. For the structure$^{\circledR}$ formed on the numbers$^{\circledR}$ c_m, I now also need their addition $c_n +^{\circledR} c_m$ as defined in section 1.4. Moreover, in order to speak about the truth$^{\circledR}$ of a formula $c_p \approx[n]\, c_q$, I also need integer multiples of the c_m, defined by repeated addition, and the notion of congruence based on it. Explicit properties of multiplication are not required, nor is an induction principle.

With the same argument as used for C_1, also C_2^* admits the induction rule (IRA).

Consider now the axiom system C_2, formulated in the language L and obtained by enlarging C_{o2} with the axioms

a_{6n}. $\quad \exists z\, \mathbb{W}\langle x \equiv n\cdot z + c_i \mid i<n\rangle \qquad$ for every n with $n>1$;

for division with remainder. Dissolving the definition a_{5n}^*, the axiom a_{6n}^* is equivalent to the L-formula

$a_{6n}^{\S}$. $\quad \mathbb{W}\langle \exists z\; x+n\cdot z \equiv y+i \mid i<n\rangle \vee \mathbb{W}\langle \exists z\; x \equiv n\cdot z+y+i \mid i<n\rangle$.

Let C_{12} be C_{o2} together with all a_{5n}^*; I shall show that under C_{12} the formula $a_{6n}^{\S}$ implies a_{6n} and that a_{6n} implies a_{6n}^*; thus all three formulas are equivalent under C_{12}. First, I specialize in $a_{6n}^{\S}$ the variable x to c_o, thus

$$\vdash^* \mathbb{W}\langle \exists z\; x+n\cdot z \equiv c_i \mid i<n\rangle \vee \mathbb{W}\langle \exists z\; x \equiv n\cdot z+c_i \mid i<n\rangle \quad .$$

But

$$\vdash^* x+n\cdot z \equiv c_i \wedge c_i<c_n \rightarrow z \equiv c_o \wedge x \equiv c_i \;,$$
$$\vdash^* x+n\cdot z \equiv c_i \rightarrow x \equiv n\cdot c_o+c_i$$
$$\vdash^* \exists z\; x+n\cdot z \equiv c_i \rightarrow \exists z\; x \equiv n\cdot z+c_i \qquad \text{for } i<n \;.$$

Hence a_{6n}^* proves $\mathbb{W}\langle \exists z\; x \equiv n\cdot z+c_i \mid i<n\rangle$ which is equivalent to a_{6n}. Conversely,

$$\vdash^* x+z_o \equiv y \wedge z_o \equiv n\cdot z+c_i \rightarrow y \equiv n\cdot z+x+i$$
$$\vdash^* x+z_o \equiv y \wedge z_o \equiv n\cdot z+c_i \rightarrow y+(n-i) \equiv n\cdot z+x+n$$
$$\vdash^* x+z_o \equiv y \wedge z_o \equiv n\cdot z+c_i \rightarrow x \approx[n]\, x+n \approx[n]\, y+(n-i)$$
$$\vdash^* x+z_o \equiv y \wedge \mathbb{W}\langle z_o \equiv n\cdot z+c_i \mid i<n\rangle \rightarrow \mathbb{W}\langle x \approx[n]\, y+(n-i) \mid i<n\rangle$$
$$\vdash^* x+z_o \equiv y \wedge \mathbb{W}\langle z_o \equiv n\cdot z+c_i \mid i<n\rangle \rightarrow \mathbb{W}\langle x \approx[n]\, y+i \mid i<n\rangle$$

since again $\vdash y+n \approx[n]\, y+0$. Analogously

$$\vdash^* x \equiv z_o+y \wedge z_o \equiv n\cdot z+c_i \rightarrow x \equiv n\cdot z+y+i$$
$$\vdash^* x \equiv z_o+y \wedge \mathbb{W}\langle z_o \equiv n\cdot z+c_i \mid i<n\rangle \rightarrow \mathbb{W}\langle x \approx[n]\, y+i \mid i<n\rangle \;.$$

Thus

$$\vdash^* (x+z_o \equiv y \vee x \equiv z_o+y) \wedge \mathbb{W}\langle z_o \equiv n\cdot z+c_i \mid i<n\rangle$$
$$\rightarrow \mathbb{W}\langle x \approx[n]\, y+i \mid i<n\rangle$$
$$\vdash^* \mathbb{W}\langle z_o \equiv n\cdot z+c_i \mid i<n\rangle$$
$$\rightarrow ((x+z_o \equiv y \vee x \equiv z_o+y) \rightarrow \mathbb{W}\langle x \approx[n]\, y+i \mid i<n\rangle)$$
$$\vdash^* \exists z\, \mathbb{W}\langle z_o \equiv n\cdot z+c_i \mid i<n\rangle$$
$$\rightarrow ((x+z_o \equiv y \vee x \equiv z_o+y) \rightarrow \mathbb{W}\langle x \approx[n]\, y+i \mid i<n\rangle)$$
$$\vdash^* \forall z_o\, \exists z\, \mathbb{W}\langle z_o \equiv n\cdot z+c_i \mid i<n\rangle$$
$$\rightarrow ((x+z_o \equiv y \vee x \equiv z_o+y) \rightarrow \mathbb{W}\langle x \approx[n]\, y+i \mid i<n\rangle)$$

$$\vdash^* (x+z_0 \equiv y \vee x \equiv z_0+y)$$
$$\qquad \rightarrow \quad (\forall z_0 \exists z \bigvee\langle z_0 \equiv n\cdot z+c_i \mid i<n\rangle \;\rightarrow\; \bigvee\langle x \approx[n]\, y+i \mid i<n\rangle)$$
$$\vdash^* \exists z_0\, (x+z_0 \equiv y \vee x \equiv z_0+y)$$
$$\qquad \rightarrow \quad (\forall z_0 \exists z \bigvee\langle z_0 \equiv n\cdot z+c_i \mid i<n\rangle \;\rightarrow\; \bigvee\langle x \approx[n]\, y+i \mid i<n\rangle).$$

But $\vdash^* \exists z_0\, (x+z_0 \equiv y \vee x \equiv z_0+y)$ by c3, hence

$$\vdash^* \forall z_0 \exists z \bigvee\langle z_0 \equiv n\cdot z+c_i \mid i<n\rangle \;\rightarrow\; \bigvee\langle x \approx[n]\, y+i \mid i<n\rangle\ ,$$

and so $a_{6n}{}^*$ is provable from (the universal closure of) a_{6n}.

It now follows that every deduction from C_2 can be expanded into a deduction from $C_2{}^*$. Hence together with $C_2{}^*$ also C_2 is consistent, and I have proved the last statement of

THEOREM 2 The axiom system C_2 is complete, decidable and consistent.

As for decidability, let $\vdash^*$ denote provability in L^* and $\vdash$ provability in L. Observe that, instead of considering provability from the open formulas C_2, C_{12} and $C_2{}^*$ I may as well use provability from their universal closures $(\forall)C_2$, $(\forall)C_{12}$ and $(\forall)C_2{}^*$. I decompose the set $G = (\forall)C_2{}^*$ into the set of the L-sentences $G_0 = (\forall)C_{02}$, the set UP of the $(\forall)a_{5n}{}^*$, and the set G_1 of the $(\forall)a_{6n}{}^*$; thus $(\forall)C_{12} = G_0 \cup UP$. For every $g = (\forall)a_{6n}{}^*$ in G_1 let g^σ be the L-sentence $(\forall)a_{6n}$. It follows from the above that $(\forall)C_{12} \vdash g \leftrightarrow g^\sigma$. Thus if $G_2{}^\sigma$ is the image of G_2 under ${}^\sigma$ then $(\forall)C_2$ is $G_0 \cup G_2{}^\sigma$.

In Chapter 2.10. I defined a translation map ${}^\pi$ which to every L^*-formula v assigns the L-formula v^π, obtained by replacing the atomic formulas with new predicate symbols by their defining L-formulas, and I observed that $UP \vdash v \leftrightarrow v^\pi$; clearly, $(a_{6n}{}^*)^\pi$ is, essentially, $a_{6n}{}^\S$. With the above definition of G, G_0 and UP, I now am in the situation of Corollary 2.10.1 which states for every L^*-sentence w:

> Every L^*-derivation of $(\forall)C_2{}^* \Longrightarrow w$ can be transformed into an L-derivation of $(\forall)C_2 \Longrightarrow w^\pi$.

For an L-sentence w then, the decision method for $C_2{}^*$ determines whether there is a derivation of $(\forall)C_2{}^* \Longrightarrow w$ or of $(\forall)C_2{}^* \Longrightarrow \neg w$, and as $w = w^\pi$ I then obtain derivations of $(\forall)C_2 \Longrightarrow w$ or of $(\forall)C_2 \Longrightarrow \neg w$.

Concerning quantifier elimination, it will be noticed that, for an L-sentence w, the deduction from $C_2{}^*$ of $w \leftrightarrow w^\#$ translates into a deduction from C_2 of $w \leftrightarrow w^{\#\pi}$, but $w^{\#\pi}$ is, in general, *not* an open L-sentence.

In concluding, let me recall that in section 1 I also mentioned the language $L^\dagger$, extending L with the definitions

$a_{5n}{}^\dagger$. $\quad n \mid y \;\leftrightarrow\; \exists z_0\, n\cdot z_0 \equiv y$.

In analogy to C_2^* then, an axiom system $C_2^\dagger$ can be obtained from C_{02}, adding the definitions $a_{5n}{}^\dagger$ and the axioms

$a_{6n}{}^\dagger$ $\quad \mathbb{V}<n|y+i \mid i<n>$.

Also $C_2^\dagger$ admits a quantifier elimination map; in formulating the necessary reductions, however, additional distinctions (referring to comparabilities $x<y$ or $x\equiv y$ or $y<x$) will have to be made as they already were included in the definition of $\approx[n]$.

4. C_2 and Induction

I observed above that C_2^* admits the induction rule

(IRA) $\quad v(c_0),\ v(x) \to v(x+1) \ \vdash \ v(y)$.

Thus also C_2 admits (IRA), because an L-formula deduced with (IRA) from C_2, hence a fortiori from C_2^*, is provable from C_2^* alone and, being an L-formula, already provable from C_2.

For the language L of C_2, or more generally for every language which has primitive or defined terms c_0 and s, I define the *induction schema* to be the family of formulas, consisting, for every L-formula v(x), of (the universal closure of) the formula

(ISA) $\quad v(c_0) \wedge \forall x\,(v(x) \to v(x+1)) \ \to \ \forall x\, v(x)$

where v(t) abbreviates rep(x,t | v). It follows from quantifier logic that (ISA) may be replaced by

(ISA') $\quad v(c_0) \wedge \forall x\,(v(x) \to v(x+1)) \ \to \ v(y)$ with y *not* free in $v(c_0)$.

An axiom system which proves (ISA) also admits (IRA). Because after generalizing the second premiss of (IRA), v(y) follows from (ISA') by modus ponens. Conversely, every instance

$$u(y) \ = \ v(c_0) \wedge \forall x\,(v(x) \to v(x+1)) \ \to \ v(y)$$

of (ISA') can be derived by (IRA), because $\vdash u(c_0)$ trivially and

$$\vdash \ \forall x(v(x) \to v(x+1)) \ \to \ (v(z) \to v(z+1))$$
$$\vdash \ (\forall x(v(x) \to v(x+1)) \ \to \ v(z)) \ \to \ (\forall x(v(x) \to v(x+1)) \ \to \ v(z+1))$$
$$\vdash \ (v(c_0) \wedge \forall x(v(x) \to v(x+1)) \ \to \ v(z)) \ \to \ (v(c_0) \wedge \forall x(v(x) \to v(x+1)) \ \to \ v(z+1))$$
$$\vdash \ u(z) \to u(z+1) \ .$$

In particular, C_2 now proves the formulas (ISA).

I define the axiom system $C_{2\mathrm{j}}$ by keeping the axioms c1. $\neg\, x<x$

c2. $x<y \wedge y<z \rightarrow x<z$

c3. $x<y \vee y<x \vee x\equiv y$

c4. $c_0<x \vee c_0\equiv x$

c4$_1$ $c_0<c_1$

c6. $x<y \rightarrow x+c_1 \lesssim y$

a3. $x+c_0\equiv x$,

omitting a1, a2, a4, b1, b2, and adding

a3'. $x+(y+c_1)\equiv(x+y)+c_1$

a4'. $c_1+y\equiv c_1+z \rightarrow y\equiv z$

b1'. $y<z \rightarrow c_1+y<c_1+z$

together with all induction axioms (ISA). Then the axioms C_{2j} are provable from C_2.

Conversely, the (omitted) axioms of C_2 are provable from C_{2j} . Because

a1. $\vdash x+(y+z) \equiv (x+y)+z$.

Induction on z. If $z=c_0$ then I use a3; in the induction step I use a3': $x+(y+(z+c_1))\equiv x+((y+z)+c_1)\equiv(x+(y+z))+c_1\equiv((x+y)+z)+c_1\equiv(x+y)+(z+c_1)$.

(j_1) $\vdash c_0+x \equiv x$.

Induction on x, making use of a3 and a3'.

(j_2) $\vdash (x+c_1)+y \equiv x+(y+c_1)$.

Induction on y. First $(x+c_1)+c_0\equiv x+c_1\equiv x+(c_0+c_1)$; in the induction step then $(x+c_1)+(y+c_1)\equiv((x+c_1)+y))+c_1\equiv(x+(y+c_1))+c_1\equiv x+((y+c_1)+c_1)$.

a2. $\vdash x+y \equiv y+x$.

Induction on y. First (j_1); then (j_2) for $x+(y+c_1)\equiv(x+y)+c_1\equiv(y+x)+c_1 \equiv y+(x+c_1)\equiv(y+c_1)+x$.

a4. $x+y\equiv x+z \rightarrow y\equiv z$.

Induction on x. First (j_1); then $(x+c_1)+y\equiv(x+c_1)+z \rightarrow c_1+(x+y)\equiv c_1+(x+z)$ by a1, a2, hence $x+y\equiv x+z$ by a4' and so $y\equiv z$ by inductive hypothesis.

b1. $y<z \rightarrow x+y<x+z$

Induction on x. First (j_1); then $y<z \rightarrow x+y<x+z$ by inductive hypothesis, but then $c_1+(x+y)<c_1+(x+z)$ by b1' whence $(x+c_1)+y<(x+c_1)+z$ by a1, a2.

b2. $y<z \rightarrow \exists x\ y+x\equiv z$.

Induction on z. In view of c4, c1 the case $z=c_0$ holds by ex absurdo quodlibet. Now $y<z+c_1 \rightarrow y\lesssim z$ as in 1.2.3. If $y\equiv z$ then $\exists x\ y+x\equiv z+c_1$ follows trivially.

If $y<z$ then b2 by inductive hypothesis. But $y+x \equiv z \rightarrow (y+x)+c_1 \equiv z+c_1$, $y+x \equiv z \rightarrow y+(x+c_1) \equiv z+c_1$. Hence also $\exists x\ y+x \equiv z \rightarrow \exists x\ y+x \equiv z+c_1$.

A more radical simplification of C_2 results if I define order from addition. So let L be a language with the constants c_0, c_1, the 2-ary operation symbol + and the predicate symbol $\equiv$. Let C_{2i} be the set of axioms

d0. $y+c_1 \equiv z+c_1 \rightarrow y \equiv z$ (variant of a4')
d1. $\neg\ x+c_1 \equiv c_0$
d2. $x+c_0 \equiv x$ (i.e. a3)
d3. $x+(y+c_1) \equiv (x+y)+c_1$ (i.e. a3')

together with all induction axioms (ISA). Employing the abbreviation $s(x)= x+c_1$, these axioms become those for a successor function and a recursive definition of addition:

$$s(y) \equiv s(z) \rightarrow y \equiv z$$
$$\neg\ s(x) \equiv c_0$$
$$x+c_0 \equiv x$$
$$x+s(y) \equiv s(x+y) .$$

Extending L with predicate symbols $\leq$ and $<$ and the defining axioms

d4 $x \leq y \leftrightarrow \exists z(z+x \equiv y)$
d5 $x<y \leftrightarrow x \leq y \wedge \neg\ x \equiv y$,

I can establish all the axioms of C_2. The details are left to the reader who may adapt the proofs to be given in Chapter 10 in a more general framework.

The ideas presented in the present and the preceding Chapter are implicit, if not explicit, in HILBERT-BERNAYS 34 . They immediately extend if not the natural numbers, but the integers are made the object of investigation, e.g. by the axiom system

c1. $\neg\ x<x$
c2. $x<y \wedge y<z \rightarrow x<z$
c3. $x<y \vee y<x \vee x \equiv y$
c4$_1$. $c_0<c_1$
c6. $x<y \rightarrow x+c_1 \leq y$

a1. $x+(y+z) \equiv (x+y)+z$
a2. $x+y \equiv y+x$
a3. $x+c_0 \equiv x$
a5. $\exists x\ y+x \equiv z$
b1. $y<z \rightarrow x+y < x+z$.

Actually, the case of integers is technically simpler because now distinctions such as "$x \leq y$ and $n|y-x$ or $y \leq x$ and $n|x-y$" do not need to be made, and so the language extension $L^{\dagger}$ with ${a_{5n}}^{\dagger}$ and ${a_{6n}}^{\dagger}$ will be the appropriate choice to use.

Introduction to the Chapters 3 - 9

Described from a narrow point of view, it is the aim of the six Chapters 4 to 9 to show that, for axiom systems of arithmetic containing, or permitting to derive, the recursion equations for addition and multiplication, (1) there can be no algorithm deciding provability from these axioms (*undecidability*), and (2) assuming the axioms to be consistent, a sentence w (of the language expressing them) can be exhibited such that neither w nor its negation can be proved from these axioms (*incompleteness*). This aim will be reached in (the Corollaries 1 and 4 of) Chapter 9.

Perceived under a wider perspective, the architecture of concepts, by which these aims will be pursued, expresses a certain interplay between logic and arithmetic:

1. Logic shall take place in a language L which, for every n in ω, contains a constant κ_n. Given an axiom system G formulated in L, hence also the notion of provability $\vdash$ from G, there are *syntactical* descriptions of relations between numbers (in particular: of sets of numbers) referring to this provability. [For instance, a subset M of ω is said to be *represented* by a formula v(x) of L if $m \in M$ implies $\vdash v(\kappa_m)$ and if *not* $m \in M$ implies $\vdash \neg v(\kappa_m)$.] Details will be discussed in Chapter 4.

2. On the other hand, there are *arithmetical* descriptions of relations between numbers, employing the notions of *recursivity*. In particular, the notion of algorithmic decidability is expressed by such concepts. Details will be discussed in Chapters 5 and 6.

3. The language L is assumed to contain countably many graphical symbols only. Employing a *coding* of these symbols (e.g. ASCII), all the syntactical objects of L can be mapped into ω, producing an *arithmetical copy* L_A which mirrors L. The relations between numbers, which in this way arise as copies of syntactical relationships, can be characterized easily by arithmetical descriptions as mentioned under 2. Details will be discussed in Chapter 7.

4. Making use of its constants κ_n, the language L can express relations between numbers. In particular, L can express relations between the codes of its syntactical objects, i.e. can speak about its own mirror, and by 3. those relations have arithmetical descriptions.

5. If the axiom system G is sufficiently strong to prove that arithmetically described relations can also be described syntactically (e.g. if recursive relations become representable) then the results about undecidability and incompleteness follow from simple combinatorial arguments about syntactically describable relations. Details will be discussed in Chapters 4 and 8.

Only in Chapter 9 examples of axiom systems G for arithmetic will be discussed which have the property mentioned in 5. Until then, the presentation in the Chapters 4 and 7-8 remains *abstract* in that G may be an arbitrary axiom system which satisfies appropriate requirements.

If the natural numbers are assumed to be a model of my axioms G then the sentence w, to be exhibited for the incompleteness result (2), will be semantically equivalent to the sentence c saying that it is not the case that the code of w belongs to the codes of provable sentences. While actually a simple consequence of a combinatorial diagonal argument, this situation may appear paradoxical if the fact that L speaks about its arithmetical copy L_A is mistaken to mean that L speaks "about itself" and that w then would state its own unprovability. Such semantical paradoxes about self-reference, exemplified in that of The Liar, have occupied philosophers since antiquity, and in Chapter 3 I shall discuss this paradox and shall use its analysis to show that, for a specified *internal* language fragment within an outer language, no definition of internal truth can be given, nor can internal provability be complete. All of Chapter 3 can be understood without previous knowledge about logic or mathematics, and its results only need that a certain technical hypothesis, connecting the interior with the exterior language, be assumed to hold. Of course, the verification of this hypothesis for concrete situations does require non-trivial mathematical efforts which are the content of the later Chapters.

The basic source for the six Chapters is the article GÖDEL 31 and its detailed elaboration in HILBERT-BERNAYS 39. Other directions of conceptual analysis, particularly of the notion of self-reference, have been developed in SMULLYAN 61 , 94, and the reader may also wish to consider SMULLYAN 92 and 93 for a technically different treatment of several details.

References

K. Gödel: Über formal unentscheidbare Sätze der Principia Mathematica und verwandter Systeme I . Monatshefte Math.Phys. 38 (1931) 173-198

D. Hilbert, P. Bernays: Grundlagen der Mathematik II . Berlin 1939

R.S. Smullyan: Theory of Formal Systems. Princeton 1961

R.S. Smullyan: Gödel's Incompleteness Theorems. Oxford 1992

R.S. Smullyan: Recursion Theory for Metamathematics. Oxford 1993

R.S. Smullyan: Diagonalization and Self-Reference. Oxford 1994

Chapter 3. Antinomies, Pseudomenos, and their Analysis

1. Antinomies

I shall talk about sentences as they are formed in ordinary language. Consider the task to establish a particular sentence A as true. The often used method of *indirect* proof then starts from the assumption that A be *not* true, and it proceeds from this assumption, by purely logical reasoning and employing further, previously made assumptions, to the result that A must be true. In the same manner, a sentence *non*A (stating that it is not the case that A) may be proven indirectly by deriving the truth of *non*A from the assumption that A be true.

There appear, however, also a few examples of sentences A for which (i) the assumption that A be true leads to the result that *non*A must be true, and (ii) the assumption that *non*A be true leads to the result that A must be true. Sentences of this kind are called *antinomical* and *antinomies*. One such example A is the sentence

This sentence is not true .

Because, it is argued,

1. let A be true	and	1. let A be not true
2. then what A says is true		2. then what A says is not true
3. A says that A is not true		3. A says that A is not true
4. hence it is true that A is not true		4. hence it is not true that A is not true
5. hence A is not true		5. hence A is true .

Having presented this argument in detail, I notice that one of its subtler points lies in identifying the truth of A with the truth of *what* A *says*. And *what* A *says* employs the pro-noun "this" which refers to the sentence A itself.

In order to further illuminate the effect of "this", I now declare explicitly that I shall use capital letters A, B, C as *names* for fixed sentences. Without knowing the particular sentence named C, I may form the sentence B

C is not true

in which I have avoided the pro-noun "this". Obviously, B will not give rise to an antinomy as long as the truth, or non-truth, of C is known *before* the name B is given to the above sentence.

Let me now *demand* that C *shall also be* the name of the sentence B. Then I have achieved the effect which the pro-noun "this" had in A : the sentence

C now *says* something about its own truth, and the antinomy arises again. But its arisal now can be understood better:

> If the truth of a sentence is the truth of what it says, then the *Principle of Genericity:*
>
> > a sentence B, saying something about sentences named C, D,... , must be assigned truth or non-truth *only after* C, D,... have been assigned truth or non-truth ,
>
> prevents the assignment of truth or non-truth to the above sentence B if it is also given the name C .

While it appears syntactically quite legitimate to formulate self referring sentences such as A, the assignment of truth values to them may prove impossible. In this sense, then, the antinomical sentence A violates genericity.

When writing the sentence B, I had tacitly suggested that a sentence named C was already present, and in that case the antinomical situation could simply be prevented by avoiding to use the same name for different things - because whichever way a given sentence named C might look as a string of graphic signs, it will be different from the sentence B using its name. This becomes particularly visible if names of sentences, as syntactical objects, are formed by enclosing the sentences into quotation marks, as it is often the practise in philosophical discussions. The name of

C is not true

then becomes "C is not true" which is clearly distinct from whichever string C may have been.

Matters do not change significantly if expressions are admitted in which the letter C is not a name of a given sentence but rather a variable for which such names may be substituted. If C is such then the expression named B becomes a *sentence form* or a *formula* which, by itself, will be neither be true nor false. If C then is replaced by the name of a sentence then a new sentence will result, but if C is replaced by the name of formula (for instance B) then the resulting expression will not be one of which truth or falseness may be asked. Again, using names formed with quotation marks, this expression resulting from B will be

"C is not true" is not true

which is not a sentence.

2. More examples and the Liar

Antinomies, based on a violation of the principle of genericity, may be formulated in ways more innocuous looking than the straightforward approach taken above. The following example, said by TARSKI 36 to have been found by Łukasiewicz, is particularly amusing. Consider the sentence A

The sentence, standing on line 3 of this page, is false .

Now the reader will see immediately that A does stand on line 3 of this page. Hence if A is true then what A says is true, and thus A must be false since this is what A says. And if A is false then what A says is false, hence it is false that this sentence, i.e. A, is false, and thus A must be true.

It is often stated that antinomical situations arise from sentences referring to themselves. But this is not quite correct as is shown by the example of *two* sentences D_0 and D_1 which arrive, D_0 before D_1, as messages in a mailbox: D_0 says

The following message is not true

and D_1 says

The previous message is true.

Now if D_0 is true then D_1 is false, meaning that D_0 is *not* true; and if D_0 is false then D_1 is true, meaning that D_0 is true. Thus I have an ***antinomical pair*** of sentences, neither of which can be attributed the truth value true or false. The situation becomes clearer if D_0 is rephrased as

D_1 is not true

and D_1 is rephrased as

D_0 is true .

For neither of these sentences then is defined by itself alone: the definition of D_0 needs that of D_1, and the definition of D_1 needs that of D_0. So again the principle of genericity is violated by such antinomical pairs (while the request that different sentences have different names is satisfied here).

Another variant of the sentence A is the sentence

This is a lie

provided I assume, for the moment at least, that everyone arguing here has an all-comprising knowledge so that, for a stated sentence, its being a lie is equivalent to its being not true. Sentences of this type have been discussed already in antiquity under the name of ψευδομενος or the ***antinomy of the Liar***; Diogenes Laertius ascribes their discovery to the Megarean *Eubulides* who must have found them before 330 b.C. For details cf. BOCHENSKI 56, § 23 and § 35, BOCHENSKI 68 p.100ff, and the extensive presentation of RÜSTOW 10. In this connection, there often is mentioned another tradition, relating the Cretan sage *Epimenides*, living in the 6th century b.C., to have stated the sentence F

All Cretans are liars ;

an indirect reference to this (with an emphatic agreement to the content of F) can be found in St. Paul's *Epist. ad Titum*, 1.12–13. Aside from the fact that Epimenides probably did not intend to formulate an antinomy, the sentence F does not give rise to one. Conclusions from F can be drawn only if, contrary to usage, a liar would be assumed to be a person lying permanently. Because in that case, if F were true, then the Cretan Epimenides, pronouncing F, must have lied himself whence F must be false. If F were false then there would be a Cretan which does not lie permanently and, therefore, utters at least one true sentence. That, of course, will reflect upon Epimenides only if I make the rather absurd assumption that Epimenides is the only Cretan. But it still will not reflect upon Epimenides' pronouncement of F. Only if I add the further, perfectly absurd assumption that F is the *only* sentence ever uttered by Epimenides, can I conclude that F is (the) one sentence of those which, stated by the one non-liar, are true.

3. Analysis: The Undefinability of Truth

The antinomies discussed above all have in common the use of names of sentences, and this in such a way that the attempt at an assignment of truth will violate the principle of genericity. What I have omitted so far is a closer analysis of the relationship between the truth of a *sentence* A and the truth of *what* A *says*. Presenting such an analysis for its own sake may appear as a superfluous task, an exercise in pedantry, as the antinomies are quite artificial and will not disturb anyone's sleep. The surprising fact, however, is that this analysis may lead to *new insights* into the notion of *truth*.

In order to carry this out, I will have to speak *about* sentences. When speaking, of course, I utter sentences myself, and if for no other reasons than those of mental hygiene, I shall distinguish between the sentences *spoken* and the sentences spoken *about*. To do so, I shall differentiate between two layers of language or, more conveniently, between two languages. There is an outer language, plain English, in which I speak and from which my spoken sentences come. And there is an inner language L, containing the sentences I speak about. It is this inner language about which I want to have complete control, and so it may be rather limited in its expressions.

There are three particular features which this inner language L will need to have in order to analyze the antinomies. The first, usually absent in natural languages (such as the outer one), is the use of *variables*: As objects to which the expressions of L refer, I shall admit not only names (of fixed things) but also variables (for which names may be substituted), so that e.g. for the name A the sentence

A is not true

would arise from the expression

x is not true

by substitution of A for the variable x. As discussed already earlier in connection with the antinomies, such an expression, containing a variable, then is not a sentence anymore (which could be true or false) but only a ***proto***-sentence or, as it is usually called, a *formula*. And the example above then shows the reason why I need such formulas: if A happens to be the name of a sentence, I now can refer to the expression into which that name is to be substituted.

The second feature, also already familiar from the antinomies, is that L will need to contain special objects which serve as *names* for its sentences. One method, mentioned already at the end of Section 1, to generate such names would be the use of quotation marks. However, the actual form of those names will not really matter here. Also, as I shall talk about arbitrary L-sentences, I will have to use *exterior* variables such as the letter s when saying

... let s be a L-sentence ... ,

and thus I shall use (the exterior functional) notation γ_s to refer to the name of s in L. Moreover, as L will have not only sentences but formulas, I shall also require the presence of names γ_u even for *formulas* u. Actually, the availability of *formulas together with names* will lie at the heart of the positive insights during the following developments.

The third, and last, particular feature of L then will be that there is a special formula in L, ***verum***(x), which does refer to L-sentences s by permitting the formation of L-sentences ***verum***(γ_s). Again, all the antinomies did use such references when ***saying*** that something be true or not true.

The precise setup now of the inner language L will be as follows.

1. *Basic* formulas of L are formed from a predicate and an object to which this predicate refers.

12. *Terms* of L are either
121. the letters x, y, z, ... , called *variables*,

122. (indexed) letters such as a_0, a_1, ... , called *constants*.

13. *Predicates* p of L need not to be specified (there may, for instance, be a predicate *blue*). Also, 1-place predicates will suffice.

14. The *basic* formulas of L are the expressions obtained from writing a predicate followed by a term put into parentheses. Thus p(x), p(a_0), ... are basic formulas of L .

2. *Formulas* of L are all basic formulas and, if *s* stands for a basic formula, the expressions *non*s . Thus *non*p(x), *non*p(a_0), ... are formulas of L.

Sentences of L are the formulas of L in which there do not occur variables (and thus occur constants only).

So much for the language L by itself. Now to the matter of *names* which I shall need not only for sentences but also for formulas:

(A) There is an injective map γ assigning to every L-formula u an L-constant γ_u (also written as $\gamma(u)$)

[here, of course, the letter u is but an (exterior) variable for L-formulas, and γ_u is an (exterior) term denoting the corresponding L-constants].

I now come to the description of *exterior truth* and *non-truth* (falseness) which are two properties that may be predicated about sentences. Let me begin by stating their *disjunctiveness*:

(DT) no sentence can be simultaneously true and non-true

which implies that the non-true sentences are not true. The stronger assumption of *complete disjunctiveness*

(CT) every sentence is either true or non-true

is equivalent to saying that the non-true sentences are precisely those which are not true; I shall protocol its uses explicitly. Observe that "if s is true then t is true" together with (DT) permits only the weak contraposition "if t is non-true then s is not true"; the full contraposition "if t is non-true then s is non-true" requires (CT).

Now, in order to discuss *interior* truth, I will assume that there is a special formula *verum*(x) in L, and I shall call *verum*(x) an *a-truth definition* if it has the property of *semantic consistency* and if it *fully reflects* exterior truth

(T_0) there is no sentence s such that both *verum*(γ_s) and *non verum*(γ_s) are simultaneously true,

(T_1) if a sentence s is true then *verum*(γ_s) is true,

(T_2) if a sentence s is non-true then *non verum*(γ_s) is true.

Observe now that the assumption (A) has been formulated so generally that no aspect of genericity is available anymore (though in all concrete cases γ still will permit arguments using genericity). In particular, the map γ may be so twisted that a sentence

f = *non verum*(a_n)

with some constant a_n has indeed as its name γ_f this very constant a_n. *If I assume* this to be the case, *then* I can reproduce for f my antinomy for the sentence A in the Section 1:

a. let f be true
b. then *non verum*(γ_f) is true, by definition of f

c. hence ***verum***(γ_f) is not true, by (T_0)
d. hence f is not true, by contraposition of (T_1)
e. hence f is non-true, by (CT) ,

a. let f be non-true
b. then ***non verum***(γ_f) is true, by (T_2)
c. but f = ***non verum***(γ_f)
d. hence f is true .

Thus even for a twisted γ, my derivation of the antinomy makes use of the assumptions that

L admits an a-truth definition, and exterior truth is completely disjunctive .

Observe that, for this result, I only needed names of sentences, not names of formulas (although the construction of names of formulas should not be - and actually is not - a harder task than that of names of sentences). - Comparing the present argumentations with those given for A in Section 1, the first one (for f being true) reaches in b. what for A was found in step 4.. The main work takes place when going on from b. to e. and establishes this: if it is externally true *what* A internally *says* about non-truth, then *what* A *says* is also externally not true. The second argumentation here does not correspond to the one given for A, because that would rather demand to argue

a. let f be non-true
b. then ***non verum***(γ_f) is non-true, by definition of f
c. hence ***non verum***(γ_f) is not true, by (DT)
d. hence f is not non-true, by contraposition of (T_2)
e. hence f is not not true, by (CT)
f. hence f is true .·

Here again step 4. is reached already in b., while the remaining arguments from b. to f. establish this: if, *what* A internally *says* about non-truth, is externally not true, then *what* A *says* is externally true.

I shall now show that, under an additional assumption (BT), the existence of a truth definition, *together* with complete disjunctiveness of exterior truth, is impossible - from which it follows that, under that additional assumption, the antinomy cannot arise. To this end, I use the notation u(x) for L-formulas containing the variable x and, in that case, if a is a constant, then I write u(a) for the formula obtained from u(x) replacing the variable x by a at all its occurrences in u(x). I choose a fixed variable x_0 and restrict myself to the consideration of formulas $u(x_0)$. For every such $u = u(x_0)$ then I have the sentence $u(\gamma_u) = u(\gamma(u))$; I denote by Δ the (exterior) function assigning the *name* of that sentence to the *name* of u:

$$\Delta(\gamma_u) = \gamma(u(\gamma_u)) .$$

I shall assume that formulas such as $v(\gamma_u)$ not only can be formed as purely syntactical objects, but that they also can be true or false, i.e. that the L-formulas $v(x)$ can be *interpreted* in the set of all constants (or at least in the set of all names γ_u) and this in such a manner that γ_u as a syntactical constant is, at the same time, interpreted by γ_u as a semantical object. What I then require is this:

(BT) For every L-formula $v(x)$ there is an L-formula $v^*(x_0)$ such that, for every name γ_u in the domain of Δ, it holds that

$v^*(\gamma_u)$ is true if, and only if, $v(\Delta(\gamma_u))$ is true.

The meaning of (BT) is clear: instead of predicating v of the complicated name $\Delta(\gamma_u)$, I may, without changing (external) truth values, predicate v^* of the simple name γ_u.

[The existence of $v^*(x_0)$ for $v(x)$ is not nearly as mysterious as it may appear at first sight: all that is necessary to assume about L is this:

(BL) The language L is a (formal) quantifier language with function symbols and the equality symbol $\equiv$. Truth of L-sentences refers to the domain of L-constants where these constants interpret themselves: a constant a satisfies a formula $v(z)$ if $v(a)$ is true. L contains a formula $d(y,z)$ such that two constants γ_u, b satisfy $d(y,z)$ if, and only if, $\Delta(\gamma_u) = b$.

Because in that case, given $v(x)$ in (BT), I define the formula $v^*(x_0)$ as

there exists x such that $(d(x_0, x)$ and $v(x))$.

Then it is equivalent that

$v^*(\gamma_u)$ is true
there exists a constant a such that $\Delta(\gamma_u) = a$ and a satisfies $v(x)$
$\Delta(\gamma_u)$ satisfies $v(x)$
$v(\Delta(\gamma_u))$ is true.

I want to mention that, up to this point, it did not matter at all that I chose v^* to contain the particular variable x_0 connected with the domain of Δ; any other variable would have done as well.

Matters can be simplified further if it is assumed that L contains a 1-ary *function symbol* δ which, on the set of names γ_u, is interpreted by the function Δ. In that case, γ_u satisfies $v(\delta(y))$ if, and only if, $\Delta(\gamma_u)$ satisfies $v(x)$; hence $v^*(x_0)$ can be defined as $v(\delta(x_0))$, because now $v^*(\gamma_u) = v(\delta(\gamma_u))$ is true if, and only if, $\Delta(\gamma_u)$ satisfies $v(x)$.]

Assuming (BT), I take a formula $w(x)$ and with $v(x) = non\, w(x)$, $p(x_0) = v^*(x_0)$ I consider the sentence $q = p(\gamma_p)$ (and now I *do need* the variable x_0):

$p(x_0) = (non\, w)^*(x_0)$,
$q = p(\gamma_p)$.

Then $\Delta(\gamma_p) = \gamma(p(\gamma_p)) = \gamma_q$, and (BT) implies that $q = (non\,w)^*(\gamma_p)$ is true if and only if $non\,w(\gamma_q)$ is true. Thus I find the equivalence

(BT_q) q is true if and only if $non\,w(\gamma_q)$ is true .

Assume now that *verum*(x) is an a-truth definition, and specialize w(x) = *verum*(x). Assume that q is true. Then *verum*(γ_q) is true by (T_1), and by (BT_q) *non verum*(γ_q) is true; this violates (T_0). Assume next that q is not true. Then *non verum*(γ_q) is true by (T_2) and thus q is true by (BT_q). Hence the sentence q can be neither true nor false, violating (CT). Thus I obtain

PROPOSITION T_0 An interior language L satisfying (A) and (BT) cannot admit an a-truth definition if exterior truth shall be completely disjunctive.

And thus the analysis of antinomies has indeed led to a positive result, viz. the undefinability of interior truth in the circumstances stated. This is a first form of TARSKI's undefinability theorem. And the failure of my derivation of the antinomy even for a twisted γ is, in the presence of (BT), just a corollary.

It will be useful to notice that a related result can be obtained if to (CT) I add the condition of *negation-complete disjunctiveness*

(GT) for every sentence s either s is true or *non* s is true

and if a formula *verum*(x) is called a *b-truth definition* if there hold (T_0), (T_1) together with

(T_3) if a sentence s is non-true then *verum*(γ_s) is non-true

instead of (T_2). Because consider again the sentence q with (BT_q); I shall show that q violates (CT). If q is true then I arrive as before at the contradiction to (T_0). If q is not true then I will show that *verum*(γ_q) violates (GT). Because in that case *verum*(γ_q) is non-true by (T_3), hence not true by (CT). But also *non verum*(γ_q) is not true, because otherwise it would be true, hence q would be true by (BT_q) and so q would violate (DT). Hence

PROPOSITION T_1 An interior language L satisfying (A) and (BT) cannot admit a b-truth definition if exterior truth shall be both completely and negation-completely disjunctive .

Let me, finally, assume (CT) together with the reductive definition of the truth of negated sentences:

non s is true if, and only if, s is non-true.

Then (GT) reduces to (CT) and (T_3) reduces to (T_2); (T_0) follows from (DT).

In concluding, I want to mention that in the literature the sentence q is often said to state *its own falseness* (and thus to be a reproduction of the antinomical sentence C from Section 1). This is quite misleading since the sentence stating q's falseness is *non verum*(γ_q) and thus is quite different from q. What *is* the case is that, by (BT_q), the latter sentence is truth-equivalent (or: semantically equivalent) to q. [This state of affairs does *not* improve even if I assume the strong form of (BL) with a function symbol δ and $v^*(x_0) = v(\delta(x_0))$. In that case, the formula $p(x_0)$ becomes *(non verum)*($\delta(x_0)$) whence $q = p(\gamma_p) =$ *(non verum)*($\delta(\gamma_p)$). But $\delta(\gamma_p)$, although semantically evaluated as γ_q, will always start with the function symbol δ and thus will be distinct from the constant γ_q in the sentence *(non verum)*(γ_q) .]

4. Analysis: Incompleteness

Insight into truth being notoriously difficult (John 18, 38), mathematics prefers to secure its knowledge by proofs from hypotheses. The undefinability of truth, discussed in the preceding Section, can be translated immediately into the undefinability of provability. To this end, I start from *one* notion of *exterior provability* and translate

true	into provable
non-true	into non-provable ,

obtaining

(DP) no sentence can be simultaneously provable and not provable,

(CP) every sentence is either provable or not provable,

as disjunctiveness and complete disjunctiveness of provability. The condition (BT) I replace by

(BP) For every L-formula v(x) there is an L-formula $v^*(x_0)$ such that, for every name γ_u, in the domain of Δ it holds that

$$v^*(\gamma_u) \text{ is provable, and only if, } v(\Delta(\gamma_u)) \text{ is provable .}$$

Instead of *verum*(x) I use an L-formula *demo*(x), describing *interior provability,* and I then obtain the conditions

(P_0) there is no sentence s such that both *demo*(γ_s) and *non demo*(γ_s) are simultaneously provable,

(P_1) if a sentence s is provable then *demo*(γ_s) is provable,
(P_2) if a sentence s is not provable then *non demo*(γ_s) is provable.

Calling *demo*(x) an ***a-provability definition*** if it has the properties (P_0), (P_1), (P_2), I obtain in place of Proposition T_0

PROPOSITION P_0 An interior language L satisfying (A) and (BP) cannot admit an a-provability definition if exterior provability shall be completely disjunctive .

While the formalism for the treatment of provability has been set up in complete parallelism to that of truth, there is a difference in the *meanings* of the corresponding assumptions. A first, more psychological one concerns (CT) and (CP): almost everyone will agree that truth should be completely disjunctive. Whereas complete disjunctiveness of provability may well be met with the objection that, while provability can be established by presenting a proof, to establish non-provability would be of a nature different *in principle*. Further, in the case of truth, the semantic consistency (T_0) might be seen as a special case of the exterior consistency (DT) if only the usual reductive definition of the truth of negated sentences were added. Whereas in the case of provability, the interior consistency (P_0) is in no ways a consequence of exterior disjunctiveness (though it obviously would follow from an exterior consistency). A more serious objection (to which my attention was drawn by A.VISSER) concerns the plausibility of the condition (P_3), sometimes called ***negative introspection***, which claims that the ***impossibility*** to prove something entails the ***possibility*** to prove something about it.

Consider now the translation of Proposition T_1. An exterior notion of provability (or, for people who do not care to carry out proofs, the L-theory collecting its hypotheses) is said to be *complete* if it satisfies

(GP) for every sentence s either s is provable or *non* s is provable .

which is nothing but the translation of (GT). Translating (T_3) into

(P_3) if a sentence s is not provable then *demo*(γ_s) is not provable

and calling *demo*(x) a ***b-provability definition*** if it has the properties (P_0), (P_1), (P_3), I obtain in place of Proposition T_1

PROPOSITION P_1 An interior language L satisfying (A) and (BP) cannot admit a b-provability definition if exterior provability shall be completely disjunctive and complete .

In presence of (CP) now, (P_1) together with (P_3) are equivalent to

(P_4) a sentence s is provable if, and only if, *demo*(γ_s) is provable .

Calling *demo*(x) a ***c-provability definition*** if it has the properties (P_0), (P_4), I obtain the following variant of (IP_1) :

PROPOSITION P_2 If an interior language L satisfies (A) and (BP) and admits a c-provability definition, and if exterior provability shall be completely disjunctive, then exterior provability cannot be complete .

Clearly, a theory supposedly giving a complete description of its field of reference should be complete. Proposition P_2 is a first form of an incompleteness *theorem*, showing that the expectation to find complete theories may be disappointed.

There is a second form of an incompleteness theorem which makes use, not only of exterior provability, but also of truth. Let there be given

an exterior notion of provability with respect to an L-theory,

an exterior notion of truth (referring e.g. to a fixed model of that theory),

an L-formula *demo*(x)

such that

(i) both exterior provability and truth are completely disjunctive,

(ii) a sentence s is true if and only if *non* s is not true,

(iii) truth satisfies (BT),

(iv) every provable sentence is true,

(v) a sentence s is provable if and only if *demo*(γ_s) is true .

PROPOSITION P_3 For an interior language L satisfying these conditions, provability is incomplete: I can exhibit a sentence q which is true but not provable .

Indeed, as q is true, *non* q is not true by (ii), hence not provable by (iv). Before producing q, let me first observe that (ii) and (v) imply

(ii') a sentence s is not true if and only if *non* s is true.

(v') a sentence s is not provable if and only if *demo*(γ_s) is not true .

The sentence q can be obtained in two ways. The first one is indirect: assume that every true sentence is also provable. Then the sentences true are the sentences provable, and the sentences not true are the sentences not provable. I then will show that *demo*(x) can serve as an a-truth definition - which in view of (iii) and (i) will contradict the undefinability of truth. The condition (T_0) follows immediately from (ii). Further, for every s which is true, hence provable, also *demo*(γ_s) is true by (v), establishing (T_1). And if s is not true, hence not provable by (iv), then *demo*(γ_s) is not true by (v') whence *non demo*(γ_s) will be true by (ii'); this establishes (T_2).

A direct construction of q proceeds along the already familiar lines. I set

$p(x_0) = (non\ demo)^*(x_0)$,
$q = p(\gamma_p)$.

In complete analogy to (BT_q), I follows from (BT) that

q is true if and only if *non demo*(γ_q) is true .

It follows from (ii') that

non demo(γ_q) is true if and only if *demo*(γ_q) is not true .

And it follows from (v') that

demo(γ_q) is not true if and only if q is not provable.

Hence

(TP_q) q is true if and only if q is not provable.

But then q must be true. For if q were not true, hence not provable by (iv), then it would be true by (TP_q) as well.

Thus the analysis of antinomies has led to a second positive result, which is a first, semantical form of GÖDEL's incompleteness theorem.

All my propositions have been established with a minimum of formalism, yet in complete rigor. But it would be a mistake to believe that now they are available for, say, philosophical discussions. Because they all depend heavily on the assumptions made, and I have *not given a single example* of an inner language L satisfying the hypotheses (BT) for truth, or (BP) or (v) for provability. Such examples will indeed be given in the following Chapters, but they will require considerably more detail, and a careful use of the machinery of mathematical logic. Speaking with precision about truth and provability of L-sentences with help of the *exterior* language is a matter of routine for everyone familiar with the basic notions of (mathematical) logic. Here, however, it will become necessary to mirror this speaking within the *inner* language L itself, setting up the predicates *verum* or *demo* in L. Whether this can be done at all, and if so under which circumstances, this at this stage is still completely open. And in particular, the verification of (BT) and of the condition (v) will, unfortunately, not be susceptible to the cavalier approach preferred up to this point.

References

J.M. Bochenski: Formale Logik. Freiburg 1956

I.M. Bochenski: Ancient Formal Logic. Amsterdam 1968

A. Rüstow: Der Lügner. Dissertation Erlangen 1908 . Leipzig 1910

A. Tarski: Der Wahrheitsbegriff in den formalisierten Sprachen. Studia Philosophica 1 (1936) 261–405

Chapter 4. Undefinability and Incompleteness, General Theory

In this Chapter I repeat the approach of Chapter 3, taking as exterior a language from mathematical logic, and as interior its arithmetical mirror obtained with help of coding functions for terms, formulas and derivations; the basic technical hypothesis now becomes the ***representability*** of certain arithmetical relations, mirroring syntactical relationships under the coding. This notion is explained in Section 4.2; the following three Sections first present TARSKI's 36 theorem on the undefinability of truth, and then incompleteness theorems of the type of GÖDEL's 31 and ROSSER's 36, differentiated in particular by their underlying semantical assumptions. Common to the proofs of these theorems is the use of fixpoint arguments as developed by Tarski in TARSKI-MOSTOWSKI-ROBINSON 51, and a simple combinatorial frame based on this idea is presented in Section 4.1.

The theorems on undefinability and incompleteness here are presented *abstractly* in that they are established under the assumptions that (1) an unspecified coding of the syntax of my language is given, (2) the language contains sufficiently many constant terms which are in correspondence with the objects serving as codes (e.g. the positive integers), and (3) a set of sentences (axioms) is given, from which the technical hypotheses on representability can be proved; these assumptions will be stated in detail at the beginning of Section 3. The technical hypotheses themselves, therefore, remain unverified; in Section 6 the special (and obviously unreasonably strong) situation that the given axioms are those of set theory will be discussed and will lead to the observation that even there it is arithmetical considerations which matter.

1. A Combinatorial Frame

By the argument of the *diagonal construction* I shall understand the following. Consider a set N and a subset R of the the set $N \times N$; for every y in N let R_y be the set of all x in N such that the pair $<x,y>$ belongs to R; let R^∇ be the set of all x in N such that $<x,x>$ is *not* in R. *Then* R^∇ *is distinct from all sets* R_y since for every y it holds that *not* $<y,y> \in R$ is equivalent to *not* $y \in R_y$.

Throughout this section, I shall consider the following situation. There are given

a (non empty) set N;

an equivalence relation A_π on N with the natural map π from N onto the partition N/A_π;

a function f from $N \times N$ into N such that

(10) $\pi(y) = \pi(y')$ implies for every x also $\pi f(x,y) = \pi f(x,y')$,
(11) for every y there exists y^* such that for all x it holds that $\pi f(f(x,x), y) = \pi f(x, y^*)$.

Replacing x in (11) by y^*, I obtain the

COMBINATORIAL FIXPOINT LEMMA

For every y: $\pi f(f(y^*,y^*), y) = \pi f(y^*, y^*)$,

and I call $f(y^*, y^*)$ the *fixpoint to* y. - Let now P be a subset of N which is closed with respect to the equivalence relation A_π. There holds the

COMBINATORIAL UNDEFINABILITY THEOREM

There is no pair of elements b, -b in N such that for every x

(20) $x \in P$ if and only if $f(x, b) \in P$,
(21) $x \in N-P$ if and only if $f(x, -b) \in P$.

In particular, there are no elements b, -b in N such that for every x

(22) there does not hold simultaneously $f(x, b) \in P$ and $f(x, -b) \in P$ (*consistency*),

(23) $x \in P$ implies $f(x, b) \in P$ and $x \in N-P$ implies $f(x, -b) \in P$ (*representability*).

Observe first that (22), (23) imply (20), (21), because if, say, f(x, b) is in P but x in not in P, then $x \in N-P$ would imply $f(x, -b) \in P$ which contradicts consistency. The proof now will be indirect; assume b and -b to exist. Define the subset R of $N \times N$ by

$\langle x,y \rangle \in R$ if and only if $f(x,y) \in P$,

which by (20) is equivalent to $f(f(x,y), b) \in P$. Then $x \in R^\nabla$ will be equivalent to each of

not $\langle x,x \rangle \in R$
not $f(x,x) \in P$
$f(f(x,x), -b) \in P$ by (21)
$f(x, (-b)^*) \in P$ by (11) and since P is closed under A_π .

For $p = (-b)^*$ this would mean that R^∇ is R_p, contradicting the diagonal argument. - This contradiction can also be obtained in the following, more explicit way. Let q be the fixpoint f(p,p) to -b. Then $q \in P$ would imply $f(q,b) \in P$ by (20), while $\pi(q) = \pi f(q,-b)$ by the Fixpoint Lemma would imply $f(q,-b) \in P$, hence $q \in N-P$ by (21). On the other side, $q \in N-P$ would imply $f(q,-b) \in P$ by (21), while $\pi(q) = \pi f(q,-b)$ would imply $f(q,-b) \in N-P$.

Assume now that there is also given a subset T of N, closed under A_π, and let F be the complement N−T. Let P be a subset of T and let b, −b be two elements in N such that

(30) $x \in P$ if and only if $f(x, b) \in T$,
(31) $f(x, b) \in F$ if and only if $f(x, -b) \in T$.

COMBINATORIAL INCOMPLETENESS THEOREM

P is a proper subset of T, and in particular the fixpoint q to −b is in T−P .

For if p is $(-b)^*$ then $f(x,p) \in T$ will be equivalent to each of

$f(x, (-b)^*) \in T$	
$f(f(x,x), -b) \in T$	by (11) and since T is closed under A_π
$f(f(x,x), b) \in F$	by (31)
not $f(f(x,x), b) \in T$	by definition of F
not $f(x,x) \in P$	by (30)

and this implies

(32) $f(x,p) \in T$ if and only if *not* $f(x,x) \in P$.

Taking x to be p, I find for $q = f(p,p)$ that

(33) $q \in T$ if and only if *not* $q \in P$.

Hence there must hold *not* $q \in P$, for $q \in P$ would imply $q \in T$ and, therefore, *not* $q \in P$. But then (33) implies also $q \in T$. – In order to see that also this proof contains the diagonal argument, define the subset R of N×N by

$\langle x,y \rangle \in R$ if and only if $f(x,y) \in P$.

Then (32) says that R^∇ is the set of all x such that $f(x,p) \in T$. Thus $R_p \subseteq R^\nabla$ because $\langle x,p \rangle \in R$ implies $f(x,p) \in P$ whence $f(x,p) \in T$. But R_p is a proper subset of R^∇ by the diagonal argument, and for some x, laying in R^∇ and not in R_p, the element $f(x,p)$ will be in T but not in P.

2. Basic Notions on Definability and Representability

This section contains, essentially, only definitions which will be fundamental in the following. I shall assume the following situation to be given:

an elementary language L with equality ,

a set G of L−sentences ,

a set N together with an injection κ from N into the set of constant (variable−less) terms of L; instead of $\kappa(n)$ I shall write κ_n , an L–structure A such that N is a subset of u(A) and every constant κ_n is interpreted in A by the element n of N .

In the following, *provability* of formula v shall mean its provability *from* G and shall be denoted by $\vdash$ v (instead of the of the usual G $\vdash$ v); in order to distinguish this from the provability from the *empty set* the latter shall be referred to as *logical* provability.

I next introduce some convenient abbreviations. If v is a formula and if $\xi = \langle x_0, \ldots, x_{k-1} \rangle$ is a sequence of variables such that $\mathrm{fr}(v) \subseteq \mathrm{im}(\xi)$ then $v(\xi)$ is the object formed from v *together* with ξ. Also, if ξ is decomposed into $\langle \xi', x_{k-1} \rangle$ with $\xi' = \langle x_0, \ldots, x_{k-2} \rangle$ then $v(\xi', x_{k-1})$ is $v(\xi)$. Given $v(\xi)$, let η be a map sending variables into terms such that $\eta(x_i) = t_i$ for $i<k$. Then $v(\tau)$ shall be the object formed from $\mathrm{sub}(\eta \mid v)$ together with $\tau = \langle t_0, \ldots, t_{k-1} \rangle$ – in particular, v(t) is sub(x,t | v) together with t. Analogously, $v(\tau', t_{k-1})$ is $v(\tau)$ for $\tau = \langle \tau', t_{k-1} \rangle$ and $\tau' = \langle t_0, \ldots, t_{k-2} \rangle$. Analogously, I consider objects $s(\xi)$ with terms s, and objects $s(\tau)$ with $h_\eta(s)$.

In particular, given $v(\xi)$ and a sequence λ in N^k, then $\kappa \cdot \lambda$ is a sequence of constant terms, and $v(\kappa \cdot \lambda)$ is formed from $\mathrm{sub}(\eta \mid v) = \mathrm{rep}(\eta \mid v)$ with the map η such that $\eta \cdot \xi = \kappa \cdot \lambda$.

Given objects $v(\xi)$ *and* $v(\tau)$, *I shall often write their formulas themselves as* $v(\xi)$ *and* $v(\tau)$, and shall employ notations such as $v(\tau) = \mathrm{sub}(\eta \mid v(\xi))$ or $v(\kappa \cdot \lambda) = \mathrm{rep}(\eta \mid v(\xi))$ which are unambiguous in the given situation.

I shall say that a sequence α in $u(A)^k$ satisfies $v(\xi)$ in the structure A if the map φ from variables into u(A) with $\varphi(x_i) = \alpha(i)$ satisfies v. For sequences λ in N^k, induction on v shows immediately : λ satisfies $v(\xi)$ if, and only if, $v(\kappa \cdot \lambda)$ is true in A .

The set S^k of all α in $u(A)^k$ which satisfy $v(\xi)$ in A is called the set *semantically defined* by v and ξ in A. Subsets M^k of N^k which are of the form $M^k = S^k \cap N^k$ for a semantically defined set S^k shall be called N-*semantically defined* by the formula $v(\xi)$. In that case, M^k consists of the λ in N^k satisfying $v(\xi)$. Hence a set M^k is

N–*semantically defined* by $v(\xi)$ if $M^k \subseteq N^k$ and if

$\lambda \in M^k$ if and only if $v(\kappa \cdot \lambda)$ is true in A .

I now introduce two more concepts, saying that a relation M^k is

formally defined by $v(\xi)$ if $M^k \subseteq N^k$ and if

$\lambda \in M^k$ if and only if $\vdash v(\kappa \cdot \lambda)$;

represented by $v(\xi)$ if $M^k \subseteq N^k$ and if

$\lambda \in M^k$ implies $\vdash v(\kappa \cdot \lambda)$, $\lambda \in N^k - M^k$ implies $\vdash \neg v(\kappa \cdot \lambda)$.

Being formally defined by a formula is notion which is quite convenient to handle technically. It is, however, quite unsatisfactory from a practical point of view: in order to conclude upon *not* $\lambda \epsilon M^k$, I have to know that $v(\kappa \cdot \lambda)$ is *not* provable, which obviously is a highly inconstructive notion. Representability avoids this inconstructiveness - at the price, occasionally, of additional technical hypotheses. - There are the following interrelations between these concepts:

If G is consistent then representability implies formal definability. Because $\vdash v(\kappa \cdot \lambda)$ then implies *not* $\vdash \neg v(\kappa \cdot \lambda)$ and thus *not* $\lambda \epsilon N^k - M^k$.

If A is a model of G then formal definability implies N-semantical definability. Because $\vdash v(\kappa \cdot \lambda)$ implies that $v(\kappa \cdot \lambda)$ is true. Conversely, if $v(\kappa \cdot \lambda)$ is not true in A then it is *not* provable and thus λ is not in M^k.

If G is the set Ths(A) of all sentences true in A then N-semantical definability, formal definability, and representability coincide.

LEMMA 1 Let $v(\xi)$ be a formula and let $\chi = \langle z_0, \ldots, z_{k-1} \rangle$ be a sequence of variables. Let id be the identity on variables, define π by $\pi(x_i) = z_i$, $i<k$, and define u_0 and u by

$$u_0 = \mathrm{tot}(\mathrm{id}, 0, \mathrm{im}(\chi) \mid v) \quad , \quad u = \mathrm{rep}(\pi \mid u_0) \quad .$$

Then u is a formula $u(\chi)$ such that, for every λ in N^k, $\vdash v(\kappa \cdot \lambda)$ if, and only if, $\vdash u(\kappa \cdot \lambda)$.

The formula u_0 differs from v in that its bound variables are guaranteed to not occur in χ. Consequently, π is free for u_0, and from $u_0 = u_0(\xi)$ there follows $u = u(\chi)$. Further, v and u_0 are provably equivalent whence also $v(\kappa \cdot \lambda) = \mathrm{rep}(\eta_1 \mid v)$ and $u_0(\kappa \cdot \lambda) = \mathrm{rep}(\eta_1 \mid u_0)$ are provably equivalent where η_1 is the map defined by $\eta_1(x_i) = \kappa \cdot \lambda(i)$. On the other hand, $u(\kappa \cdot \lambda)$ is $\mathrm{rep}(\eta_2 \mid u)$ where η_2 is the map defined by $\eta_2(z_i) = \kappa \cdot \lambda(i)$. Since π is free for u_0, it follows as in Lemma 13.6 that

$$\mathrm{rep}(\eta_2 \mid u) = \mathrm{rep}(\eta_2 \mid \mathrm{rep}(\pi \mid u_0)) = \mathrm{rep}(\eta_2 \cdot \pi \mid u_0) \quad .$$

But $\eta_2(\pi(x_i)) = \eta_2(z_i) = \kappa \cdot \lambda(i) = \eta_1(x_i)$, and so $h_2 \cdot \pi$ and η_1 coincide on the free variables of u_0. Thus $u(\kappa \cdot \lambda) = \mathrm{rep}(\eta_2 \mid u) = \mathrm{rep}(\eta_1 \mid u_0)$ is provably equivalent to $v(\kappa \cdot \lambda)$.

It follows from this observation that, when talking about formal definability or representability of a relation M^k, I always can choose the formula $v(\xi)$ such that the variables ξ belong to a sequence given in advance - and, in particular, do *not* belong to a certain other (finite) set of variables.

I consider functions $f = f^k$ such that $\mathrm{def}(f^k) \subseteq N^k$ and $\mathrm{im}(f^k) \subseteq N$, and if $\mathrm{def}(f^k)$ is a proper subset of N^k then I call f^k ***partial***, otherwise ***total***. A function f is uniquely determined by its ***graph*** G(f), namely the set of all $\langle \xi, f(\xi) \rangle$ with $\xi \epsilon \mathrm{def}(f)$. G(f) being a (k+1)-ary relation, it thus is explained what it means that G(f) is defined or represented by a formula

$v(\xi,x_k)$, and synonymously then I may say that f itself is defined or represented by such formula. I shall now introduce a farther reaching notion, calling a (possibly partial) function f *functionally represented* by $v(\xi,x_k)$ if

$$\lambda\epsilon\text{def}(f) \text{ implies } \vdash v(\kappa\cdot\lambda, x_k) \longleftrightarrow x_k \equiv \kappa_{f(\lambda)} .$$

In that case also $\vdash v(\kappa\cdot\lambda,x)\wedge v(\kappa\cdot\lambda,y) \rightarrow x\equiv y$ by equality logic from $\vdash v(\kappa\cdot\lambda,x) \longleftrightarrow x\equiv \kappa_{f(\lambda)}$ and $\vdash v(\kappa\cdot\lambda,y) \longleftrightarrow y\equiv \kappa_{f(\lambda)}$.

If f is functionally represented by $v(\xi,x_k)$ then I can conclude that G(f) is represented by this formula provided I know that f is total and that $n\neq m$ implies $\vdash \neg\kappa_n\equiv \kappa_m$ for all n,m . Because in that case $<\lambda,a>$ does not belong to G(f) then $a\neq f(\lambda)$ implies $\vdash \neg\ \kappa_a\equiv \kappa_{f(\lambda)}$ whence $\vdash \neg v(\kappa\cdot\lambda,\kappa_a)$.

A (possibly partial) function f is said to be *represented by a term* $t(\xi)$ if

$$\lambda\epsilon\text{def}(f) \text{ implies } \vdash t(\kappa\cdot\lambda)\equiv \kappa_{f(\lambda)} ;$$

in that case, f is functionally represented by the formula $t(\xi)\equiv x_k$ as follows from the equality axioms. I observe the

LEMMA 2 If f is functionally represented by $v(\xi,x_k)$ then (making use of classical logic) I can expand L conservatively by a term which represents f .

To this end, I choose an element o in N such that the constant term κ_o is available. Writing x and y instead of x_k, x_{k+1} and employing the abbreviations

$$\begin{aligned}
u(\xi) &= \forall x\forall y(v(\xi,x)\wedge v(\xi,y) \rightarrow x\equiv y)\\
a_0(\xi,x) &= \exists y(v(\xi,y)\wedge u(\xi)) \wedge v(\xi,x)\\
a_1(\xi,x) &= \neg\exists y(v(\xi,y)\wedge u(\xi))\wedge x\equiv \kappa_o\\
a(\xi,x) &= a_0(\xi,x) \vee a_1(\xi,x) ,
\end{aligned}$$

I deduce

$$\begin{array}{l}
v(\xi,y)\wedge u(\xi) \vdash v(\xi,y)\wedge u(\xi)\\
v(\xi,y)\wedge u(\xi) \vdash \exists y(v(\xi,y)\wedge u(\xi)) \qquad v(\xi,y)\wedge u(\xi) \vdash v(\xi,y)\\
\qquad v(\xi,y)\wedge u(\xi) \vdash \exists y(v(\xi,y)\wedge u(\xi))\wedge v(\xi,y)\\
\qquad v(\xi,y)\wedge u(\xi) \vdash (\exists y(v(\xi,y)\wedge u(\xi))\wedge v(\xi,y))) \vee a_1(\xi,y)\\
\qquad v(\xi,y)\wedge u(\xi) \vdash \exists x((\exists y(v(\xi,y)\wedge u(\xi))\wedge v(\xi,x)))\\
\qquad\qquad\qquad \vee\ a_1(\xi,x))\\
\qquad v(\xi,y)\wedge u(\xi) \vdash \exists x(a_0(\xi,x) \vee a_1(\xi,x))\\
\qquad \exists y(v(\xi,y)\wedge u(\xi)) \vdash \exists x(a_0(\xi,x) \vee a_1(\xi,x))\\
\qquad \exists y(v(\xi,y)\wedge u(\xi)) \vdash \exists x a(\xi,x)
\end{array}$$

and

$$\begin{array}{l}
\neg\exists y(v(\xi,y)\wedge u(\xi)) \vdash \neg\exists y(v(\xi,y)\wedge u(\xi))\wedge \kappa_o\equiv \kappa_o\\
\neg\exists y(v(\xi,y)\wedge u(\xi)) \vdash a_0(\xi,x) \vee (\neg\exists y(v(\xi,y)\wedge u(\xi))\wedge \kappa_o\equiv \kappa_o)\\
\neg\exists y(v(\xi,y)\wedge u(\xi)) \vdash \exists x(a_0(\xi,x) \vee (\neg\exists y(v(\xi,y)\wedge u(\xi))\wedge x\equiv \kappa_o))
\end{array}$$

$$\neg\exists y(v(\xi,y)\wedge u(\xi)) \vdash \exists x(a_0(\xi,x) \vee a_1(\xi,x))$$
$$\neg\exists y(v(\xi,y)\wedge u(\xi)) \vdash \exists x a(\xi,x) .$$

But $\vdash \exists y(v(\xi,y)\wedge u(\xi)) \vee \neg\exists y(v(\xi,y)\wedge u(\xi))$ by tertium non datur; hence

(1) $\vdash \exists x a(\xi,x)$.

I next wish to show

(2) $\vdash a(\xi,x) \wedge a(\xi,y) \to x\equiv y$

and since

$$\vdash a(\xi,x)\wedge a(\xi,y) \to (a_0(\xi,x)\wedge a_0(\xi,y)) \vee (a_0(\xi,x)\wedge a_1(\xi,y)) \vee (a_1(\xi,x)\wedge a_0(\xi,y)) \vee (a_1(\xi,x)\wedge a_1(\xi,y))$$

this follows from

$\vdash (\forall x\forall y(v(\xi,x)\wedge v(\xi,y) \to x\equiv y)) \to (v(\xi,x)\wedge v(\xi,y) \to x\equiv y)$
$\vdash (\forall x\forall y(v(\xi,x)\wedge v(\xi,y) \to x\equiv y) \wedge v(\xi,x)\wedge v(\xi,y)) \to x\equiv y$
$\vdash u(\xi) \wedge v(\xi,x) \wedge v(\xi,y) \to x\equiv y$
$\vdash \exists y(u(\xi)\wedge v(\xi,y)) \wedge v(\xi,x)\wedge v(\xi,y) \to x\equiv y$
$\vdash a_0(\xi,x) \wedge \exists y(u(\xi)\wedge v(\xi,y))\wedge v(\xi,y) \to x\equiv y$
$\vdash a_0(\xi,x) \wedge a_0(\xi,y) \to x\equiv y$

$\vdash a_1(\xi,x)\to x\equiv \kappa_0$
$\vdash a_1(\xi,x) \wedge a_1(\xi,y) \to x\equiv \kappa_0\wedge y\equiv \kappa_0$
$\vdash a_1(\xi,x) \wedge a_1(\xi,y) \to x\equiv y$ by equality logic ,

$\vdash a_0(\xi,x) \to \exists y(v(\xi,y)\wedge u(\xi))$
$\vdash a_1(\xi,y) \to \neg\exists y(v(\xi,y)\wedge u(\xi))$
$\vdash a_0(\xi,x) \wedge a_1(\xi,y) \to x\equiv y$ by ex absurdo quodlibet ,

$\vdash a_1(\xi,x) \wedge a_0(\xi,y) \to x\equiv y$ analogously .

It follows from (1) and (2) that my language L can be expanded conservatively by a term $t(\xi)$ such that

$$\vdash a(\xi,x) \longleftrightarrow t(\xi)\equiv x .$$

Assume now that $\lambda\epsilon\mathrm{def}(f)$. Since $v(\xi,x)$ functionally represents f, $\lambda\epsilon\mathrm{def}(f)$ implies

$\vdash v(\kappa\cdot\lambda,x)\wedge v(\kappa\cdot\lambda,y) \to x\equiv y$
$\vdash u(\kappa\cdot\lambda)$
$\vdash u(\kappa\cdot\lambda)\wedge \kappa_{f(\lambda)}\equiv \kappa_{f(\lambda)}$
$\vdash u(\kappa\cdot\lambda)\wedge v(\kappa\cdot\lambda,\kappa_{f(\lambda)})$
$\vdash \exists y(u(\kappa\cdot\lambda)\wedge v(\kappa\cdot\lambda,y)) \wedge v(\kappa\cdot\lambda,\kappa_{f(\lambda)})$
$\vdash a_0(\kappa\cdot\lambda,\kappa_{f(\lambda)})$
$\vdash a(\kappa\cdot\lambda,\kappa_{f(\lambda)})$
$\vdash t(\kappa\cdot\lambda)\equiv \kappa_{f(\lambda)}$.

This concludes the proof of Lemma 2 .

I next mention an illustrative application. Let p and q be two distinct elements of N. For every subset M of N^k, I define its *characteristic* function χ_M from N^k into N by $\chi_M(\lambda) = p$ if $\lambda \epsilon M$ and $\chi_M(\lambda) = q$ otherwise. I find

LEMMA 3 If M is represented by $u(\xi)$ then χ_M is functionally represented by the formula $v(\xi,x_k) = (u(\xi) \wedge x_k \equiv \kappa_p) \vee (\neg u(\xi) \wedge x_k \equiv \kappa_q)$.

If χ_M is functionally represented by $v(\xi,x_k)$ and if $\vdash \neg\, \kappa_p \equiv \kappa_q$ then M is represented by $v(\xi,\kappa_p)$.

Let M be represented by $u(\xi)$ and assume $\chi_M(\lambda) = p$; it follows from $\lambda \epsilon M$ that $\vdash u(\kappa\cdot\lambda)$, hence $\vdash u(\kappa\cdot\lambda) \wedge \kappa_p \equiv \kappa_p$ and thus $\vdash v(\kappa\cdot\lambda,\kappa_p)$. It follows from the equality axioms that $\vdash x_k \equiv \kappa_p \rightarrow ((v(\kappa\cdot\lambda,\kappa_p) \rightarrow v(\kappa\cdot\lambda,x_k))$, hence $\vdash x_k \equiv \kappa_p \rightarrow v(\kappa\cdot\lambda,x_k)$. Now $u \rightarrow (v \rightarrow x_k \equiv \kappa_p)$ is a tautology in view of the form of v; hence $\vdash u(\kappa\cdot\lambda)$ implies $\vdash v(\kappa\cdot\lambda,x_k) \rightarrow x_k \equiv \kappa_p$. If $\chi_M(\lambda) = q$ then I may argue analogously since also $\neg u \rightarrow (v \rightarrow x_k \equiv \kappa_q)$ is a tautology. Assume next that χ_M is functionally represented by $v(\xi, x_k)$, and let λ be in M; then $\chi_M(\lambda) = p$ and $\vdash v(\kappa\cdot\lambda,x_k) \leftrightarrow x_k \equiv \kappa_p$ imply $\vdash v(\kappa\cdot\lambda,\kappa_p)$. If λ is not in M then $\vdash v(\kappa\cdot\lambda,x_k) \leftrightarrow x_k \equiv \kappa_q$ implies $\vdash v(\kappa\cdot\lambda,\kappa_p) \leftrightarrow \kappa_p \equiv \kappa_q$ whence also $\vdash \neg v(\kappa\cdot\lambda,\kappa_p)$.

Closer connections between representability and functional representability of f^k can be established if L contains, besides $\equiv$, another 2-ary predicate symbol $\lesssim$, if N is countably infinite, and thus may be assumed right away to be the set ω of natural numbers, and if G proves for every n in ω

(MINn) $\vdash y \lesssim \kappa_n \rightarrow (y \equiv \kappa_0 \vee y \equiv \kappa_1 \vee \dots \vee y \equiv \kappa_n)$

(CMPn) $\vdash \neg\, y \lesssim \kappa_n \rightarrow \kappa_n \lesssim y$.

I express this situation by saying that ω is *provably initial* and shall assume if for the remainder of this section.

LEMMA 4 Let $v(\xi,x)$ be an L-formula, and let λ be in ω^k, n in ω. Assume that

$\vdash v(\kappa\cdot\lambda,\kappa_n)$ and for every i: if i<n then $\vdash \neg v(\kappa\cdot\lambda,\kappa_i)$.

Define $u(\xi,x) = v(\xi,x) \wedge \forall z((v(\xi,z) \wedge \neg x \equiv z) \rightarrow x \lesssim z)$ where z is a new variable.

Then $\vdash u(\kappa\cdot\lambda,x) \leftrightarrow x \equiv \kappa_n$.

Observe that

$$\vdash \forall z((v(\kappa\cdot\lambda,z) \wedge \neg x \equiv z) \rightarrow x \lesssim z) \rightarrow ((v(\kappa\cdot\lambda,\kappa_n) \wedge \neg x \equiv \kappa_n) \rightarrow x \lesssim \kappa_n)$$
$$\vdash u(\kappa\cdot\lambda,x) \rightarrow ((v(\kappa\cdot\lambda,\kappa_n) \wedge \neg x \equiv \kappa_n) \rightarrow x \lesssim \kappa_n)$$
$$\vdash u(\kappa\cdot\lambda,x) \wedge \neg x \equiv \kappa_n) \rightarrow x \lesssim \kappa_n$$

hence by (MINn)

$$\vdash u(\kappa\cdot\lambda,x) \wedge \neg x \equiv \kappa_n \rightarrow (x \equiv \kappa_0 \vee \ldots \vee x \equiv \kappa_{n-1} \vee x \equiv \kappa_n)$$
$$\vdash u(\kappa\cdot\lambda,x) \wedge \neg x \equiv \kappa_n \rightarrow (x \equiv \kappa_0 \vee \ldots \vee x \equiv \kappa_{n-1}) \ .$$

The hypotheses for $i<n$ imply $\vdash x \equiv \kappa_i \rightarrow \neg v(\kappa\cdot\lambda,x)$ whence

$$\vdash u(\kappa\cdot\lambda,x) \wedge \neg x \equiv \kappa_n \rightarrow \neg v(\kappa\cdot\lambda,x)$$
$$\vdash u(\kappa\cdot\lambda,x) \wedge v(\kappa\cdot\lambda,x) \rightarrow x \equiv \kappa_n$$
$$\vdash u(\kappa\cdot\lambda,x) \rightarrow x \equiv \kappa_n \ .$$

Conversely, by (CMPn) and (MINn) also

$$\vdash \neg \kappa_n \leq z \rightarrow z \leq \kappa_n$$
$$\vdash \neg \kappa_n \leq z \rightarrow (z \equiv \kappa_0 \vee \ldots \vee z \equiv \kappa_{n-1} \vee z \equiv \kappa_n) \ .$$

But the hypotheses for $i<n$ also imply $\vdash z \equiv \kappa_i \rightarrow \neg v(\kappa\cdot\lambda,z)$ whence

$$(z \equiv \kappa_0 \vee \ldots \vee z \equiv \kappa_{n-1}) \rightarrow \neg v(\kappa\cdot\lambda,z)$$
$$\vdash \neg \kappa_n \leq z \rightarrow (\neg v(\kappa\cdot\lambda,z) \vee z \equiv \kappa_n)$$

hence

$$\vdash (v(\kappa\cdot\lambda,z) \wedge \neg z \equiv \kappa_n) \rightarrow \kappa_n \leq z$$
$$\vdash \forall z((v(\kappa\cdot\lambda,z) \wedge \neg \kappa_n \equiv z) \rightarrow \kappa_n \leq z) \ .$$

Now $\vdash v(\kappa\cdot\lambda,\kappa_n)$ implies $\vdash u(\kappa\cdot\lambda,\kappa_n)$, and therefore $\vdash x \equiv \kappa_n \rightarrow u(\kappa\cdot\lambda,x)$.

LEMMA 5 A (possibly partial) function f^k which is representable is also functionally representable :

If $v(\xi,x)$ represents $G(f^k)$ and if $u(\xi,x)$ is defined as in Lemma 4, then $u(\xi,x)$ functionally represents f^k, and $u(\xi,x)$ also represents $G(f^k)$.

It follows from Lemma 4 that $u(\xi,x)$ functionally represents f^k. If $\langle\lambda,a\rangle \in G(f^k)$ then $\vdash v(\kappa\cdot\lambda, \kappa_a)$ by assumption, hence $\vdash u(\kappa\cdot\lambda, \kappa_a)$ by Lemma 4 and equality logic. If $\langle\lambda,a\rangle$ is not in $G(f^k)$ then $\vdash \neg v(\kappa\cdot\lambda, \kappa_a)$ by assumption, hence $\vdash \neg u(\kappa\cdot\lambda, \kappa_a)$ by definition of u.

It may be noticed, as an additional observation, that if G secures $\vdash x \leq y \wedge y \leq x \rightarrow x \equiv y$ then for the formula $u(\xi,x)$ of Lemma 4 there holds $\vdash u(\xi,x) \wedge u(\xi,y) \rightarrow x \equiv y$.

3. Undefinability

In this section, I shall assume the following data to be given:

an elementary language L with equality,

a set G of L–sentences ,

a set N together with an injection κ from N into the set of constant terms of L; instead of $\kappa(n)$, I shall write κ_n ,

three injective maps g_T, g_F, g_S which have as ranges three pairwise disjoint subsets of N; g_T is defined on the set of all L–terms, g_F on the set of all L–formulas, g_S on the set of all L–derivations ,

an L–structure A such that N is a subset of u(A) and every constant κ_n is interpreted in A by the element n of N .

As already stated in the last section, *provability* shall always mean *provability from* G. There is the assumption that N is given within a structure A which on N induces relations and functions; it will be used when referring to the representability (or definability) of such relations and functions. Naturally, to speak about their representability, or about their formal definability, does suppose these particular objects to be available; it does, however, make *no* assumptions about the interpretability of the predicate and function symbols occurring in the sentences of G, and it does *not* imply anything about the question of whether these sentences are valid in A or not. Consequently, those semantical assumptions about N (and A), which I shall call *external* ones, should be distinguished carefully from possible model theoretical hypotheses, concerning the interpretation and validity of the formulas from G in the structure A.

For a term t, a formula v and a derivation S, the image under g_T, g_F, g_S will be written also as $\ulcorner t \urcorner$, $\ulcorner v \urcorner$, $\ulcorner S \urcorner$; for an element n in N, belonging to the range of one of these maps g_T, g_F, g_S, the unique original shall be denoted as $[\![n]\!]$.

The maps $\kappa \cdot g_T$ and $\kappa \cdot g_F$ I shall write as γ. For instance, for any formula v then γ_v is $\kappa_{\ulcorner v \urcorner}$.

I now choose a fixed variable x_0 and define a function *rep* from $N \times im(g_F)$ into N by

$$rep(r,k) = \ulcorner rep(x_0, \kappa_r \mid [\![k]\!]) \urcorner \ ;$$

hence $v(x_0) = [\![k]\!]$ implies $rep(r,k) = \ulcorner v(\kappa_r) \urcorner$. There holds the first *Fixpoint Lemma*

LEMMA 6 Assume that *rep* is functionally representable. Let $v = v(x)$ be a formula, and if the chosen variable x_0 is free in v then let x be x_0.

Then I can construct a formula w such that $fr(w) = fr(v) - \{x\}$ and $\vdash w \longleftrightarrow v(\gamma_w)$.

Let $REP(x',y',z)$ be a formula which functionally represents *rep* and such that z does not occur in v and is distinct from x_0. I then set

$$v^*(x_0) = \exists z\,(REP(x_0,x_0,z) \wedge v(z)) .$$

I want to show that

(ϵ) if $rep(r,r) = m$ then $\vdash v^*(\kappa_r) \longleftrightarrow v(\kappa_m)$.

If $rep(r,r) = m$ then the functional representability

$$\vdash REP(\kappa_r,\kappa_r,z) \longleftrightarrow z \equiv \kappa_m$$

implies by equality logic

$$\vdash \exists z\,(REP(\kappa_r,\kappa_r,z) \wedge v(z)) \longleftrightarrow \exists z\,(z \equiv \kappa_m \wedge v(z)) .$$

Here the left side is $v^*(\kappa_r)$, and $\vdash \exists z(z \equiv \kappa_m \wedge v(z)) \longleftrightarrow v(\kappa_m)$ proves (ϵ). I now define

$$p = \ulcorner v^*(x_0) \urcorner$$
$$w = rep(x_0,\kappa_p \mid v^*)$$
$$q = \ulcorner w \urcorner$$

whence $rep(p,p) = q$. Thus w does what was required.

LEMMA 6a If *rep* is represented by a term then the formula w may be chosen to have the same quantifier degree as has v; hence if v is open then so will be w.

Let $REP(v',y')$ be a term representing *rep*. I define for $v(x)$

$$v^*(x_0) = sub(x, REP(x_0,x_0) \mid v) = v(REP(x_0,x_0)) .$$

If $rep(r,r) = m$ then $\vdash REP(\kappa_r,\kappa_r) \equiv \kappa_m$ implies by equality logic

$$\vdash v(REP(\kappa_r,\kappa_r)) \longleftrightarrow v(\kappa_m) .$$

If x_0 is not bound in v (which will be the case if v is open) then

$$v(REP(\kappa_r,\kappa_r)) = rep(x,REP(\kappa_r,\kappa_r) \mid v) =$$
$$= rep(x_0,\kappa_r \mid sub(x,REP(x_0,x_0) \mid v)) = v^*(\kappa_r) ;$$

in the general case, induction on the structure of v gives by pure logic

$$\vdash rep(x,REP(\kappa_r,\kappa_r) \mid v) \longleftrightarrow rep(x_0,\kappa_r \mid sub(x,REP(x_0,x_0) \mid v)) .$$

Hence I have in any case

(ϵ) $\vdash \mathrm{rep}(x_0,\kappa_r \mid v^*) \longleftrightarrow \mathrm{rep}(x,\kappa_m \mid v)$,

and so I can proceed as before.

The following second Fixpoint Lemma requires a weak model theoretical hypothesis, viz.

> L contains a binary predicate $\leq$ interpreted in A by a relation $\leq^A$;
> in A the relation $\equiv$ is interpreted by equality ,
> if $n \in N$ and $a \leq^A n$ then $a \in N$.

If this hypothesis holds I shall say that N is *semantically* $\leq$ *-descending* in the structure A.

LEMMA 7 Assume that *rep* is N-semantically definable and that N is semantically $\leq$ -descending. Then I can still find a formula w such that $\mathrm{fr}(w) = \mathrm{fr}(v) - \{x\}$ and $w \longleftrightarrow v(\gamma_w)$ is true in A.

If G is the set Ths(A) of sentences true in A then provability and truth in A coincide. If REP(x',y',z) N-semantically defines *rep*, let REP'(x',y',z) be the formula

$$\mathrm{REP}(x',y',z) \wedge \forall z' \, (\mathrm{REP}(x',y',z') \rightarrow z \leq z')$$

and define $v^*(x_0)$ with REP'(x',y',z) instead of REP(x',y',z). In order to repeat the argument from the proof above, it will suffice to show that

if $rep(r,k) = m$ then $\mathrm{REP}'(\kappa_r,\kappa_k,z) \longleftrightarrow z \equiv \kappa_m$ is true in A.

Now $rep(r,k) = m$ implies the truth of $\mathrm{REP}(\kappa_r,\kappa_k,\kappa_m)$. If a in A satisfies $\mathrm{REP}'(\kappa_r,\kappa_k,z)$ then it also satisfies $\mathrm{REP}(\kappa_r,\kappa_k,z)$, and for every b in A then <a,b> satisfies $\mathrm{REP}(\kappa_r,\kappa_k,z') \rightarrow z \leq z'$. Taking b to be κ_m, the truth of $\mathrm{REP}(\kappa_r,\kappa_k,\kappa_m)$ implies $a \leq^A m$ whence $a \in N$ by my semantical hypothesis. Since the formula REP(x',y',z) N-semantically defines *rep*, there now follows $rep(r,k) = a$, and as *rep* is a function, this implies $a = m$. Consequently, $\mathrm{REP}'(\kappa_r,\kappa_k,z) \rightarrow z \equiv \kappa_m$ is true in A. The reverse implication is true in A because if a satisfies $z \equiv \kappa_m$ then $a = m$.

I now shall view the data, stated at the beginning of this section, as a *model* of the combinatorial frame of Section 1. For a *first interpretation*, I consider the set M_0 of all formulas which either are sentences or contain the variable x_0 as free. Let N_0 be the image of M_0 under g_F.

The equivalence relation A_π on N , I define as the set of all pairs <n,m> such that

$$n = m \quad \text{or} \quad (\langle n,m\rangle \in N_0 \times N_0 \text{ and } \vdash [\![n]\!] \longleftrightarrow [\![m]\!]) .$$

The function f from N×N into N , I define by

$$f(r,n) = rep(r,n) \quad \text{if n in } N_0$$
$$f(r,n) = \circledcirc \quad \text{else, where } \circledcirc \text{ is fixed in } N-N_0 .$$

For $y = \ulcorner v \urcorner$ then $f(r,y) = \ulcorner rep(x_0,\kappa_r | v)\urcorner$. Evidently, the combinatorial condition (10) holds. In the Fixpoint Lemma 6, I determined for every $v(x_0)$ in M_0 the formula $v^*(x_0)$; I now define for every n in N the element n^* by

$$n^* = \ulcorner [n]^* \urcorner \quad \text{for n in } N_0 , \qquad n^* = n \quad \text{otherwise.}$$

The statement (ϵ), derived in Lemma 6 under the assumption $rep(r,r) = m$, now gives $\pi f(r,\ulcorner v\urcorner^*) = \pi f(f(r,r),\ulcorner v\urcorner)$ for $\ulcorner v\urcorner$ in N_0; replacing $\ulcorner v\urcorner$ by an element from $N-N_0$, I obtain the trivial identity $\pi(\circledcirc) = \pi(\circledcirc)$. Thus also the combinatorial condition (11) holds, and the statement of the Combinatorial Fixpoint Lemma coincides with that of Lemma 6.

For a *second interpretation*, let M_0 be the set of all formulas which freely contain at most the variable x_0. Let N_0 be the image of M_0 under g_F. The equivalence relation $A_\pi{}^\dagger$ (together with the map $\pi^\dagger$) I define as the set of all pairs $\langle n,m \rangle$ such that

$$n = m \quad \text{or} \quad ([\![m]\!],[\![n]\!] \text{ both are sentences and } \vdash [n] \leftrightarrow [m]) .$$

The function f and the elements n^* I define as for the first interpretation. Again the conditions (10) and (11) hold.

Let now P be the image under g_F of the set of provable formulas, let P_s be the image under g_F of the set of all provable sentences. There holds the *Undefinability Theorem*

THEOREM 1 Assume that *rep* is functionally representable. Then not both sets P and $-P$, as well as not both sets P_s and $-P_s$, can be formally definable. If G is consistent then neither P nor P_s is representable.

Because consider the first interpretation. Then both P and $-P$ are closed under A_π. If P and $-P$ were formally definable, they were so by formulas $v(x_0)$ and $w(x_0)$. Setting $b = \ulcorner v \urcorner$, $-b = \ulcorner w \urcorner$, the conditions (20), (21) for the Combinatorial Undefinability Theorem for frames would hold, and this theorem states that this cannot happen. If P were representable, then the consistency of G would imply that both P and $-P$ are formally definable, and again the Combinatorial Undefinability Theorem shows that this situation cannot occur. Next, consider the second interpretation. Then both P_s and $-P_s$ are closed under $A_\pi{}^\dagger$, and so I may argue as I just did for P.

Let T be the image under g_F of the set of all formulas true in A; let T_s be the image under g_F of the set of all sentences true in A. Taking G to be Ths(A), there holds the

COROLLARY 1 Assume that *rep* is N-semantically definable and that N is semantically $\leq$-descending. Then neither T nor T_s is N-semantically definable.

4. Incompleteness

I shall continue by assuming the same data to be given which were listed at the beginning of the last section; I also shall use the notations introduced there. I define on N a relation *ded* :

> *ded* shall consist of all pairs $<n,m>$ such that $[n]$ is a formula and $[m]$ is a derivation of $[n]$ from G.

Under assumptions to be discussed in Chapter 8, there will exist a formula $DED(x_0,y)$ which - at least - defines *ded* semantically in N. Employing such a formula, I can form the ***provability predicate***

$$BEW(x_0) = \exists y \; DED(x_0,y) .$$

Evaluating it in the structure A, the variable y will range over all elements of u(A), not only over those of N. In connection with semantical arguments, therefore, I shall make the additional assumption

> if $DED(x_0,y)$ is satisfied by a,b in A then a,b belong to N ;

this amounts to a semantical description of N, contained in $DED(x_0,y)$, which it is easy to realize in suitable examples.

For the following theorem I shall use two semantical hypotheses:

(H_0) The notion of *truth in* A is defined for sentences $DED(\kappa_n,\kappa_m)$, $\neg DED(\kappa_n,\kappa_m)$, $BEW(\kappa_n)$, $\neg BEW(\kappa_n)$. In particular, for each sentence s of this type either s or $\neg s$ is true, and if $BEW(\kappa_n)$ is true then there exists some b in A such that n,b satisfy $DED(x_0,y)$. If s and $s \leftrightarrow s'$ are true then s' is true.

(H_1) If w is semantically equivalent in A to a sentence $\neg BEW(\kappa_n)$ then $\vdash w$ implies that w is true in A .

Obviously, (H_1) certainly will hold if A is assumed to be a model of G. The following is the *Model Theoretical Incompleteness Theorem* :

THEOREM 2 Assume that *rep* is N-semantically definable and that N is semantically $\leq$-descending. Assume that *ded* is N-semantically defined by $\mathbf{DED}(x_0,y)$. Assume (H_0), (H_1). Then it holds that

(i) the fixpoint w to $\neg\,\mathbf{BEW}(x_0)$ is a sentence unprovable and true in A.

(ii) If A is a model of G then P and P_s are proper subsets of T and T_s.

I begin the proof by showing that P is N-semantically defined by $\mathbf{BEW}(x_0)$, i.e. that

(A0) $\ulcorner v \urcorner \epsilon P$ if and only if $\mathbf{BEW}(\gamma_v)$ is true in A.

For let $[m]$ be a derivation of v. Then $<\ulcorner v \urcorner, m>$ is in *ded*, and N-semantical definability of this relation implies that $\mathbf{DED}(\gamma_v, \kappa_m)$ is true in A; hence also $\mathbf{BEW}(\gamma_v)$ is true in A. Conversely, if the sentence $\exists y\, \mathbf{DED}(\gamma_v, y)$ is true in A, then (H_0) implies that there exists m in u(A) satisfying $\mathbf{DED}(\gamma_v, y)$. Then m is in N and thus interprets κ_m in A whence $\mathbf{DED}(\gamma_v, \kappa_m)$ is true in A. Hence N-semantical definability implies that $<\ulcorner v \urcorner, m>$ is in *ded*; thus $[m]$ is a derivation of v, and $\ulcorner v \urcorner$ is in P. – For the rest of the proof I now use the sentence w such that

(AF) $w \longleftrightarrow \neg\,\mathbf{BEW}(\gamma_w)$ is true in A,

which exists by Lemma 7. I proceed in two steps:

(A1) $\vdash w$ does not hold.

Because otherwise

(A10) $\vdash w$ whence $\mathbf{BEW}(\gamma_w)$ would be true in A by (A0), and
(A11) from $\vdash w$ there would follow that w is true in A by (H_1), hence $\neg\mathbf{BEW}(\gamma_w)$ would be true in A by (AF), and this contradicts (H_0).

(A2) w is true in A.

For $\ulcorner w \urcorner$ is not in P by (A1), hence $\mathbf{BEW}(\gamma_w)$ is not true by (A0) whence $\neg\,\mathbf{BEW}(\gamma_w)$ is true in A by (H_0), and thus also w is true because of (AF).

This argument proves (i), and this implies (ii) if provable formulas are true. I observe that, under this assumption, the proper inclusion of P_s in T_s follows already from (A0) since T cannot be N-semantically definable after the corollary to the undefinability theorem. I also observe that (i) may be obtained as a special case of the Combinatorial Incompleteness Theorem if I describe the present situation in the following ***third interpretation***. Its only difference from the *first* interpretation, given in the preceding section, is the use of a different equivalence relation A_π which shall consist of all pairs $<n,m>$ such that

$n = m$ or ($<n,m> \in N_0 \times N_0$ and $[n] \leftrightarrow [m]$ is true in A) .

The function f is defined as before; again (10) holds, and (11) follows from (AF). The set T is closed under A_π ; for b = ⌜BEW(x_0)⌝ and $-b$ = ⌜¬BEW(x_0)⌝ there follows (30) from (A0), and (31) holds since $[f(x,b)]$ and $[f(x,-b)]$ always are sentences.

Theorem 2 depends essentially on the semantical hypotheses (H_0), (H_1). While (H_0) is external, the hypothesis (H_1) is model theoretical since it connects provability from G and truth in A. (H_1) is used in the proof of (A1) for the sentence w, and although the syntactical form of this sentence is known from Corollary 1 of the fixpoint lemma, it cannot be excluded that its provability indeed requires the axioms in G. Since incompleteness is a purely syntactical concept, concerning alone the notion of provability from G, there arises the task to search for other incompleteness theorems which do not require model theoretical hypotheses.

I shall begin with some more definitions. The set G is said to be

ω-consistent if
for every formula u(y): if for all m in N $\vdash u(\kappa_m)$, then *not* $\vdash \neg\forall y\, u(y)$;

ω-incomplete if
there exists a formula u(y): for all m in N: $\vdash u(\kappa_m)$, and *not* $\vdash u(y)$.

A set which is ω-consistent is also consistent, because if u(y) is the formula $y \equiv y$ then $\vdash u(y)$ implies $\vdash u(\kappa_m)$ for all m in N, whence ω-consistency shows that not *all* sentences are derivable.

The definition of ω-incompleteness does not require quantifiers; if quantifier logic is available then an ω-consistent set, which is ω-incomplete, will also be incomplete per se. Because if u(y) is the formula for which ω-incompleteness prevails, then both *not* $\vdash \forall y\, u(y)$ and *not* $\vdash \neg\forall y u(y)$. If L is only an *open* language then the usual notion of incompleteness is inadequate and *must* be replaced by that of ω-incompleteness: if A, for instance, is a structure on the set of natural numbers, if L permits it to speak about addition and multiplication, and if u(y) is the formula $y+y \equiv y \cdot y$, then neither u(y) nor $\neg u(y)$ belongs to the set Th(A) of formulas true in A which, understandably, I would wish to consider as complete.

The following is GÖDEL's *Incompleteness Theorem* :

THEOREM 3 Assume that *rep* is functionally representable and that *ded* is represented by DED(x_0,y) .

If G is consistent then it is ω-incomplete; if G is ω-consistent then it is incomplete.

Here the second statement will follow immediately from the first one. The proof of the first statement I start with the observation

(B0) if $\vdash$ v then $\vdash$ BEW(γ_v) ,

because if [m] is a derivation of v then $\vdash$ DED(γ_v, κ_m) follows from the representability of *ded*, hence also $\vdash$ BEW(γ_v) . Further, let w be the sentence such that

(BF) $\vdash$ w $\leftrightarrow$ $\neg$BEW(γ_w)

which exists by the fixpoint lemma. I proceed again in two steps:

(B1a) $\vdash$ w does not hold .

Because otherwise

(B10) $\vdash$ w whence $\vdash$ BEW(γ_w) by (B0) , and
(B11) from $\vdash$ w there would follow $\vdash$ $\neg$BEW(γ_w) by (BF) , and this contradicts the consistency of G .

(B1b) *not* $\vdash$ $\neg$DED(γ_w,y) .

Because $\vdash$ $\neg$DED(γ_w,y) would imply $\vdash$ $\forall$y $\neg$DED(γ_w,y) , and thus
$\vdash$ $\neg$BEW(γ_w) . But then w would be provable by (BF) .

(B2) for all m in N: $\vdash$ $\neg$DED(γ_w,κ_m) .

Because as w has no derivation [m], none of <⌜w⌝,m> is in *ded*, and thus my claim follows from the representability of *ded*.

Thus the set G is ω-incomplete for u(y) = $\neg$DED(γ_w,y). – More generally, G also is ω-incomplete for every formula $\neg$DED(κ_n,y) such that n is in N–P and *not* $\vdash$ $\neg$BEW(κ_n): taking w to be [n], the statements (B1b) and (B2) remain in effect (and the use of (BF) at the end of (B1b) disappears). Instead of applying (BF), the existence of such n in N–P can be concluded from the undefinability theorem saying that P is not representable if G is consistent; because of (B0) this can only be the case if such an n exists.

It will be observed that the proof was carried out completely parallel to that of Theorem 2; this was done in order to make visible the different effects of the different hypotheses. The content of (B2) corresponds to that part of (A2) where from *not* $\vdash$ w the falsity of BEW(γ_w) is concluded (i.e. the contraposition of the argument from right to left in (A0)). It will be a useful illustration of the notion of ω-incompleteness to compare this in detail.

In the model theoretic situation, I concluded in (A2) from *not* $\vdash$ w that no pair <⌜w⌝,m> is in *ded*; hence for every m the sentence DED(γ_w,κ_m) is *not* true; this corresponds exactly to (B2). From the assumption that any b satisfying DED(γ_w,y) must be in N, it then followed that there is no b in A satisfying this formula, hence BEW(γ_w) was false and $\neg$BEW(γ_w) was true.

Phrased differently, all sentences $\neg\mathrm{DED}(\gamma_w,\kappa_m)$ were true and the sentence $\forall y\neg\mathrm{DED}(\gamma_w,y)$, equivalent to $\neg\mathrm{BEW}(\gamma_w)$, was also true. In the present, syntactical situation though I do not know more than (B2) since the sentence $\forall y\neg\mathrm{DED}(\gamma_w,y)$, provably equivalent to w by (BF), cannot be proved by (B1a).

Of course, I know from the definitions that (1) truth of a universally quantified sentence $\forall x u(x)$ follows from the truth of all of its substitution instances, provided there are constant terms naming all objects of the universe under consideration, whereas (2) provability of $\forall x u(x)$ requires the uniformity of a proof of u(y) with a free variable y. An axiom system which is ω-incomplete presents an example in which the notions of truth and provability for such sentences actually differ. Speaking in the language of models, truth refers to one of the models of G, while provability refers to all of them; the appearance of *non standard* models, discussed for arithmetic in Chapter 1.16 and known since SKOLEM [23], is another, model theoretical means to illustrate this distinction.

The following corollary shows that I actually do not need quantifiers but only constants in order to produce the phenomenon of ω-incompleteness:

COROLLARY 1 Let L be an open language and strengthen the assumptions such that *rep* now is representable by a term. If G is consistent in the sense that there is no sentence provable simultaneously with its negation, then G will be ω-incomplete.

Because it then follows from Lemma 6a that there is a formula w(y) such that

(BF') $\vdash w(y) \leftrightarrow \neg\mathrm{DED}(\gamma_w,y)$.

I then vary the preceding proof as follows:

(B1a') $\vdash$ w does not hold .

Because otherwise

(B10') $\vdash$ w with a derivation **[**m**]** whence $\vdash \mathrm{DED}(\gamma_w,\kappa_m)$, and
(B11') from $\vdash$ w there would follow $\vdash \neg\mathrm{DED}(\gamma_w,y)$ by (BF') , thus a fortiori $\vdash \neg\mathrm{DED}(\gamma_w,\kappa_m)$, contradicting the consistency of G.

(B1b') *not* $\vdash \neg\mathrm{DED}(\gamma_w,y)$.

(B2') for all m in N: $\vdash \neg\mathrm{DED}(\gamma_w,\kappa_m)$.

The two incompleteness theorems are related in still a further way which concerns the *content* of the sentences that establish incompleteness. If G is ω-consistent then it is incomplete for the fixpoint w from (BF), hence also for the sentence

$c = \forall y \neg DED(\gamma_w, y)$

which is classically equivalent to $\neg BEW(\gamma_w)$. Making now semantical assumptions about the structure A, it becomes possible to speak about the *content* of this sentence – which coincides with that of w. Because assume the assumptions of the model theoretical Theorem 2 to hold; then it follows that w and, therefore, c is true. But *that means per definitionem* that for every m in N also $\neg DED(\gamma_w, \kappa_m)$ is true, i.e. that w has *no derivation*. Consequently, the fact stated in Theorem 2 (viz. that w is an unprovable, but true sentence) becomes the *content* of the (true but) unprovable sentence in Theorem 3.

Another noteworthy consequence of GÖDEL's incompleteness theorem is the appearance of axiom systems G which are consistent and, at the same time, are

ω-inconsistent:
there exists a formula u(y): for all m in N: $\vdash u(\kappa_m)$, and $\vdash \neg \forall y\, u(y)$.

To see this, I start from the assumptions of GÖDELs theorem and, again, work with the fixpoint w to $\neg BEW(x_0)$. If G" is obtained from G by adding the sentence $\neg w$, then

(B3) G" is consistent.

Because otherwise every sentence could be derived from G", thus in particular w. Hence there would be a finite number of sentences $g_0, \ldots, g_{r-1}$ in G such that w could be derived from them together with $\neg w$. But as these are sentences, the deduction rule would lead to a derivation of $\neg w \rightarrow w$ from $g_0, \ldots, g_{r-1}$, i.e. from G. Since $(\neg a \rightarrow a) \rightarrow a$ is a tautology, it would follow that also w can be derived from G, and this contradicts (B1a). – On the other hand, G" is ω-inconsistent, viz.

(B4) for all m in N: $G'' \vdash \neg DED(\gamma_w, \kappa_m)$ and $G'' \vdash \neg \forall y \neg DED(\gamma_w', y)$

because that, which can be derived from G, *a fortiori* can be derived from G", and the latter sentence in (B4) is provably equivalent to $\neg w$ by (BF). – If, in addition, I add the semantical hypotheses of the model theoretical Theorem 2, then the sentence $\neg w$ is false in A, and thus A will not be a model of the consistent set G".

Choosing a different fixpoint in a suitable manner, it becomes possible to state a third, *Semantical Incompleteness Theorem* which produces an unprovable, true sentence alone from *external* semantical assumptions. To this end I consider the function *ne* such that

$ne(\ulcorner v \urcorner) = \ulcorner \neg v \urcorner$

and use, besides (H_0), the hypotheses

(H_2) If w is provably equivalent to a sentence $BEW(\kappa_n)$ then w is true (or not true) in A simultaneously with that sentence. .

(H_3) If $\mathrm{DED}(\kappa_n,\kappa_m)$ is true in A then $\vdash \mathrm{DED}(\kappa_n,\kappa_m)$.

THEOREM 4 Assume that *rep* is functionally representable and that *ded* is represented by $\mathrm{DED}(x_0,y)$; assume that *ne* is represented by a term $\mathrm{NE}(x)$. Assume (H_0), (H_2), (H_3), and let w be the fixpoint to $\mathrm{BEW}(\mathrm{NE}(x_0))$.

If G is consistent then $\neg$w is an unprovable, true sentence.

By definition of w, I have

$$\vdash w \longleftrightarrow \mathrm{BEW}(\mathrm{NE}(\gamma_w)) .$$

If w were true then so would be $\mathrm{BEW}(\mathrm{NE}(\gamma_w))$ by (H_2). Hence by (H_0) also $\mathrm{DED}(\mathrm{NE}(\gamma_w),\kappa_m)$ would be true for some m, and then (H_3) would lead to $\vdash \mathrm{DED}(\mathrm{NE}(\gamma_w),\kappa_m)$. But *ded* is represented by $\mathrm{DED}(x,y)$, and thus the pair $<ne(\ulcorner w \urcorner),m>$ would be in *ded*; hence $[m]$ would be a derivation of $\neg$w. But $\vdash \mathrm{BEW}(\mathrm{NE}(\gamma_w))$ would imply $\vdash$ w , and this would contradict the consistency of G. Thus w is a false sentence, and so $\neg$w is true. If $\neg$w were provable then a derivation $[m]$ of $\neg$w would give $<ne(\ulcorner w \urcorner), m> \in ded$, hence $\vdash \mathrm{DED}(\mathrm{NE}(\gamma_w),\kappa_m)$ and thus $\vdash \mathrm{BEW}(\mathrm{NE}(\gamma_w))$. Consequently, also w would be provable, and this once more would contradict the consistency of G.

Comparing (H_1) and (H_2), I begin by observing that the sentence $\neg \mathrm{BEW}(\kappa_n)$ in (H_1) is of a higher complexity than the sentence $\mathrm{BEW}(\kappa_n)$ in (H_2). Further, in (H_2) I deal with provable instead of semantical equivalence, and in examples, to be considered later, such provability will depend alone on certain elementary arithmetical assumptions but not on the full content of the axioms in G. Therefore, also the semantical equivalence in (H_2), concerning truth in A, will follow already from these elementary arithmetical assumptions, and will avoid possible model theoretical hypotheses about G. As for the new hypothesis (H_3), it will be seen that the sentences $\mathrm{DED}(\kappa_n,\kappa_m)$ are of a particularly simple kind, and thus it will become possible to conclude also the provability claimed in (H_3) from those elementary, external-semantical assumptions.

5. Incompleteness, continued

Under the hypotheses made so far, the syntactical methods lead to the result that a consistent set G is ω-incomplete. I now shall strengthen the hypotheses concerning G in order to be able to conclude that a consistent set G actually is incomplete. So I shall assume the same data L and N as before, but in addition also that ω be ***provably initial***:

L contains, besides $\equiv$, an other 2–ary predicate symbol written $\leq$,

N is countably infinite, and thus shall be assumed right away to be the set ω of natural numbers,

G proves for every n in ω the formulas

(MINn) $\vdash y \leq \kappa_n \rightarrow (y \equiv \kappa_0 \vee y \equiv \kappa_1 \vee \ldots \vee y \equiv \kappa_n)$

(CMPn) $\vdash \neg\, y \leq \kappa_n \rightarrow \kappa_n \leq y$.

I now can state ROSSER's *Incompleteness Theorem* :

THEOREM 5 Assume that the functions *rep* and *ne* are representable and that *ded* is represented by $\mathrm{DED}(x_0,y)$.

If G is consistent then G is incomplete.

The proof proceeds in two steps, the first of which is preparatory only. I consider the relation *neded* of all $<n,m>$ such that $<ne(n),m> \in ded$. By Lemma 5, I find a formula $\mathrm{NE}(x,x_1)$ functionally representing *ne*; I define the formula $\mathrm{NEDED}(x,y)$ as

$$\exists x_1\, (\mathrm{NE}(x,x_1) \wedge \mathrm{DED}(x_1,y)) \,.$$

I want to show that *neded* is represented by NEDED. Given n, let q abbreviate $ne(n)$. If $<n,m>$ is in *neded* then $\vdash \mathrm{DED}(\kappa_q,\kappa_m)$ and $\vdash \mathrm{NE}(\kappa_n,\kappa_q)$ imply $\vdash \mathrm{NEDED}(\kappa_n,\kappa_m)$. On the other hand, if $<n, m>$ does not belong to *neded* then $\vdash \neg\mathrm{DED}(\kappa_q,\kappa_m)$ implies

$$\vdash x_1 \equiv \kappa_q \rightarrow \neg\mathrm{DED}(x_1,\kappa_m) \,.$$

As $\mathrm{NE}(x,x_1)$ functionally represents *ne*, it follows from $ne(n) = q$ that

$$\vdash \mathrm{NE}(\kappa_n,x_1) \rightarrow x_1 \equiv \kappa_q$$

whence now

$$\vdash \mathrm{NE}(\kappa_n,x_1) \rightarrow \neg\mathrm{DED}(x_1,\kappa_m)$$

$$\vdash \forall x_1\, (\mathrm{NE}(\kappa_n,x_1) \rightarrow \neg\mathrm{DED}(x_1,\kappa_m))$$

$$\vdash \forall x_1\, \neg(\mathrm{NE}(\kappa_n,x_1) \wedge \mathrm{DED}(x_1,\kappa_m))$$

$$\vdash \neg\, \exists x_1\, (\mathrm{NE}(\kappa_n,x_1) \wedge \mathrm{DED}(x_1,\kappa_m))$$

and thus $\vdash \neg\mathrm{NEDED}(\kappa_n,\kappa_m)$.

I now introduce technical tool by defining, for every natural number m, two relations pe_m and qu_m both of arity m+2: let $<j, k_0,\ldots, k_m>$ be in pe_m, respectively in qu_m, if there exists i with $i \leq m$ such that $<j,k_i>$ is in *ded* , respectively is in *neded*. I then find that pe_m and qu_m are represented by the formulas

$$P_m(x_0, y_0, \ldots, y_m) = \mathrm{DED}(x_0,y_0) \vee \ldots \vee \mathrm{DED}(x_0,y_m)$$

and

$$Q_m(x_0, y_0, \ldots, y_m) = \mathrm{NEDED}(x_0,y_0) \vee \ldots \vee \mathrm{NEDED}(x_0,y_m) .$$

Because if $<j, k_0, \ldots, k_m>$, say, is *not* in pe_m then none of $<j, k_i>$ is in *ded*, and thus for every i with $i \leq m$ there holds $\vdash \neg \mathrm{DED}(\kappa_j, \kappa(k_i))$. – Now if G is assumed to be consistent, then pe_m and qu_m will also be *formally defined* by P_m and by Q_m – and it is this fact alone which will be needed (although it could not have been shown, say for pe_m, if *ded* only had been assumed to be formally definable by DED: the provability of a disjunction does not imply the provability of one of its members!).

Now the proof can begin. I define

$$v(x_0) = \forall y\, (\mathrm{DED}(x_0,y) \rightarrow \exists z\, (z \leq y \wedge \mathrm{NEDED}(x_0,z))) ,$$

and by the Fixpoint Lemma I find the sentence w such that $\vdash w \longleftrightarrow v(\gamma_w)$. I shall show that neither w nor $\neg w$ is provable.

Assume $\vdash w$. There then would be an m such that $<\ulcorner w \urcorner, m>$ is in *ded*, hence $\vdash \mathrm{DED}(\gamma_w, \kappa_m)$. By choice of w, there also would hold $\vdash v(\gamma_w)$ where $v(\gamma_w)$ may be written as $\forall y\, u(\gamma_w, y)$ with

$$u(\gamma_w,y) = \mathrm{DED}(\gamma_w,y) \rightarrow \exists z\, (z \leq y \wedge \mathrm{NEDED}(\gamma_w,z)) .$$

Now $\vdash \forall y\, u(\gamma_w,y)$ implies $\vdash u(\gamma_w,\kappa_m)$. Together with $\vdash \mathrm{DED}(\gamma_w,\kappa_m)$ this would give

$$\vdash \exists z\, (z \leq \kappa_m \wedge \mathrm{NEDED}(\gamma_w,z))$$

by modus ponens. Applying (MINn) to the first formula in the conjunction, I would find

$$\vdash \exists z\, ((z \equiv \kappa_0 \vee \ldots \vee z \equiv \kappa_m) \wedge \mathrm{NEDED}(\gamma_w,z))$$

which would imply $\vdash Q_m(\gamma_w, \kappa_0, \ldots, \kappa_m)$. As qu_m is formally defined by Q_m, there now would exist an i with $i \leq m$ such that $< ne(\ulcorner w \urcorner), i>$ were in *ded*. But then $[ne(\ulcorner w \urcorner)]$ would have a derivation, and since $[ne(\ulcorner w \urcorner)] = \neg [\ulcorner w \urcorner] = \neg w$, this would contradict the consistency of G.

Assume now $\vdash \neg w$. There then would be an m such that $< ne(\ulcorner w \urcorner), m>$ is in *ded*, hence $\vdash \mathrm{NEDED}(\gamma_w,\kappa_m)$ and, a fortiori, $\vdash \neg y \leq \kappa_m \rightarrow \mathrm{NEDED}(\gamma_w,\kappa_m)$. Applying (CMPn), there would follow

$$\vdash \neg y \leq \kappa_m \rightarrow (\kappa_m \leq y \wedge \mathrm{NEDED}(\gamma_w,\kappa_m))$$

$$\vdash \neg y \leq \kappa_m \rightarrow \exists z\, (z \leq y \wedge \mathrm{NEDED}(\gamma_w,z))$$

$$\vdash \neg \exists z\, (z \leq y \wedge \mathrm{NEDED}(\gamma_w,z)) \rightarrow y \leq \kappa_m .$$

Since $\vdash w \longleftrightarrow v(\gamma_w)$ holds, I also would have $\vdash \neg v(\gamma_w)$ which I can write as

$$\vdash \neg \forall y\, (\mathrm{DED}(\gamma_w,y) \rightarrow \exists z\, (z \leq y \wedge \mathrm{NEDED}(\gamma_w,z)))$$

$$\vdash \exists y\, (\mathrm{DED}(\gamma_w,y) \wedge \neg \exists z\, (z \leq y \wedge \mathrm{NEDED}(\gamma_w,z))) .$$

Together with the previously established derivability, this would lead to

$$\vdash \exists y\ (\mathrm{DED}(\gamma_w,y) \wedge y \leq \kappa_m)$$

and then (MINn) would give

$$\vdash \exists y\ (\mathrm{DED}(\gamma_w,y) \wedge (y \equiv \kappa_0 \vee \ldots \vee y \equiv \kappa_m))$$

which implies $\vdash P_m(\gamma_w, \kappa_0, \ldots, \kappa_m)$. As pe_m is formally defined by P_m, there then would be an i with $i \leq m$ for which $<\ulcorner w \urcorner, i>$ were in *ded* . But then w would have a derivation, contradicting the consistency of G .

The following remarks about this proof are important:

(1) The proof *does* implicitly use assumptions on the connection between formal provability and the structure of the set N: the functions *rep* and *ne* and the relation *ded* are assumed to be representable, and the relations pe_m, qu_m, representable in consequence of these assumptions, are used in the proof.

(2) Apart from these hypotheses about representability, the arguments in the proof do *not* use explicit semantic assumptions about N , i.e. about ω. The predicate symbols $\equiv$ and $\leq$ of L do *not* need to be interpreted, in particular, they do *not* need to be interpreted by the identity and by the usual $\leq$-relation between numbers - just as the constant terms κ_n in the references to (MINn) and (CMPn) do *not* need to be interpreted by the objects in N . That *which is* employed of the natural numbers is alone that they are used to count (a) formulas and (b) the constant terms κ_n (and for this reason the index set N of those constants may be chosen as ω itself). These two facts are required in order to define the formulas (MINn) which contain a disjunction of n+1 members depending on the index n of κ_n . While there is no semantical interpretation of the formulas (MINn), there *are* operations with the formulas P_m, Q_m, but they do not use the constant terms κ_m, much less so an interpretation of the κ_m by, say, their indices m in N .

(3a) Let me now assume that the given structure A is a model of G , and let me also make the assumption, mentioned at the beginning of the preceding section, that elements a,b satisfying $\mathrm{DED}(x_0,y)$ in A already belong to N . Then it follows that $v(\gamma_w)$ is true in A . Because $\mathrm{DED}(\gamma_w,\kappa_m)$ is false for every m in N since w is not provable. Hence $u(\gamma_w,\kappa_m)$ is true for every such m , and since in $v(\gamma_w) = \forall y\ u(\gamma_w,y)$ the variable y needs to run only over N , also $v(\gamma_w)$ will be true.

Let me assume, additionally, that the predicate symbol $\leq$ of L *is* interpreted in A by the usual $\leq$-relation on the set $N = \omega$. Then the true, but not provable sentence $v(\gamma_w)$ says that, for every derivation [m] of the sentence w (provably equivalent to $v(\gamma_w)$) there exists a derivation of $\neg w$ with a code smaller than m . Again, therefore, the fact that w is an unprovable sentence of a particular shape now becomes the content of the true, but unprovable sentence $v(\gamma_w)$.

(3b) The formula $u(\gamma_w, \kappa_m)$ is of a particularly simple form; besides the formulas DED and NEDED it also contains a quantifier $\exists z$ whose effect, however, is bounded by $z \leq k_m$. In later examples, formulas of this type will reveal themselves as being *primitive recursive*. This will have the consequence that, in order to evaluate $u(\gamma_w, \kappa_m)$ semantically, I do not need A to be a model of full arithmetic: rather it will suffice to know of such primitive recursive sentences that, if provable, they are true in A.

(4) I continue to consider the sentences $v(\gamma_w)$ and $\neg v(\gamma_w)$ under the assumptions in (3a). From G, I form the set G' by adjoining the first and the set G" by adjoining the second of these sentences. Then both G' and G" are consistent, as follows with the same argument which in the preceding section was used to derive (B3); in case of G' I here use the tautology $(a \to \neg a) \to \neg a$. In analogy to (B4) then also here G" will be ω-inconsistent. Making the assumption (3a), A still is a model of G', but not a model of G". Can it be shown that now the relation *ded'* determined by G', respectively the relation *ded"* determined by G", is representable (for G' respectively G"), then also G', respectively G", will be incomplete again.

6. Set Theory - an Example

The general theorems established in this Chapter were concerned with a set G of sentences - an axiom system - which was not further specified, though it had to have certain properties. Of these, the only ones not evident were those requiring representability (of the syntactical functions and relations *rep*, *ded* and *ne*). For representability requires to provide proofs, and proofs can be provided only if the axioms in G are sufficiently strong. Now every mathematician knows that everything he ever does can be phrased in the language, and can be derived from the axioms, of set theory. Taking G to be such axiom system should, therefore, provide an easy example. And indeed it does - as long as I restrict myself to simple semantical considerations.

So I consider a language L, at least containing the predicate symbols $\equiv$ and ϵ, and a set G of L-sentences forming an axiom system for set theory - it will not matter which particular system I choose. And for the sake of simplicity, let me assume that I am given a model A of G. My first task will be to find the set N and the maps g_T, g_F, g_S. Now I do have infinitely many L-formulas, and as g_F shall be injective, also N will have to be infinite. But then I need the names κ_n for the n from N; as they are not yet *in* L, I will need a definitorial expansion of L which introduces these constant terms κ_n - meaning that I need a formal definition in L for every single one of them. It is this request for definitions which excludes the most simple-minded choice of N as u(A).

The second most simpleminded choice is that in which N consists of natural numbers of A. There is an obvious set theoretical definition of the empty set, which I choose as $\kappa_0{}^A$, and with the set theoretical successor map s^A, defined by $s^A(x) = x \cup \{x\}$, I then take $\kappa_{n+1}{}^A$ to be $s^A(\kappa_n{}^A)$. If s is the function symbol corresponding to s^A, I then have for κ_{n+1} the defining axiom $\kappa_{n+1} \equiv s(\kappa_n)$. Taking N to be the set of all these $\kappa_n{}^A$, I obtain an infinite set of natural numbers in A. Of course, the set N will, in general, *not* be the set of *all* natural numbers of A; its members are called the *standard numbers* of A. Also, (exterior) induction (on the n of $\kappa_n{}^A$) shows that N is semantically ϵ-descending in A. From the recursive definitions of the usual arithmetical operations it follows immediately that, e.g., $\kappa_a{}^A + \kappa_b{}^A = \kappa_{a+b}{}^A$ holds in A; hence the set N is closed under these operations.

Having decided to use A-numbers as elements of N, it remains to set up the maps g_T, g_F, g_S, and this is easy for the first two of them. For already in Chapter 1.5 I have pointed out several arithmetical realizations of term algebras (with at most countably many operations). For both the algebra of terms and the algebra of formulas of my language L for set theory, such arithmetical realizations can be copied within the set N of standard numbers of A, and so g_T and g_F become obvious. As for g_S, for modus ponens calculi I need an arithmetical coding of sequences of formulas, and for sequent calculi I need to repeat this in order to code trees of such sequences; again, this is easily achieved by coding sequences of numbers such that they become the sequences of exponents in the prime factor decomposition of their code.

Questions of definability demand a more careful consideration. I shall denote by ϵ_A the element relation of the model A; I shall denote, for every a in A, by M(a) the subset of u(A) consisting of all b in A such that $b \epsilon_A a$; thus M(a) is the *externalization* of a, i.e. the external set which externally contains the objects of A being internally the elements of a. If ω_A is the object in A which, in A, is the A-set of natural numbers, then $M(\omega_A)$ *is* semantically definable by the usual formula describing the set of natural numbers; the subset N of $M(\omega_A)$, however, will in general be not definable. Consider now, say, the range $im(g_T)$ of the function g_T. It consists of the $\kappa_n{}^A$ for the numbers n which code terms. As does everything in the mathematician's world, these numbers n form a well defined set which can be described by a set theoretical formula. In the model A this formula will semantically define a set T^A, but if non-standard numbers are present in A then they will also be present in T^A, and so $im(g_T)$ will be the intersection $N \cap T^A$. Consequently, $im(g_T)$ will be N-semantically definable by that set theoretical formula. In the same way, $im(g_F)$ will be N-semantically definable as $N \cap F^A$ with F^A semantically defined in A.

Let me now discuss the substitution function *rep*. The construction, from a term t and formula v, of the formula $w = rep(x_0, t \mid v)$ can be described by a set theoretical (recursive) function rep_0. The set theoretical formula describing rep_0 determines in A a corresponding function rep_0 from $T^A \times F^A$ to F^A. to $im(g_F)$. Further, there is a function *co* assigning to n the code $\ulcorner \kappa_n \urcorner$ of κ_n, and *rep* then is nothing but the superposition $rep_0(co(-),-)$. Consequently, also *rep* has a definition by a set theoretical formula, and in A, again, that formula semantically defines a function from $\omega_A \times F^A$ to F^A which, intersected with N×N×N, results in *rep*. Thus also *rep* is N-semantically defined.

For the relation *ded*, the situation is the same. The relationship between a formula and a derivation of that formula (viewed either as a sequence of formulas or as a tree of sequents) has a set theoretical description, and thus *ded* is again N-semantically defined.

Consequently, the model theoretical incompleteness theorem holds for this example: if the axioms G of set theory have a model A then the fixpoint sentence w is true in A but is not provable. (Observe that the incompleteness theorem itself might not need the full assumption that a model exists; here, however, I have used this assumption in order to secure the required N-definabilities.) It is well known that there are set theoretical sentences w (e.g. the axiom of choice or the continuum hypothesis) which are independent from the axioms of set theory and, in contrast to the fixpoint sentence, have an *obvious* set theoretical *meaning*. The independent sentence w of the incompleteness theorem, however, is *uniform* in that it will be independent in all other examples as well.

Up to this point, the study of my example was a fairly easy matter. The assumption that A was a model of set theory was used in order to form the object ω_A (but not in order to form external subsets of A such as T^A and F^A) and in order to conclude from the set theoretical (recursive) definitions of functions such as rep_0, and from the set theoretical descriptions of formulas and deductions, that the function *rep* and the relation *ded* were N-definable. Turning now from semantical definability to syntactical representability, this situation changes drastically and for the worse. While the statement that an object $\kappa_n{}^A$ satisfies a formula v(x) may be dealt with rather abstractly (viz. by a straightforward induction over the subformulas of v(x)), the statement that $v(\kappa_n)$ is provable will, in general, demand that an actual proof be carried out. Keeping in mind the appearance of ω-inconsistency, it cannot be excluded that there is a proof of $\exists y\ v(\kappa_n,y)$, although for no m the sentence $v(\kappa_n,\kappa_m)$ may be provable. Actually, this is the syntactical counterpart to the semantical appearance of non-standard numbers which could be handled quite easily. But worse still, representability of *ded*, say, requires not only a proof of $DED(\kappa_n,\kappa_m)$ in case $<n,m>$ is in *ded* , but also a proof of $\neg DED(\kappa_n,\kappa)$ if $<n,m>$ is not in *ded*. Such is the paradox of syntactical incompleteness that, in order to provide a sentence w for which neither w nor $\neg$w is provable, I first will have to show a kind of *local completeness*, viz. representability, for some auxiliary formulas such as DED !

Clearly, formulas answering this immodest demand will have to be of a special, particularly simple structure – and to exhibit this will be one of the tasks ahead.

Let me illustrate this type of reasoning, required for proofs of representability, by looking at the simplest facts from arithmetic as they are needed already for the coding of terms and formulas. Representability of addition and multiplication (so to speak the *tables* for addition and multiplication) will follow by straightforward induction from the defining axioms of the κ_n and the recursion equations describing these two operations. As for the relation of divisibility, I shall avoid the formula $v_0(x,y) = \exists z(y \equiv x \cdot z)$ and rather use the formula $v_1(x,y) = \exists z(z \leq y \wedge y \equiv x \cdot z)$. For if m divides n such that $n = m \cdot i$ then also $\kappa_i \leq \kappa_n \wedge \kappa_n \equiv \kappa_m \cdot \kappa_i$ is provable from the multiplication tables, hence also $v_1(\kappa_m, \kappa_n)$ is provable. If, however, m does not divide n then representability of multiplication makes

$$(z \equiv \kappa_0 \vee z \equiv \kappa_1 \vee \ldots \vee z \equiv \kappa_n) \rightarrow \neg \kappa_n \equiv \kappa_m \cdot z$$

provable. Assuming now the provability of

$$\text{(MINn)} \qquad z \leq \kappa_n \rightarrow (z \equiv \kappa_0 \vee z \equiv \kappa_1 \vee \ldots \vee z \equiv \kappa_n)$$

also $\neg(z \leq \kappa_n \wedge \kappa_n \equiv \kappa_m \cdot z)$ becomes provable whence $\forall z \ \neg(z \leq \kappa_n \wedge \kappa_n \equiv \kappa_m \cdot z)$ and, therefore, $\neg v_1(\kappa_m, \kappa_n)$ will be provable. Making use of divisibility, I find that the set of all primes is representable. For it again suffices to define n as prime if for all m *such that* $1 < m \leq n$ it holds that m does not divide n: in this situation, the formula (MINn) can be put to use as before. By now, the reader can conclude that, quite generally, it will be *bounded quantification* which produces representable relations from representable ones (as long as they take place between natural numbers). It will, therefore, be no surprise that attention will be paid to the quantifications occurring in the definitions of the relations (and functions) expected to be representable.

At this stage, my considerations are taking a turn from set theory towards arithmetic. Concerning syntactical representability, the power of set theory does not appear to be of use, and so I now shall change from the study of axiom systems for set theory to that of axioms for arithmetic. The structure A then will right away have as underlying set the natural numbers $N = \omega$, and based now on arithmetical axioms I will have to carry out the program of developing a suitable family of representable arithmetical relations which, in the end, will lead to a representable *rep* and a representable *ded*. For this development, a particular technical problem comes from the fact that the various syntactical notions, such as terms, formulas, deductions, were defined recursively, and while it has been known since Dedekind that recursive definitions produce set theoretically definable objects, it is much less obvious how a recursively defined arithmetical function may be defined by an explicit formula of an arithmetical language. It is this the reason why the study of recursive functions and recursive relations will play a prominent rôle in the following Chapters.

References

J.B.Rosser: Extensions of some theorems of Gödel and Church. J.Symb.Logic 1 (1936) 87–91

Th.Skolem: Einige Bemerkungen zur axiomatischen Begründung der Mengenlehre. In: 5. skandinaviska matematikerkongressen i Helsingfors 1922. Helsinki 1923, 217–232

A.Tarski, A.Mostowski, R.M.Robinson: Undecidable Theories. Amsterdam 1953

Chapter 5. Elementary and Primitive Recursive Functions

The theorems on undefinability and incompleteness have so far been studied *in abstracto*, viz. for languages L, assumed to admit coding functions g_T, g_F g_S, and for axiom systems G, assumed to make representable, or definable, certain functions and relations coded with help of these functions, such as as *sub* and *ded*. In the following Chapters, I shall choose the set N, in which to code the language L, to be the set ω, and I shall study axiom systems G for (fragments of) arithmetic; for them I shall have to verify those assumptions to hold.

The task to exhibit ω-valued coding functions for a language L amounts to the re-construction of an arithmetical copy L_A of L, and the study of the coded functions and relations *sub* and *ded* requires to re-build these syntactical objects within that arithmetization L_A. Usually, most syntactical constructions proceed by recursion on the complexity of terms or formulas, where recursion is understood in some general, or set-theoretical sense. In an arithmetical copy L_A, recursion will be number theoretical, and various syntactical objects will arise as functions under recursion from simpler ones. Since it will be my aim to prove the representability of certain complicated objects, there will be many situations in which it has to be shown that representability is preserved under number theoretical recursion. Recalling the definition of representability by some formula, it should be plausible that this problem will not easily be susceptible to a direct attack.

It is this the reason to separate the following work into two parts. In a first step, to be performed in this Chapter and the next one, I shall study arithmetical functions and relations generated by number theoretical, or *primitive recursion*. In particular, I shall arrive at several characterizations of classes of functions closed under primitive recursion which, however, can be formulated without employing the notion of primitive recursion itself – instead some other concepts will be used which will be easier to handle when later investigating representability by formulas. This then will be carried out in the second step, beginning with Chapter 8. Before that, the arithmetization L_A will be constructed in Chapter 7.

The autonomous study of *recursive functions and relations* is, therefore, aimed to prepare the tools for the logical investigation of axioms systems for arithmetic. There are other, important aspects under which to study recursive functions, connected with the notion of computability, which cannot be discussed here.

The present Chapter is devoted to the study of *primitive recursively* closed classes of functions. They are typically defined by requesting them to contain a certain number of *initial* functions and to be *closed* under certain constructions, such as superposition or primitive recursion, leading from given functions to new ones. Various of their properties can be proven already for rather weak closure conditions, leading to the preparatory study of *simply closed* and of *elementarily closed* function classes. The principal results state that the primitive recursively closed classes are closed also under *course of values recursion*, and that the elementarily closed classes are closed under *bounded* primitive and *bounded* course of values recursion. The study of elementarily closed classes is of particular interest, because it will turn out in Chapter 7 that for most languages L an arithmetization L_A can be obtained already with *elementary* functions, i.e. with functions from the smallest elementarily closed class.

1. Basic Constructions

I shall consider the set ω of natural numbers. For $k>1$, let ω^k be the set of all k-tuplets α such that $\alpha = <\alpha_0, \ldots, \alpha_{k-1}>$ with members α_i in ω. If α is in ω^k and α' is $<\alpha_0, \ldots, \alpha_{k-2}>$ in ω^{k-1} then I shall write α sometimes as $<\alpha', \alpha_{k-1}>$, and related notations will be used analogously. Further, I set ω^1 to be ω itself. Subsets of ω^k, $k>0$, are called k-ary *relations*, and are denoted as R^k or S^k &c. By $-R^k$ I denote the complement of R^k in ω^k. The 2-ary identity relation = on ω^2 I shall often denote as E^2. When talking about a *map defined* on a natural number n, I shall assume that n is the *set* of numbers $0, 1, \ldots, n-1$.

By a k-ary function f^k, I understand a function defined on all of ω^k and with values in ω . The 1-ary successor function s is defined by $s(x) = x+1$. Throughout this Chapter, I shall distinguish between a function f^k and its *graph* $G(f^k)$, i.e. the (k+1)-ary relation consisting of all $<\alpha, a>$ such that $f^k(\alpha) = a$; thus functions here will be viewed as objects in their own right. The *characteristic function* χ_R of a relation $R = R^k$ is the k-ary function such that $\chi_R(\alpha) = 1$ for $\alpha \in R$ and $\chi_R(\alpha) = 0$ otherwise. There are two types of particularly simple functions: for every n in ω and every $k>0$ the

constant function c^k_n , defined by $c^k_n(\alpha) = n$ for every α ,

and for every $k>0$ and every $i<k$ the

projection function p^k_i , defined by $p^k_i(\alpha) = \alpha(i)$.

For functions f^k and $g^m_0, \ldots, g^m_{k-1}$, I define the

superposition $f^k \circ <g^m_0, \ldots, g^m_{k-1}>$

as the function h^m such that $h^m(\alpha) = f^k(g^m_0(\alpha), \ldots, g^m_{k-1}(\alpha))$. In case $k=1$, $f^k \circ <g^m_0>$ reduces to the composition $f^k \circ g^m_0$ of the two functions. For a relation $R = R^k$ and a sequence $g^m_0, \ldots, g^m_{k-1}$ of functions I define the

superposition $R^k[g^m_0, \ldots, g^m_{k-1}]$

as the relation S^m with the characteristic function $\chi_R \circ <g_0^m, \ldots, g_{k-1}^m>$; thus

$\alpha \in S^m$ if and only if $<g_0^m(\alpha), \ldots, g_{k-1}^m(\alpha)> \in R^k$.

I now shall list a number of basic constructions on relations and functions.

(CC1) *Transformation of Variables*. Let π be a map from k into m.

$g^m = f^k \circ <p_{\pi(0)}^m, \ldots, p_{\pi(k-1)}^m>$ is the function arising under *transformation of variables by* π from f^k.

$S^m = R^k[p_{\pi(0)}^m, \ldots, p_{\pi(k-1)}^m]$ is the relation arising under *transformation of variables by* π from R^k.

If α is in ω^m then $g^m(\alpha) = f^k(\alpha\pi)$, and $\alpha \in S^m$ if, and only if, $\alpha\pi \in R^k$. Special cases are:

Permutation of arguments: π is a bijection from k onto k.

Addition of empty (dummy) arguments: π is an order preserving injection from k into a larger number m. E.g. if π is the identity from k into k+1 then $g^{k+1}(\alpha,a) = f^k(\alpha)$, and $<\alpha,a> \in S^{k+1}$ if, and only if, $\alpha \in R^k$.

Contraction of identical arguments: π is a surjection from k onto a smaller number m. E.g. if $k = m+1$ and $\pi(i) = i$ for $i \leq m$, $\pi(m) = m-1$ then $g^m = f^k \circ <p_0^m, \ldots, p_{m-1}^m, p_{m-1}^m>$. If $\beta = \alpha\pi$ is in R^k and α is in S^m then $\beta(m) = \alpha\pi(m) = \alpha\pi(m-1) = \beta(m-1)$, and S^m consists of those α which arise from those β in R^k [the subset of R^k consisting of those β will be constructed in (CC4)].

(CC2) *Specification or contraction of a constant argument*. Let n be in ω and assume $i \leq k$.

$g^k = f^{k+1} \circ <p_0^k, \ldots, p_{i-1}^k, c_n^k, p_i^k, \ldots, p_{k-1}^k>$ is the function arising from f^{k+1} *by specifying to n the i-th argument*;

if $k = 2$ it is also written as $f^2(n,-)$ or $f^2(-,n)$. The relation S^k arising from R^{k+1} *by contracting to n the i-th argument* is that whose characteristic function arises that way from the characteristic function of R^{k+1}; is is the set of all $<\alpha_0, \ldots, \alpha_{i-1}, \alpha_{i+1}, \ldots, \alpha_k>$ with $\alpha \in R^{k+1}$ and $\alpha_i = n$.

(CC3) *Keeping an argument constant.* Let n be in ω and assume $i \leq k$. The relation S^{k+1}, arising from R^{k+1} *by keeping constant at* n *the i-th argument*, is the set of all α in R^{k+1} such that $\alpha_i = n$. Then $S^{k+1} = R^{k+1} \cap E^2[p_i^{k+1}, c_n^{k+1}]$.

(CC4) *Identification of arguments.* Let i and j be such that $i < j \leq k$. The relation S^{k+1}, arising from R^{k+1} by identifying the i-th argument with the j-th argument is the set of all α in R^{k+1} such that $\alpha_i = \alpha_j$. Then $S^{k+1} = R^{k+1} \cap E^2[p_i^{k+1}, p_j^{k+1}]$.

(CC5) *Bounded Quantification:* The four bounded quantifications of R^{k+1} are the (k+1)-ary relations $\exists_{\le}R^{k+1}$, $\forall_{\le}R^{k+1}$, $\exists_{<}R^{k+1}$, $\forall_{<}R^{k+1}$, defined as

$<\alpha,a> \in \exists_{\le}R^{k+1}$ if, and only if
there exists b such that $b \le a$ and $<\alpha,b> \in R^{k+1}$

$<\alpha,a> \in \forall_{\le}R^{k+1}$ if, and only if,
for every b: if $b \le a$ then $<\alpha,b> \in R^{k+1}$

$<\alpha,a> \in \exists_{<}R^{k+1}$ if, and only if,
$a>0$ and there exists b such that $b<a$ and $<\alpha,b> \in R^{k+1}$

$<\alpha,a> \in \forall_{<}R^{k+1}$ if, and only if,
$a>0$ and for every b: if $b<a$ then $<\alpha,b> \in R^{k+1}$.

The *internally bounded quantifications,* defined for every i such that $i<k$, are the k-ary relations $\exists^{i}_{\le}R^{k+1}$, $\forall^{i}_{\le}R^{k+1}$, $\exists^{i}_{<}R^{k+1}$, $\forall^{i}_{<}R^{k+1}$, defined by

$\alpha \in \exists^{i}_{\le}R^{k+1}$ if, and only if,
there exists b such that $b \le \alpha(i)$ and $<\alpha,b> \in R^{k+1}$

&c. They arise as $(\exists_{\le}R^{k+1})[p^k_0, \ldots, p^k_{k-1}, p^k_i]$ &c. Conversely, bounded quantifications can be reduced to internally bounded quantifications since e.g. $\forall_{\le}R^{k+1} = \forall^{k}_{\le}(R^{k+1}[p^{k+2}_0, \ldots, p^{k+2}_{k-1}, p^{k+2}_{k+1}])$.

(CC6) *Bounded Minimization:* The bounded minimizations of R^{k+1} are the two (k+1)-ary functions $\mu_{\le}R^{k+1}$ and $\mu_{<}R^{k+1}$, defined by

$\mu_{\le}R^{k+1}(\alpha,a)$ is the smallest b such that $b \le a$ and $<\alpha,b> \in R^{k+1}$,
provided such b exists, and is a+1 otherwise

$\mu_{<}R^{k+1}(\alpha,a)$ is the smallest b such that $b<a$ and $<\alpha,b> \in R^{k+1}$,
provided such b exists, and is a otherwise.

The *internally bounded minimizations*, defined for every i such that $i<k$, are the k-ary functions $\mu^{i}_{\le}R^{k+1}$ and $\mu^{i}_{<}R^{k+1}$, defined by

$\mu^{i}_{\le}R^{k+1}(\alpha)$ is the smallest b such that $b \le \alpha(i)$ and $<\alpha,b> \in R^{k+1}$,
provided such b exists, and is $\alpha(i)+1$ otherwise

&c. Thus $\mu^{i}_{\le}R^{k+1} \circ <p_0^k, \ldots, p^k_{k-1}, p^k_i>$ and $\mu^{i}_{<}R^{k+1} \circ <p_0^k, \ldots, p^k_{k-1}, p^k_i>$.

(CC7) *General Minimization*. The (general or unbounded) minimization of R^{k+1} is the (k+1)-ary relation νR^{k+1} consisting of all $<\alpha,a>$ in R^{k+1} for which there is no b such that $b<a$ and $<\alpha,b> \in R^{k+1}$.

If $<\alpha,a>$ and $<\alpha,a'>$ are in νR^{k+1} then $a=a'$. Hence is νR^{k+1} graph of a partial function, contained in R^{k+1}, which is denoted as μR^{k+1}. This will be a total function, defined on all of ω^k, if, and only if, for every α there exists (at least one) a such that $<\alpha,a> \in R^{k+1}$; in that case the relation R^{k+1} is called *full*.

Observe that $\mu_< R^{k+1}$, $\mu_\leq R^{k+1}$ are always total functions, and they are of arity k+1 while μR^{k+1} is of arity k.

For any - possibly partial - function f^k with the graph $G(f^k)$ there holds $f^k = \mu G(f^k)$.

The following formulas express certain of my constructions in terms of others of them; they will be used at later stages of the development.

(C1) $\forall_\leq R^{k+1} = -\exists_\leq -R^{k+1}$ and $\exists_\leq R^{k+1} = -\forall_\leq -R^{k+1}$.

(C2) $\forall_< R^{k+1} = -E^2[p_k^{k+1}, c_0^{k+1}] \cap -\exists_< -R^{k+1}$ and
$\exists_< R^{k+1} = -E^2[p_k^{k+1}, c_0^{k+1}] \cap -\forall_< -R^{k+1}$.

It is $<\alpha,a>$ *not* in $\exists_< -R^{k+1}$ if, and only if, $a = 0$ or if there is *no* b such that $b < a$ and $<\alpha,b> \epsilon -R^{k+1}$, i.e. $<\alpha,b> \epsilon R^{k+1}$ for every $b < a$.

(C3) $\forall_\leq R^{k+1} = (\forall_< R^{k+1} \cup E^2[p_k^{k+1}, c_0^{k+1}]) \cap R^{k+1}$ and
$\exists_\leq R^{k+1} = \exists_< R^{k+1} \cup R^{k+1}$.

(C4) $\forall_\leq R^{k+1} = (\forall_< R^{k+1})[p_0^{k+1},\ldots,p_{k-1}^{k+1}, s\circ p_k^{k+1}]$ and
$\exists_\leq R^{k+1} = (\exists_< R^{k+1})[p_0^{k+1},\ldots,p_{k-1}^{k+1}, s\circ p_k^{k+1}]$.

It is $<\alpha,a>$ in the first right hand set if, and only if, $<\alpha, s(a)> \epsilon \forall_< R^{k+1}$, i.e. if, and only if,

($s(a) \neq 0$) and $<\alpha,b> \epsilon R^{k+1}$ for every $b < s(a)$, i.e. for every $b \leq a$.

(C5) $\forall_< R^{k+1} =$
$\exists^k_\leq (G(s)[p_{k+1}^{k+2}, p_k^{k+2}] \cap (\forall_\leq R^{k+1})[p_0^{k+2},\ldots, p_{k-1}^{k+2}, p_{k+1}^{k+2}])$.

The set T^{k+2}, standing under the quantifier $\exists^k_\leq$, contains

all $<\alpha,a,c>$ such that $<c,a> \epsilon G(s)$ and $<\alpha,c> \epsilon \forall_\leq R^{k+1}$.

Hence $\exists^k_\leq T^{k+2}$ is the set of

all $<\alpha,a>$ such that there exists $c \leq a$: $a = s(c)$ and $<\alpha,c> \epsilon \forall_\leq R^{k+1}$.

(C6) $\exists_< R^{k+1} = (-E^2)[\mu_< R^{k+1}, p_k^{k+1}]$ and
$\exists_\leq R^{k+1} = (-E^2)[\mu_\leq R^{k+1}, s\circ p_k^{k+1}]$.

It is $<\alpha,a>$ in $\exists_< R^{k+1}$ if, and only if, $\mu_< R^{k+1}(\alpha,a) \neq a$.

(C7) $\exists_< R^{k+1} = (-G(\mu_< R^{k+1}))[p_0^{k+1},\ldots, p_k^{k+1}, p_k^{k+1}]$.

(C8) $\forall_< R^{k+1} = -E^2[p_k^{k+1}, c_0^{k+1}] \cap E^2[\mu_< -R^{k+1}, p_k^{k+1}]$ and
$\forall_\leq R^{k+1} = E^2[\mu_\leq -R^{k+1}, s\circ p_k^{k+1}]$.

It is $<\alpha,a>$ in $\forall_< R^{k+1}$ if $a \neq 0$ and if there is no b such that $b < a$ and $<\alpha,b> \in -R^{k+1}$; this is equivalent to $a \neq 0$ and $\mu_< -R^{k+1}(\alpha,a) = a$.

(C9) $$\nu R^{k+1} = R^{k+1} \cap (\forall_< -R^{k+1} \cup E^2[p_k^{k+1}, c_0^{k+1}]) .$$

(C10) $$\begin{aligned} G(\mu_< R^{k+1}) = {} & (E^2[p_k^{k+2}, p_{k+1}^{k+2}] \cap -\exists_< R^{k+1}[p_0^{k+2},\dots, p_k^{k+2}]) \cup \\ & (\exists_< R^{k+1}[p_0^{k+2},\dots, p_k^{k+2}] \cap R^{k+1}[p_0^{k+2},\dots, p_{k-1}^{k+2}, p_{k+1}^{k+2}] \cap \\ & (E^2[p_{k+1}^{k+2}, c_0^{k+2}] \cup \forall_< -R^{k+1}[p_0^{k+2},\dots, p_{k-1}^{k+2}, p_{k+1}^{k+2}])) . \end{aligned}$$

It is $<\alpha,a,b>$ in $G(\mu_< R^{k+1})$ if, and only if,

($a = b$ and not $<\alpha,a> \in \exists_< R^{k+1}$) or
($<\alpha,a> \in \exists_< R^{k+1}$ and $<\alpha,b> \in R^{k+1}$ and ($b = 0$ or not $<\alpha,b> \in \exists_< R^{k+1}$)).

(C11) $$S^{k+2} = R^{k+1}[p_0^{k+2},\dots, p_{k-1}^{k+2}, p_{k+1}^{k+2}] \cup E^2[p_k^{k+2}, p_{k+1}^{k+2}] \text{ is full, and } G(\mu_< R^{k+1}) = \nu S^{k+2} .$$

It consists S^{k+2} out of all $<\alpha,a,c>$ such that

(1) $<\alpha,c> \in R^{k+1}$ or $a = c$.

Hence for every such $<\alpha,a>$ there is a c such that $<\alpha,a,c>$ in S^{k+2}, i.e. S^{k+2} is full. There holds

(2) $<\alpha,a,c> \in \nu S^{k+2}$ if, and only if, $<\alpha,a,c> \in S^{k+2}$ and for all $b < c$: *not* $<\alpha,a,b> \in S^{k+2}$.

Consider the case $a = c$. Then $<\alpha,a,a> \in S^{k+2}$, and so the following statements become equivalent:

$<\alpha,a,a> \in \nu S^{k+2}$
for all $b < a$ *not* $<\alpha,a,b> \in S^{k+2}$ — by (2)
for all $b < a$ *not* $<\alpha,b> \in R^{k+1}$ — by (1), since $b < a$ implies $b \neq a$
$\mu_< R^{k+1}(\alpha,a) = a$.

Next consider the case $a \neq c$. Then $<\alpha,a,c> \in S^{k+2}$ and $<\alpha,c> \in R^{k+1}$ are equivalent by (1), and so the following statements become equivalent:

$<\alpha,a,c> \in \nu S^{k+2}$
$<\alpha,c> \in R^{k+1}$ and for all $b < c$: not $<\alpha,a,b> \in S^{k+2}$ — by (2)
$<\alpha,c> \in R^{k+1}$ and for all $b < c$: $b \neq a$ and not $<\alpha,a,b> \in S^{k+2}$ — since $<\alpha,a,a>$ is in S^{k+2}
$<\alpha,c> \in R^{k+1}$ and $c < a$ and for all $b < c$: not $<\alpha,b> \in R^{k+1}$ — by (1) and since $c \neq a$
$\mu_< R^{k+1}(\alpha,a) = c$.

(C12) $$G(\mu_\leq R^{k+1}) = (G(\mu_< R^{k+1}))[p_0^{k+2},\dots, p_{k-1}^{k+2}, s \circ p_k^{k+2}, p_{k+1}^{k+2}] .$$

I now observe some easy consequences:

(D1) If g^m (respectively S^m) arises from f^k (respectively R^k) under transformation of variables by it π and if h^n (respectively T^n) arises from g^m (respectively S^m) under transformation of variables by σ then h^n (respectively T^n) arises from f^k (respectively R^k) under transformation of variables by $\sigma\pi$.

Because $g^m(\alpha) = f^k(\alpha\pi)$ and $h^n(\beta) = g^m(\beta\sigma)$ imply $h^n(\beta) = f^k(\beta\sigma\pi)$, and $\alpha \epsilon S^m$ is equivalent to $\alpha\pi\epsilon R^k$ and $\beta\epsilon T^n$ is equivalent to $\beta\sigma\epsilon S^m$, hence also equivalent to $\beta\sigma\pi\epsilon R^k$.

A class **F** of functions, or a class **R** of relations, is called *trivially closed* if for each of its elements it contains also those which arise from it under transformation of variables.

A class **F** of functions is closed under superposition if together with f^k and $g_0^m, \ldots, g_{k-1}^m$ also $f^k \circ \langle g_0^m, \ldots, g_{k-1}^m \rangle$ is in **F**.

Let **R** be trivially closed and let **F** be a class of functions. Then **R** is called *closed under superposition* with functions from **F** if **R** contains the relations $R^k[g_0^m, \ldots, g_{k-1}^m]$ for every R^k in **R** and for every sequence of functions $g_0^m, \ldots, g_{k-1}^m$ which are either projections p_i^m or arise under transformation of variables from functions in **F**. (This includes in particular the case that **F** contains one function only.) The trivially closed classes **R** are those closed under superposition with projections, because transforming the variables of a projection gives a projection again.

If **R** is trivially closed and closed under superposition with the functions c_n^1 then the constructions (CC2) do not lead out of **R**.

The class **R** is called *positively-Boolean* closed if, together with R^k and S^k also $R^k \cup S^k$ and $R^k \cap S^k$ belong to **R**; it is called *Boolean* closed if, moreover, together with R^k also the complement $-R^k$ of R^k in ω^k belong to **R**.

(D2) Let **R** be trivially closed, closed under superposition with the functions c_n^1, and also positively-Boolean closed; let E^2 be in **R**. Then **R** contains ω^k and every finite subset of ω^k.

Observe first that $\omega^k = E^2[p_0^k, p_0^k]$ is in **R**. By (CC3) then **R** contains, for every β in ω^k and $i < k$, the set $M(\beta, i)$ consisting of all α in ω^k such that $\alpha_i = \beta_i$. Hence **R** contains the intersection of the k sets $M(\beta, i)$ which is $\{\beta\}$. Hence also every non-empty finite subset subset of ω^k is in **R**, and the empty subset I obtain as $\{\beta\} \cap \{\gamma\}$ for two distinct β and γ.

In the following, I shall consider classes **F** of functions (of varying arities). To every **F** I shall associate the class

R(**F**) of all relations R which have their characteristic function χ_R in **F**.

To every class **R** of relations I shall associate the class

F(**R**) of all functions which have their graphs in **R**.

A class **R** is, in general, *not* closed under superposition with functions from **F(R)**. For the following observations I shall assume throughout that **R** is trivially closed.

(D3) If **R** is closed under superposition with functions from **F(R)** then **F(R)** is closed under superposition.

For $<\alpha,a> \epsilon G(f^k \circ <g_0^m,\ldots, g_{k-1}^m>)$ and $<g_0^m(\alpha),\ldots,g_{k-1}^m(\alpha),a> \epsilon G(f^k)$ are equivalent. Thus

$$G(f^k \circ <g_0^m,\ldots, g_{k-1}^m>) =$$
$$G(f^k)[g_0^m \circ <p_0^{m+1},\ldots, p_{m-1}^{m+1}>,\ldots, g_{k-1}^m \circ <p_0^{m+1},\ldots, p_{m-1}^{m+1}>, p_m^{m+1}] .$$

If $g_0^m,\ldots, g_{k-1}^m$ are in **F(R)** then every $g_i^m \circ <p_0^{m+1},\ldots, p_{m-1}^{m+1}>$ is in **F(R)** since its graph arises from $G(g_i^m)$ by adding a dummy argument: it contains $<\alpha,b,a>$ if, and only if, $<\alpha,a>$ is in $G(g_i^m)$.

(D4) If E^2 is in **R** then the p_i^k are in **F(R)**

since $G(p_i^k) = E^2[p_i^{k+1}, p_k^{k+1}]$.

(D5) If **R** is positively-Boolean closed then **F(R)** is closed under *definition by cases*: Let $R_0,\ldots, R_{n-1}$ be k-ary relations in **R(F)** such that every α in ω^k belongs to precisely one R_i. Let $r_0,\ldots, r_{n-1}$ be k-ary functions in **F(R)**. Then **F(R)** contains the function f^k defined by

$$f^k(\alpha) = r_i(\alpha) \text{ for } \alpha \epsilon R_i .$$

For every i<n it follows from (CC1) that **R** contains the set S_i of all $<\alpha,a>$ such that $\alpha \epsilon R_i$, and $G(f^k)$ is the union of the sets $G(r_i) \cap S_i$.

(D6) If **R** is closed under superposition with c_1^1 then **R(F(R))** $\subseteq$ **R** .

Because $R \epsilon$ **R(F(R))** means that $G(\chi_R)$ is in **R**, and from $G(\chi_R)$ I obtain R by contracting the last argument 1 with the construction (CC2).

(D7) Let **R** be positively-Boolean closed and closed under superposition with c_0^1, c_1^1, assume also that E^2 is in **R**. Then **R** = **R(F(R))** holds if, and only if, **R** is closed under forming complements.

Assume first that **R** is closed under complements; I shall show that $R \epsilon$ **R** implies $G(\chi_R) \epsilon$ **R**. Because **R** contains the set of all $<\alpha,a>$ such that $\alpha \epsilon R$, and the construction (CC3), keeping the last argument constant at 1, shows that **R** contains the set of all $<\alpha,1>$ such that $\alpha \epsilon R$. The same argument shows that **R** contains the set of all $<\alpha,0>$ such that *not* $\alpha \epsilon R$; thus the union of these two sets belongs to **R**, and that is $G(\chi_R)$. Conversely, if $G(\chi_R) \epsilon$ **R** then the construction (CC2), contracting the last argument to 1 or to 0, produces both R and $-$R in **R**.

2. Simple Functions

The *simply closed* classes of functions are defined by closure conditions which are met by many examples to be studied in the following; there will be a smallest simply closed class, called the class of *simple* functions.

Let *sg* be the 1-ary *signum function*, defined by $sg(0) = 0$ and $sg(n) = 1$ for $n \neq 0$.

A class **F** of functions is called *simply closed* if

(FSF0) **F** contains the *simple initial functions*: c_1^1, p_i^k, $+$, $\cdot$, *sg* and the characteristic functions of the relations $<$ and E^2

(FSF1) **F** is closed under superposition

(FSF2) If R^{k+1} is in **R**(**F**) then $\mu_< R^{k+1}$ is in **F** .

The elements of the *smallest* simply closed class **FSF** I call the *simple* functions; they belong to every simply closed class. The elements of **R**(**FSF**) I call *simple* relations.

The condition (FSF2), referring to the class **R**(**F**) of relations, can be replaced by the condition

(FSF2$_0$) If f^{k+1} is in **F** then $\mu^*_< f^{k+1}$ is in **F**

which avoids such reference; here $\mu^*_< f^{k+1}$ is the (k+1)-ary function defined by

$$\mu^*_< f^{k+1}(\alpha,a) = \begin{cases} b & \text{if } f^k(\alpha,b) \neq 0 \text{ and } b<a \text{ and } f^k(\alpha,c) = 0 \text{ for all } c<b \\ a & \text{else} \end{cases}$$

First (FSF2$_0$) follows from (FSF2) since $\chi_R = sg \circ \chi_R$ implies $\mu_< R = \mu^*_<(\chi_R)$ for any relation R. Conversely, (FSF2) follows from (FSF2$_0$) because given f^{k+1}, define R^{k+1} by $<\alpha,b> \in R^{k+1}$ if, and only if, $f^{k+1}(\alpha,b) \neq 0$; hence $\mu_< R^{k+1}$ is $\mu^*_< f^{k+1}$. But the characteristic function of R^{k+1} is $sg \circ f^{k+1}$, and together with f^{k+1} also this function is in **F** by (FSF1), hence $R^{k+1} \in$ **R**(**F**), and so $\mu^*_< f^{k+1}$ is in **F**.

From now on let **F** be a simply closed class. Alone from (FSF1) there follows

(FS0) **R**(**F**) is closed under superposition with functions from **F**

because $S = R[g_0^m \ldots, g_{k-1}^m]$ implies $\chi_S = \chi_R \circ <g_0^m \ldots, g_{k-1}^m>$.

(FS1) If f^k is in **F** then the graph $G(f^k)$ is in **R**(**F**)

because $G(f^k) = E^2[f^k \circ <p_0^{k+1}, \ldots, p_{k-1}^{k+1}>, p_k^{k+1}>]$. Another formulation is

(FS1$_R$) If **F** is simply closed then $\mathbf{F} \subseteq \mathbf{F}(\mathbf{R}(\mathbf{F}))$.

(FS2) **F** is closed under *definition by cases*: Let $R_0,\dots, R_{n-1}$ be k-ary relations in **R(F)** such that every α in ω^k belongs to precisely one R_i. Let $r_0,\dots, r_{n-1}$ be k-ary functions in **F**. Then **F** contains the function f^k defined by $f^k(\alpha) = r_i(\alpha)$ for $\alpha \in R_i$.

Because let χ_i be the characteristic function of R_i. Then $f^k(\alpha)$ becomes $\Sigma\langle r_i(\alpha)\cdot\chi_i(\alpha) \mid i<n\rangle$, and I obtain f^k as t_{n-1} in $\langle t_j \mid j<n\rangle$ where $t_0 = \cdot\circ\langle r_0, \chi_0\rangle$ and $t_{j+1} = +\circ\langle t_j, \cdot\circ\langle r_{j+1}, \chi_{j+1}\rangle\rangle$.

(FS3) **F** contains the successor function s, contains all constant functions c_n^k, and for R^{k+1} in **R(F)** also $\mu_{\leq}R^{k+1}$ is in **F**.

Because $s = +\circ\langle p_0^1, c_1^1\rangle$, and for $n \geq 1$ there holds $c_{n+1}^1 = s\circ c_n^1$, $c_n^k = c_n^1 \circ p_0^k$; the case c_0^1 will be covered following (FS7). Finally, $\mu_{\leq}R^{k+1}$ is the superposition $\mu_{<}R^{k+1}\circ\langle p_0^{k+1},\dots, p_{k-1}^{k+1}, s\circ p_k^{k+1}\rangle$.

Observe that by (CC6) now also $\mu^i_{<}R^{k+1}$ and $\mu^i_{\leq}R^{k+1}$ will belong to **F**.

Let f^k be a function whose graph $G(f^k)$ is in **R(F)**. In general, I *cannot* conclude that f^k belongs to **F**. However:

(FS4) If $G(f^k)$ is in **R(F)** and if there exists a function h^k in **F** which majorizes f^k (i.e. $f^k(\xi) \leq h^k(\xi)$ for all ξ in ω^k) then f^k is in **F**.

Since $R^{k+1} = G(f^k)$ is in **R(F)**, the (k+1)-ary function $\mu_{\leq}R^{k+1}$ is in **F**, hence also the k-ary function $(\mu_{\leq}R^{k+1})\circ\langle p_0^k,\dots, p_{k-1}^k, h^k\rangle$. But this is f^k, because its value at α is $(\mu_{\leq}R^{k+1})(\alpha, h^k(\alpha))$, and that is the smallest number a (existing because of $f^k(\alpha) \leq h^k(\alpha)$) such that $a \leq h^k(\alpha)$ and $\langle\alpha, a\rangle \in R^{k+1}$, i.e. $a = f^k(\alpha)$.

(FS5) **R(F)** is closed under unions of relations of the the same arity.

Because $\chi_{R\cup S} = sg\circ+\circ\langle\chi_R, \chi_S\rangle$.

(FS6) The function of *bounded subtraction* $\dot{-}$, defined by $x\dot{-}y = x-y$ for $x \geq y$ and $x\dot{-}y = 0$ otherwise, is simple.

Because let R be the set of all $\langle x,y,z\rangle$ such that

$$y+z = x \quad \text{or} \quad x<y$$

which belongs to **R(F)** as $G(+)[p_1^3, p_2^3, p_0^3] \cup \langle[p_0^3, p_1^3]$. Then $\mu^o_{\leq}R(x,y)$ is the smallest z such that $z \leq x$ and $R(x,y,z)$ provided such z exists. But such z *does* exist, and it is $x\dot{-}y$. Hence $x\dot{-}y = \mu^o_{\leq}R(x,y)$.

(FS7) The 1-ary *predecessor function cs*, defined as $cs(n) = n-1$ for $n>0$ and $cs(0) = 0$, is simple.

It is $cs = \dot{-}\circ\langle p_0^1, c_1^1\rangle$. – Now also $c_0^1 = cs\circ c_1^1$ belongs to **F**. Observe further that in (FSF0) the function *sg* may be replaced by c_0^1 since it can be defined from c_0^1 and c_1^1 by cases, making use of the fact that $\{0\} = E^2(p_0^1, c_0^1)$ and $\omega = E^2(p_0^1, p_0^1)$ are in **R**(**F**).

(FS8) The 1-ary *cosignum function csg*, defined as $csg(0) = 1$ and $csg(n) = 0$ for $n > 0$, is simple.

It is $csg(x) = 1 \dot- x$, i.e. $csg = \dot- \circ \langle c_1^1, p_0^1 \rangle$.

(FS9) **R**(**F**) is Boolean closed.

Because $\chi_{-R} = csg \circ \chi_R$, and so (FS5) implies closure under intersections since $R \cap S = -(-R \cup -S)$. - Now (D7) implies that $\mathbf{R}(\mathbf{F}) = \mathbf{R}(\mathbf{F}(\mathbf{R}(\mathbf{F})))$.

(FS10) **R**(**F**) is closed under bounded quantifications.

This follows for $\exists_<$, $\exists_\leq$ from (C6), for $\forall_\leq$ from (C1), for $\forall_<$ from (C2).

(FS11) The functions $min(x,y) = x \dot- (x \dot- y)$, $max(x,y) = (x+y) \dot- min(x,y)$ and $|x-y| = (x \dot- y) + (y \dot- x)$ are simple.

(FS12) The functions $qu(x,y)$ of the *quotient* and $mod(x,y)$ of the remainder are simple where

$$y = qu(x,y) \cdot x + mod(x,y) \text{ for all } x,y,\ 0 \leq mod(x,y) < x \text{ for } x > 0,$$
$$mod(0,y) = y,\ qu(0,y) = 0 .$$

Observe first that **R**(**F**) contains the relation $\leq$ as union of $<$ and E^2. Let R^3 be the set of all $\langle x,y,q \rangle$ such that $xq \leq y < x(q+1)$. Being intersection of the superpositions $\leq [\cdot \circ \langle p_0^3, p_2^3 \rangle, p_1^3]$ and $< [p_1^3, \cdot \circ \langle p_0^3, s \circ p_2^3 \rangle]$, R^3 then is in **R**(**F**). Hence $f = \mu^1{}_\leq R^3$ is in **F**. For $x > 0$ there exists q such that $q \leq y$ and $\langle x,y,q \rangle \in R^3$, viz. the integer quotient (in particular $q = 0$ for $y = 0$) which so is the value $f(x,y)$. For $x = 0$ there are no y,q such that $\langle 0,y,q \rangle \in R^3$; hence always $f(0,y) = y+1$. Since **R**(**F**) contains 2-ary relations, it contains ω^2 by (F9) and then also $A = \omega^2 \cap E^2[p_0^2, c_0^2]$. Consequently, the set A of all $\langle 0,y \rangle$ is in **R**(**F**). Now I define *qu* by cases: $qu(x,y) = f(x,y)$ if $\langle x,y \rangle \in \omega^2 - A$, and $qu(0,y) = 0$ else. Thus *qu* is simple by (FS2), hence so is $mod = \dot- \circ \langle p_1^2, \cdot \circ \langle qu, p_0^2 \rangle \rangle$, i.e. $mod(x,y) = y \dot- qu(x,y) \cdot x$.

(FS13) The relation $div(x,y)$ of *divisibility* with the characteristic function $csg \circ mod$ is simple.

Observe that $div(x,0)$ for all x and *not* $div(0,y)$ for all $y > 0$ by definition of *mod* .

(FS14) The function *esq* of the *integer square root* is simple: $esq(x)^2 \leq x$ and $(esq(x)+1)^2 > x$.

The set R^2 of all $\langle x, y \rangle$ such that $(y+1)^2 > x$ is in **R**(**F**), hence $f = \mu^0{}_\leq R^2$ is in **F**. Since $(x+1)^2 > x$, the value $f(x)$ is the smallest y such that $y \leq x$ and $(y+1)^2 > x$.

(FS15) The bijective pairing function

$$cau(x,y) = (x+y)(x+y+1)/2 + x \ .$$

from ω^2 onto ω is simple, as are the 1-ary component functions cro_0 and cro_1 of the function uac inverse to cau satisfying $z = cau(cro_0(z), cro_1(z))$.

This function, counting the matrix ω^2 along its diagonals from their upper right to their lower left, was used by CAUCHY 21, p. 542 (8), in order to define the terms of the product of two infinite series. The proof that the above definition gives indeed a bijection from ω^2 onto ω, can be left to the reader. Then cau is in **F** since $cau(x,y) = qu(2, (x+y)(x+y+1)) + x$. Hence $G(cau)$ is in $\mathbf{R}(\mathbf{F})$ and so is the set R^3 of all $<z,x,y>$ such that

(c) $\quad z = cau(x,y)$.

Since $y \leq z$ the set $R^2 = \exists^o_{\leq} R^3$ contains all $<z,x>$ which satisfy (c) for some y. Since also $x \leq z$, and since cau is a bijection, it follows that $cro_0 = \mu^o_{\leq}\exists^o_{\leq}R^3$ and, analogously, $cro_1 = \mu^o_{\leq}\exists^o_{\leq}S^3$ for S^3 obtained from R^3 by permuting x and y. – A more practical approach to cro_0 and cro_1 makes use of the fact that (c) is equivalent to $2z = (x+y)(x+y+1) + 2x$ and to

$$(x+y)(x+y+1) + 2x < (x+y)(x+y+1) + 2(x+y+1) = (x+y+1)(x+y+2) \ .$$

Thus $x+y$ is the unique number s satisfying

(a) $\quad s(s+1) \leq 2z < (s+1)(s+2)$,

whence x becomes $z - (^1/_2) \cdot s \cdot (s+1)$ and y becomes $s-x$. But (a) implies $s^2 \leq 2z < (s+2)^2$ whence $s \leq esq(2z) < s+2$ and $esq(2z)-1 \leq s \leq esq(2z)$. Thus the function sending z into s can be defined by cases as either $esq(2z) \dot{-} 1$ or $esq(2z)$, depending on which of these two numbers satisfies (a).

(FS16) The bijection cau^k from ω^k onto ω, defined as $cau^1 = p^1_0$ and as

$$cau^k(\alpha,a) = cau(cau^{k-1}(\alpha), a)$$

for $k>1$, is simple, as are the k component functions cro^k_i of the inverse uac_k to cau^k, defined with $uac_1 = p^1_0$ for $k>1$ as

$$uac_k(x) = <uac_{k-1}(cro_0(x)), cro_1(x)> \ .$$

Induction shows that cau_k is in **F**, that $uac_k \circ cau^k$ is the identity on ω^k and $cau^k \circ uac_k$ is the identity on ω. For $k = 2$ this was (FS15), and induction gives

$$\begin{aligned} uac_k(cau^k(\alpha,a)) &= uac_k(cau(cau^{k-1}(\alpha), a)) \\ &= <uac_{k-1}(cro_0(cau(cau^{k-1}(\alpha), a))), cro_1(cau(cau^{k-1}(\alpha), a))> \\ &= <uac_{k-1}(cau^{k-1}(\alpha), a> = <\alpha,a> \end{aligned}$$

and

$$\begin{aligned} cau^k(uac_k(x)) &= cau^k(uac_{k-1}(cro_0(x)), cro_1(x)) \\ &= cau(cau^{k-1}(uac_{k-1}(cro_0(x))), cro_1(x)) = cau(cro_0(x), cro_1(x)) = x \end{aligned}$$

The component functions cro_i^k can be defined as $cro_0^2 = cro_0$, $cro_{k-1}^k = cro_1$ and, for $k>2$, $cro_i^k = cro_i^{k-1} \circ cro_0$ for $i<k-1$. Because it then follows from $uac_{k-1}(x) = <cro_i^{k-1}(x) \mid i<k-1>$ that

$$\begin{aligned} uac_k(x) &= <uac_{k-1}(cro_0(x)),\ cro_1(x)> \\ &= <<cro_i^{k-1}(cro_0(x)) \mid i<k-1>,\ cro_1(x)> = <cro_i^k(x) \mid i<k> . \end{aligned}$$

(FS17) If f^{k+1} is in **F** then so is the function $max f^{k+1}$ of *bounded maximization* defined as

$$max f^{k+1}(\alpha,n) = \max<f^{k+1}(\alpha,i) \mid i \leq n> .$$

The set R^{k+2} of all $<\alpha,y,n>$ such that $f^{k+1}(\alpha,y) \geq f^{k+1}(\alpha,n)$ is in **R(F)**, hence also the set $\forall_\leq R^{k+2}$ of all $<\alpha,y,n>$ such that $f^{k+1}(\alpha,y) \geq f^{k+1}(\alpha,i)$ for all $i \leq n$. Hence also the set S^{k+2} of all $<\alpha,n,y>$ such that $<\alpha,y,n> \in \forall_\leq R^{k+2}$ is in **R(F)**, and $\max<f^{k+1}(\alpha,i) \mid i \leq n>$ is $\mu^k_\leq S^{k+2}(\alpha,n)$.

If g^k is in **F**, if R is a finite subset of ω^k, and if f^k and g^k coincide outside of R, then f^k is in **F**. Because together with R also $-R$ is in **R(F)**, further every singleton $\{\alpha\}$ for $\alpha \in R$ is in **R(F)**, and f^k may be defined by cases from g^k and the functions c_a^k with $a = f^k(\alpha)$.

The simplicity of simple functions becomes apparent in their being *polynomially bounded*:

(FS18) For every f^k in **FSF** there are numbers a and q such that for every ξ in ω^k there holds

$$f^k(\xi) \leq a \cdot \max(\xi)^q .$$

This holds for the initial functions listed in (FSF0) (with $q=2$ for $\cdot$, $a=2$ for $+$, and $q=a=1$ in all other cases); it holds with $q=a=1$ for every function $\mu_< R$. If a, q are known for f^k and if a(i), q(i) are known for functions g_i^m, $i<k$, then with $p = \max<q(i) \mid i<k>$ and $b = \max<a(i) \mid i<k>$ I conclude from

$$g_i^m(\xi) \leq a(i) \cdot \max(\xi)^{q(i)} \leq b \cdot \max(\xi)^p$$

and

$$f^k(g_i^m(\xi), \dots, g_i^m(\xi)) \leq a \cdot \max<g_i^m(\xi) \mid i<k>^q \leq a \cdot (b \cdot \max(\xi)^p)^q$$

upon

$$f^k \circ <g_i^m, \dots, g_i^m>(\xi) \leq (a \cdot b^q) \cdot \max(\xi)^{p \cdot q} .$$

Thus, in particular, the function 2^x cannot be simple.

3. Elementary Functions

For every function f^{k+1} I define the $(k+1)$-ary *bounded summation* Σf^{k+1} and the $(k+1)$-ary *bounded multiplication* Πf^{k+1} by

$$\Sigma f^{k+1}(\alpha,0) = f^{k+1}(\alpha,0)\ ,\quad \Sigma f^{k+1}(\alpha, n+1) = \Sigma f^{k+1}(\alpha,n) + f^{k+1}(\alpha, n+1)$$

$$\Pi f^{k+1}(\alpha,0) = f^{k+1}(\alpha,0)\ ,\quad \Pi f^{k+1}(\alpha, n+1) = \Pi f^{k+1}(\alpha,n) \cdot f^{k+1}(\alpha, n+1)$$

and write them as

$$\Sigma f^{k+1}(\alpha,n) = \Sigma\langle f^{k+1}(\alpha,i) \mid i \leq n\rangle\ ,\quad \Pi f^{k+1}(\alpha,n) = \Pi\langle f^{k+1}(\alpha,i) \mid i \leq n\rangle\ .$$

Anyone familiar with the use of Σ and Π will want to read the latter expressions as definitions; the recursive definition though is required in order to actually prove certain statements about Σf^{k+1} and Πf^{k+1}.

A class **F** of functions is called *elementarily closed* if it contains the *elementary initial functions*

$$c_1^1\ ,\ p_i^k\ ,\ +\ ,\ \cdot\ ,\ \dot{-}$$

and is closed with respect to superposition and the formation of bounded summations and bounded multiplications: together with f^{k+1} also Σf^{k+1} and Πf^{k+1} shall be in **F**. The members of the *smallest* elementarily closed class **FEF** are called the *elementary functions*, and the relations in **R(FEF)** are called the *elementary relations*. Thus the elementary functions belong to every elementarily closed class. The following functions are elementary:

The successor function s and the constant functions c_n^k. Proof as in (FS3).

The predecessor function *cs*. Proof as in (FS7).

The signum function and the cosignum function. Proof for *csg* as in (FS8); then $sg = csg \circ csg$.

The characteristic functions of the 2-ary relations $<$, $\leq$, $=$ (or E^2). Because $\chi_<(x,y) = sg(y \dot{-} x)$, $\chi_\leq(x,y) = sg(y \dot{-} s(x))$, $\chi_=(x,y) = \chi_\leq(x,y) \cdot \chi_\leq(y,x)$ (and $\chi_=(x,y) = sg|x-y|$) .

From now on let **F** be an elementarily closed class. Then **F** is is closed under definitions by cases and **R(F)** is Boolean closed; the proofs from (FS2) and (FS9) remain in effect.

(FE0) If f^{k+1} is in **F** then so are the functions of *strictly bounded summation* and *multiplication*

$$\Sigma\langle f^{k+1}(\alpha,i) \mid i < n\rangle\ ,\quad \Pi\langle f^{k+1}(\alpha,i) \mid i < n\rangle$$

where

$$\Sigma\langle f^{k+1}(\alpha,i) \mid i < 0\rangle = 0\ ,$$
$$\Sigma\langle f^{k+1}(\alpha,i) \mid i < n+1\rangle = \Sigma\langle f^{k+1}(\alpha,i) \mid i < n\rangle + f^{k+1}(\alpha,n)$$

$$\Pi\langle f^{k+1}(\alpha,i) \mid i < 0\rangle = 1\ ,$$
$$\Pi\langle f^{k+1}(\alpha,i) \mid i < n+1\rangle = \Pi\langle f^{k+1}(\alpha,i) \mid i < n\rangle \cdot f^{k+1}(\alpha,n)\ .$$

Because together with f^{k+1} and Σf^{k+1} also

$$g = \Sigma f^{k+1} \circ \langle p_0^{k+1}, .., p_{k-1}^{k+1}, cs \circ p_k^{k+1}\rangle$$

is in $\mathbf{F}$, and $g(\alpha,n) = \Sigma\langle f^{k+1}(\alpha,i) \mid i \leq n-1\rangle = \Sigma\langle f^{k+1}(\alpha,i) \mid i<n\rangle$ if $n>0$. Since $\mathbf{R}(\mathbf{F})$ is Boolean closed, it follows from (D2) that it contains the set M of all α in ω^{k+1} such that $\alpha(k) = 0$, and so $\Sigma\langle f^{k+1}(\alpha,i) \mid i<n\rangle$ may be defined by cases: $\Sigma\langle f^{k+1}(\alpha,i) \mid i<n\rangle(\alpha) = 0$ for $\langle\alpha,n\rangle \epsilon M$, and otherwise $\Sigma\langle f^{k+1}(\alpha,i) \mid i<n\rangle(\alpha) = g(\alpha)$. The case of products is analogous.

(FE1) An elementarily closed $\mathbf{F}$ is simply closed.

I only have to verify (FSF2). So let R^{k+1} be in $\mathbf{R}(\mathbf{F})$ and define g and f by $g(\alpha,i) = \Pi\langle csg \circ \chi_R(\alpha,j) \mid j \leq i\rangle$ and $f(\alpha,n) = \Sigma\langle g(\alpha,i) \mid i<n\rangle$; they both are in $\mathbf{F}$. Now $g(\alpha,i) = 1$ if, and only if, for all $j \leq i$, the pair $\langle\alpha,j\rangle$ is not in R^{k+1}. Starting with 0, every such i contributes 1 to the sum $f(\alpha,n)$whence $f = \mu_< R^{k+1}$.

Inspection of the above proofs will show that all which was used about the function $\dot{-}$ was that it made available *csg*. Thus in the list of elementary initial function, the function $\dot{-}$ may be replaced by *csg*.

(FE2) $x^y = \Pi\langle p_0^2(x,i) \mid i<y\rangle$ and $y! = \Pi\langle p_1^2(x,s(i)) \mid i<y\rangle$ are elementary functions.

Because the first definition, say, indeed implies $x^0 = 1$ and $x^{y+1} = x^y \cdot x$. In particular, for every n also $n^y = (x^y)\circ\langle c_n^1, p_0^1\rangle$ is elementary, hence $n^{g(y)}$ is in $\mathbf{F}$ for every g in $\mathbf{F}$. Thus 2^y is an elementary function which is not simple.

The functions $b_0 = p_0^1$, $b_{n+1}(y) = 2^{b_n(y)}$ for $n = 0,1,\ldots$ are called the *elementary scaling* functions. It has been shown by BERECZKI 50 that for *every* elementary function f^k there exists a number n such that, for every ξ in ω^k, there holds $f^k(\xi) \leq b_n(\max(\xi))$ (cf. e.g. FELSCHER 93). This can be used in order to show that the 2-ary function

$$g(n,y) = b_n(y)$$

is *not* elementary. Because otherwise also the function $f(x) = g(x,x)$ would be so, and thus there would be an m such that $f(x) \leq b_m(x)$. Now $y<2^y$ implies $b_n(y)<b_{n+1}(y)$ for every n, and so I would arrive at the contradiction

$$f(m+1) \leq b_m(m+1) < b_{m+1}(m+1) = g(m+1) = f(m+1) .$$

A number p is (multiplicatively) *irreducible* if $p>1$ and if $p = a \cdot b$ implies that a or b is 1. I observed in Chapter 1.16, during the discussion of Full Arithmetic, statement (a2), that the irreducible numbers are the *prime* numbers: if p is irreducible and divides $c \cdot d$ then it either divides c or d.

(FE3) The set of all prime numbers is (simple and therefore) elementary.

Because let R^2 be the set of all $\langle y,x\rangle$ such that *div*(x,y) and $1<x$. It is simple by (FS13), and so $M = \exists^\circ_< R^2$ is simple. But y is in M if it is not irreducible, and so y is prime if $y>1$ and y is in the complement of M.

(FE4) The function *prino* is elementary which gives as *prino*(n) the n-th prime number p_n .

Because let f be the characteristic function of the set of prime number which is elementary. The condition

$$f(y) = 1 \text{ and } x+1 = \Sigma\langle f(i) \mid i \lesssim y\rangle$$

characterizes the set R^2 of $\langle x,y\rangle$ such that y is the x-th prime number p_x (beginning with $p_0 = 2$); thus R^2 is elementary and is the graph of the function *prino*. Now I shall be able to apply (FS4) once I can find an elementary function h such that $prino(x) \lesssim h(x)$. It is known since Euclid that the product $1+p_0 \cdots p_{x-1}$ is divisible by some p_z with $x \lesssim z$; from $p_0 = 2^1$ and $1+2+\cdots+2^x < 2^{x+1}$ there follows by mathematical induction

$$p_{x+1} \lesssim 1+p_0 \cdots p_x < 1+2^{2^{x+1}} , \quad p_{x+1} \lesssim 2^{2^{x+1}} .$$

So $p_x < h(x) = 2^{g(x)}$ for $g(x) = 2^x$. [There actually holds $p_x \lesssim 2^{x+1}$ as follows from the validity *Bertrand's conjecture*, stating that for every n there exists a prime number p such that $n < p \lesssim 2n$, cf. e.g. NIVEN-ZUCKERMAN 60, p.185]

(FE5) The function *exp* is elementary, where $exp(x,0) = exp(x,1) = 0$, and for $y>1$ the number $exp(x,y)$ is the largest exponent with which *prino*(x) divides y .

Let R^3 consist of all $\langle x,y,z\rangle$ such that $y>1$, $div(prino(x)^z, y)$ and not $div(prino(x)^{z+1}, y)$. Define *exp* by cases as c_0^2 for $y<2$ and as $\mu^1_{\lesssim} R^3$ for $y>1$.

(FE6) The function *len* is elementary, where $len(0) = len(1) = 0$, and for $y>1$ the number $len(y)$ is $x+1$ for the largest x such that *prino*(x) divides y.

Let R^3 be the set of $\langle x,y,z\rangle$ such that $y>1$, $div(prino(x), y)$ and not $div(prino(x+z+1), y)$. R^3 is elementary, hence so is $R^2 = (\forall^1_{\lesssim} R^3)[p^2_1, p^2_0]$ containing all $\langle y,x\rangle$ such that $\langle x,y,z\rangle \in R^3$ for all $z \lesssim y$. Define *len* by cases as c_0^1 for $y<2$ and as $1+\mu^0_{\lesssim} R^2$ for $y>1$.

For every $n>0$ the function $f^n(\alpha) = \Pi\langle prino(i)^{\alpha(i)} \mid i<n\rangle$ is an elementary injection from ω^n into ω; for every x the sequence $\alpha = \langle exp(i,x) \mid i<n\rangle$ satisfies $f^n(\alpha) = x$. For $k \neq n$, however, f^n and f^k will have overlapping ranges, e.g. $f^3(1,0,0) = f^2(1,0)$. This mishap can be avoided by using the elementary functions g^n defined as

$$g^n(\alpha) = \Pi\langle prino(i)^{1+\alpha(i)} \mid i<n\rangle$$

for $n \geq 1$. A number x then belongs to the range $im(g^n)$ if $n = len(x)$ and if $div(prino(i), x)$ for all $i < len(x)$. The set *seq* of numbers x belonging to the union of these ranges is characterized as

$$x>1 \text{ and for all } i<len(x): div(prino(i), x) .$$

The set R^3 of all <x,y,z> with $x>1$, $len(x)=y$ and $div(prino(z),x)$ is elementary; hence so is the set $R^2 = \forall^1_< R^3$ of all <x,y> with <x,y,z> in R^3 for all $z<y$. But *seq* is $R^2[p^1_0, len \circ p^1_0]$; thus also *seq* will be elementary. The numbers in *seq* may be called *elementary sequence numbers.*

The usual treatment of prime factor decompositions freely uses *sequences* of prime factors and *sequences* of the exponents with which these prime factors occur. Doing so is easy if the set theoretical notion of arbitrary sequences is admitted; shall that be avoided then the following approach may be taken. Let me recall the observation

(a3) if x,y are coprime and x divides $b \cdot y$ then x also divides b

proved in Chapter 1.16 when discussing Full Arithmetic. I next observe

(a8) If p,q are different primes then p^n and q^m are coprime.

Consider an irreducible divisor r of p^n. Being irreducible, r is prime. If $n=1$ then $r=p$ since p itself is irreducible. If $n>1$ then r divides $p^{n-1} \cdot p$, hence r divides p^{n-1} or p, and induction on n now gives $r=p$. Thus p is the only irreducible divisor of p^n. Likewise, q is the only irreducible divisor of q^m. But any common divisor of p^n, q^m, different from 1, would contain an irreducible one. Hence p^n, q^m are coprime.

(a9) If p,q are different primes and if q^n divides $a \cdot p^m$ then q^n divides a.

This follows from (a8) together with (a3). – I now introduce two elementary functions *car* and *cdr*:

$$\text{if } x>1 \text{ then } car(x) = prino((len(x)\dot{-}1)^{exp(len(x)\dot{-}1,\, x)}$$
$$\text{if } x \leq 1 \text{ then } car(x) = 0\,,$$

$$\text{if } x>1 \text{ then } cdr(x) = qu(car(x), x)\,, \quad \text{if } x \leq 1 \text{ then } cdr(x) = 0\,.$$

If $x>1$ then $prino(len(x)-1)$ *does* divide x, hence

(a10) If $x>1$ then $car(x)>0$ and $x = cdr(x) \cdot car(x)$.

Since $prino(len(x)-1)^{exp(len(x)-1,\, x)\,+1}$ does *not* divide x by definition of $exp(len(x)-1, x)$, the prime $prino(len(x)-1)$ does *not* divide $cdr(x)$. As any prime dividing $cdr(x)$ will also divide x, in particular the largest prime $prino(len(cdr(x))-1)$ dividing $cdr(x)$, there follows

(a11) If $len(x)=1$ then $cdr(x)=1$; if $len(x)>0$ then $cdr(x)<x$ and $len(cdr(x)) \leq len(x)-1$.

It follows from (a10) that $exp(i, cdr(x)) \leq exp(i,x)$ for any i; it follows from (a9) that, for $i \neq len(x)-1$, any number $prino(i)^n$ dividing x also divides $cdr(x)$. Hence

(a12) If $x>1$ then for all $i<len(x)-1$: $exp(i, cdr(x)) = exp(i,x)$

whence in particular

(a13) If $x \epsilon seq$ and $len(x) > 1$ then $cdr(x) \epsilon seq$ and $len(x) = 1 + len(cdr(x))$.

It also follows from (a10) that

(a14) If $x>0$, $y>0$, $len(x) = len(y)$, $car(x) = car(y)$, $cdr(x) = cdr(y)$ then $x = y$

and if x,y are in *seq* then the hypothesis $len(x) = len(y)$ may be omitted.

(a15) If $exp(j, x) > 0$ then $j < len(cdr(x))$ or $j = len(x)-1$.

Because if $prino(j)$ divides x then $x > 1$ whence (a10) implies that it either divides $cdr(x)$ or $car(x)$; in the latter case (a8) implies $j = len(x)-1$. – My main result now is

(a16) If $x>0$ then $x = \Pi < prino(i)^{exp(i,x)} \mid i < len(x) >$,
if $x>1$ then $cdr(x) = \Pi < prino(i)^{exp(i,x)} \mid i < len(x)-1 >$.

This holds for $x = 1$ by definition of strictly bounded multiplication since $len(1) = 0$. For $x>1$, I proceed by order induction, and as $cdr(x) < x$, I have by inductive hypothesis

$$cdr(x) = \Pi < prino(i)^{exp(i,\, cdr(x))} \mid i < len(cdr(x)) >$$
$$= \Pi < prino(i)^{exp(i,x)} \mid i < len(cdr(x)) > ,$$

making use of (a12). This is the special case for $h = 0$ of

if $0 \leq h < len(x) - len(cdr(x))$ then
$$cdr(x) = \Pi < prino(i)^{exp(i,x)} \mid i < len(cdr(x)) + h > .$$

Applying induction, I assume that $h+1 < len(x) - len(cdr(x))$ whence (a15) gives $exp(len(cdr(x))+h+1, x) = 0$ and thus

$$\Pi < prino(i)^{exp(i,x)} \mid i < len(cdr(x))+h+1 >$$
$$= \Pi < prino(i)^{exp(i,x)} \mid i < len(cdr(x))+h > \cdot prino(len(cdr(x))+h)^{exp(len(cdr(x))+h,\, x)}$$
$$= \Pi < prino(i)^{exp(i,x)} \mid i < len(cdr(x))+h > .$$

This concludes the induction proof whence, in particular,

$$cdr(x) = \Pi < prino(i)^{exp(i,x)} \mid i < len(x)-1 >$$

which is the second part of (a15). Making use of this representation of $cdr(x)$, (a10) implies

$$x = \Pi < prino(i)^{exp(i,x)} \mid i < len(x)-1 > \cdot prino((len(x)-1)^{exp(len(x)-1,\, x)}$$
$$= \Pi < prino(i)^{exp(i,x)} \mid i < len(x) >$$

again by definition of strictly bounded multiplication.

Every prime factor p of x must be of the form *prino*(i) with $i < len(x)$. Hence the representation of x in (a16) contains p, and does so with the maximal exponent with which p divides x – by definition of *exp*(i,x). In this sense, it may be called the canonical prime factor decomposition of x.

Consider now functions φ and ψ and a positive number n, and assume that $\varphi(i) < \varphi(i+1)$ for all $i < n-1$ and $\psi(i) > 0$ for all $i < n$. In informal mathematical reasoning, the reader may call the number

$$x = \Pi < prino(\varphi(i))^{\psi(i)} \mid i < n >$$

the confluent product of *the sequence* $< prino(\varphi(i)) \mid i < n >$ with *the sequence* $< \psi(i) \mid i < n >$ of exponents, and he will agree that it thus is divided by these primes with these exponents and, in view of uniqueness of factorization, is divided by no other primes. Here the statement

(a17) The primes *prino*(j) dividing $x = \Pi < prino(\varphi(i))^{\psi(i)} \mid i < n >$ are precisely the $prino(\varphi(i))$ for $i < n$, and then $exp(\varphi(i), x) = \psi(i)$.

is proved by induction on n. If $n = 1$ then $x = prino(\varphi(0))^{\psi(0)}$ and (a17) is trivial as different primes are coprime. If $n > 1$ then, by definition of strictly bounded multiplication,

$$x = \Pi < prino(\varphi(i))^{\psi(i)} \mid i < n-1 > \cdot\, prino(\varphi(n-1))^{\psi(n-1)}$$

where I shall abbreviate the first factor as b. Any prime p dividing x must divide one of these two factors. If p divides the second one then p is $prino(\varphi(n-1))$; if p divides b, then I apply the inductive hypothesis to b and obtain that p is one of $prino(\varphi(i))$ with $i < n-1$. So it is at most the $prino(\varphi(i))$ for $i < n$ which divide x. But $prino(\varphi(n-1))$ does divide x, and as, again by inductive hypothesis for b, it is not one of the primes dividing b, there follows $exp(\varphi(n-1), x) = \psi(n-1)$. It follows from (a9) that the numbers $prino(\varphi(i))$ with $i < n-1$ divide b and x with same exponents and, once more by inductive hypothesis for b, these exponents are the $\psi(i)$.

4. Primitive Recursive Functions

Let k be such that $k > 1$ and let g^k, r^{k+2} be two functions, or let k be 1 and let a be a number and r^2 be a function. A function f^{k+1} is said to be defined *by primitive recursion from* g^k, r^{k+2}, respectively from g, r^2 if, for all of its arguments, there hold the *recursion schemata*

$$f^{k+1}(\alpha,0) = g^k(\alpha) \qquad \text{respectively} \qquad f^1(0) = a$$

$$f^{k+1}(\alpha, n+1) = r^{k+2}(\alpha, n, f^{k+1}(\alpha,n)) \qquad\qquad f^1(n+1) = r^2(n, f^1(n)).$$

The schema for 1-ary functions f^1 can be reduced to the general one if the constant functions c^1_a and superpositions are available: if f^1 is defined by a and r^2, and if f^2 is defined by c^1_a and $r^3 = r^2 \circ <p^3_1, p^3_2>$, then f^1 becomes $f^2 \circ <p^1_0, p^1_0>$.

A class F of functions is said to be *primitive recursively closed* if it contains the primitive recursive initial functions

$$s,\ c^1_0\ ,\ p^k_i$$

and if it is closed under superposition and primitive recursion. The members of the *smallest* primitive recursively closed class **FPF** are called the *primitive recursive functions*, and the relations in **R(FPF)** are called the *primitive recursive relations*. A primitive recursively closed class is closed under *simple recursion*: if g^k and r^{k+1} are in F then so is the function f^{k+1} satisfying

$$f^{k+1}(\alpha,0) = g^k(\alpha)$$

$$f^{k+1}(\alpha, n+1) = r^{k+1}(\alpha, f^{k+1}(\alpha,n))$$

since it arises from g^k and $r^{k+2} = r^{k+1} \circ <p^{k+2}_0, \dots, p^{k+2}_{k-1}, p^{k+2}_{k+1}>$ by primitive recursion.

Every primitive recursively closed class **F** *also is elementarily closed.* To see this, observe that the following functions are primitive recursive:

The constant functions c^1_n . Because I have c^1_0 and s, and $c^1_n = s \circ c^1_{n-1}$.

The predecessor function *cs* with $cs(0) = 0$, $cs(n+1) = p^2_0(n, cs(n))$.

The function $\dot{-}$ with the definition $x \dot{-} (n+1) = cs(x \dot{-} n)$, i.e.
$\dot{-}(x,0) = p^1_0(x)$, $\dot{-}(x,n+1) = cs \circ p^2_1(x, \dot{-}(x,n))$.

The functions $+(x,y) = x+y$ and $\cdot(x,y) = x \cdot y$. Because I have
$+(x,0) = p^1_0(x)$, $+(x, n+1) = s \circ p^3_2(x,n, +(x,n))$ and
$\cdot(x,0) = c^1_0(x)$, $\cdot(x, n+1) = + \circ <p^3_0, p^3_2>(x,n, \cdot(x,n))$.

It remains to verify that **F** is closed under bounded summation and multiplication. But Σf^{k+1} can be defined recursively by $\Sigma f^{k+1}(\alpha,0) = f^{k+1}(\alpha,0)$, $\Sigma f^{k+1}(\alpha, n+1) = r(\alpha,n, \Sigma f^{k+1}(\alpha,n))$ where $r(\alpha,n,x) = x + f^{k+1}(\alpha,s(n))$; the case of Πf^{k+1} is analogous. In particular, elementary functions and relations are also primitive recursive. – A primitive recursively closed class **F** also is closed under primitive recursive definition *by cases*:

Let $R_0, \dots, R_{m-1}$ be (k+1)-ary relations in **R(F)** such that every α in ω^{k+1} belongs to precisely one of the R_i; let $r_0, \dots, r_{m-1}$ be (k+2)-ary functions in **F**, and let g^k be in **F** . Then **F** contains the function f^{k+1} such that

$$f^{k+1}(\alpha,0) = g^k(\alpha),\ f^{k+1}(\alpha, n+1) = r_i(\alpha, n, f^{k+1}(\alpha,n)) \text{ for } <\alpha,n> \epsilon R_i .$$

Because $f^{k+1}(\alpha, n+1) = \Sigma < r_i(\alpha,n, f^{k+1}(\alpha,n)) \cdot \chi_{R(i)}(\alpha,n) \mid i<m >$ leads to a primitive recursive definition.

A finite sequence of functions f^{k+1}_j, $j<m$, is said to be defined by *simultaneous recursion* from sequences g^k_j, r^{k+m+1}_j, $j<m$, of functions if for every $j<m$

$$f_j^{k+1}(\alpha,0) = g_j^k(\alpha)$$
$$f_j^{k+1}(\alpha, n+1) = r_j^{k+m+1}(\alpha,n, f_0^{k+1}(\alpha,n),\dots, f_{m-1}^{k+1}(\alpha,n)) \ .$$

The functions cro_0, cro_1, for instance, could be defined simultaneously by

$$cro_0(0) = 0 \ , \quad cro_1(0) = 0 \ ,$$
$$cro_0(n+1) = (sg \circ cro_1(n)) \cdot (1+cro_0(n)) \ ,$$
$$cro_1(n+1) = (csg \circ cro_1(n)) \cdot (1+cro_0(n))) + (sg \circ cro_1(n)) \cdot (cro_1(n) \dot{-} 1)$$

though computing them by this schema would be quite time consuming.

A primitive recursively closed class **F** is also closed with respect to simultaneous recursion. Because **F** contains the two functions

$$g^k(\alpha) = cau^m (g_0^k(\alpha),\dots, g_{m-1}^k(\alpha)) \ ,$$
$$r^{k+2}(\alpha,n,x) = cau^m (r_0^{k+m+1}(\alpha,n, cro_0^m(x),\dots, cro_{m-1}^m(x)), \dots , \\ r_{m-1}^{k+m+1}(\alpha,n, cro_0^m(x),\dots, cro_{m-1}^m(x)))$$

and if f^{k+1} is defined by primitive recursion from g^k and r^{k+2} then the functions f_j^{k+1} become $cro_j^m \circ f^{k+1}$. Because $f_j^{k+1}(\alpha,0) = g_j^k(\alpha) = cro_j^m(g^k(\alpha)) = cro_j^m(f^{k+1}(\alpha,0))$ by definition of g^k, and if $f_j^{k+1}(\alpha,n) = cro_j^m(f^{k+1}(\alpha,n))$ has been proven then also

$$\begin{aligned} cro_j^m(f^{k+1}(\alpha,n)) &= cro_j^m(r^{k+2}(\alpha,n,f^{k+1}(\alpha,n))) \\ &= r_j^{k+m+1}(\alpha,n,cro_0^m(f^{k+1}(\alpha,n)),\dots, cro_{m-1}^m(f^{k+1}(\alpha,n))) \\ &= r_j^{k+m+1}(\alpha,n, f_0^{k+1}(\alpha,n),\dots, f_{m-1}^{k+1}(\alpha,n)) \\ &= f_j^{k+1}(\alpha, n+1) \ . \end{aligned}$$

The schema of *extended* simultaneous recursion requires for every $0<j<m$

$$f_j^{k+1}(\alpha,0) = g_j^k(\alpha)$$
$$f_0^{k+1}(\alpha, n+1) = r_0^{k+m+1}(\alpha,n, f_0^{k+1}(n,n), \dots , f_{m-1}^{k+1}(\alpha,n))$$
$$f_j^{k+1}(\alpha, n+1) = r_j^{k+m+1}(\alpha,n, f_0^{k+1}(\alpha, n+1), \dots , \\ f_{j-1}^{k+1}(\alpha, n+1), f_j^{k+1}(n,n), \dots , f_{m-1}^{k+1}(\alpha,n)) \ .$$

It reduces to the earlier one since for $j<m$

$$f_0^k(\alpha,1) = r_0^{k+m+1}(\alpha,n, g_0^k(\alpha), \dots , g_{m-1}^k(\alpha))$$
$$\dots\dots\dots$$
$$f_j^k(\alpha,1) = r_j^{k+m+1}(\alpha,n, f_0^k(\alpha,1), \dots , f_{j-1}^k(\alpha,1), g_j^k(\alpha), \dots , g_{m-1}^k(\alpha))$$

can be defined explicitly, and if now the two functions

$$g^k(\alpha) = cau^{m+m-1}(g_0^k(\alpha),\dots, g_{m-1}^k(\alpha), f_0^k(\alpha,1),\dots, f_{m-1}^k(\alpha,1))) \ ,$$
$$r^{k+2}(\alpha,n,x) = cau^{m+m-1}(cro_m^{m+m-1}(x), \dots , cro_{m-1}^{m+m-1}(x) \ , \\ r_0^{k+m+1}(\alpha,n, cro_0^{m+m-1}(x), \dots , cro_{m-1}^{m+m-1}(x)) \ , \ \dots , \\ r_j^{k+m+1}(\alpha,n, cro_m^{m+m-1}(x), \dots , \\ cro_{m+j-2}^{m+m-1}(x), cro_j^{m+m-1}(x),\dots, cro_{m-1}^{m+m-1}(x)), \dots ,$$

$$r^{k+\mathbf{m}+1}_{m-1}(\alpha,n,\ cro^{m+\mathbf{m}-1}_{m}(x),\ldots,\ cro^{m+\mathbf{m}-1}_{m+\mathbf{m}-2}(x),\ cro^{m+\mathbf{m}-1}_{m-1}(x))\ )$$

are used to define f^{k+1} by primitive recursion then the functions f^{k+1}_j become $cro^{\mathbf{m}+m-1}_j \circ f^{k+1}$ since $f^{k+1}_j(\alpha,n) = cro^{m+\mathbf{m}-1}_{m+j-1} \circ f^{k+1}(\alpha,n)$ holds for $n>0$.

Let P be a 2-ary function satisfying $P(m,x)<P(m+1,x)$ and such that for *every* primitive recursive function f^k there exists a number m such that, for every ξ in ω^k, there holds $f^k(\xi) \leq P(m, \max(\xi))$. Then P *cannot* be primitive recursive. Because otherwise also the function $f(x) = P(x,x)$ would be so, and hence there would be an m such that $f(x) \leq P(m,x)$. Thus I would arrive at the contradiction

$$f(m+1) \leq P(m, m+1) < P(m+1, m+1) = f(m+1) = f(m+1)\ .$$

Simplifying a 3-ary function invented by ACKERMANN 29 , PETER 35 has exhibited such a function defined by the recursion equations

$$P(0,n) = n+1\ ,\quad P(m+1,0) = P(m,1)\quad ,\ P(m+1,n+1) = P(m, P(m+1,n))$$

(cf. e.g. FELSCHER 93). Thus *primitive* recursion does not cover all functions which are defined by recursion equations.

5. Elementary Arithmetization

The function f assigning to n the nth Fibonacci number can be defined by

$$f(0) = 0\ ,\quad f(1) = 1\ ,\quad f(n+1) = f(n-1) + f(n)\ \text{ for } n>1\ .$$

Then f is primitive recursive since $f = cro^3_2 \circ g^3$ for the primitive recursive function g^3 defined as

$$g^3(0) = cau^3(0,0,0)\ ,\quad g^3(1) = cau^3(0,0,1)\ ,$$

$$g^3(n+1) = cau^3(cro^3_1(n),\ cro^3_2(n),\ cro^3_1(n) + cro^3_2(n))\ .$$

Matters are different if a function f is defined at n+1 not only with help of f(n) and f(n−1), but with help of preceding values $f(t_0(n)), f(t_1(n)), \ldots$ making use of (primitive recursive) functions t_i of which I still know that $t_i(n) \leq n$ but whose values $t_i(n)$ may vary erratically between 0 and n: Given 1-ary functions $t_0, \ldots, t_{p-1}$ satisfying $t_j(x) \leq x$ for all x; a function f^{k+1} is said to be defined by recursion *with regression functions* t_j from the functions g^k, v^{k+p+1} if

$$\text{(WR)}\qquad \begin{aligned} &f^{k+1}(\alpha,0) = g^k(\alpha)\\ &f^{k+1}(\alpha, n+1) = v^{k+p+1}(\alpha, n, f^{k+1}(\alpha, t_0(n)), \ldots, f^{k+1}(\alpha, t_{p-1}(n))\ )\ . \end{aligned}$$

In that case, the computation of $f^{k+1}(\alpha, n+1)$ requires to have available all preceding values $f^{k+1}(\alpha, n)$. The technique of arithmetization, to be discussed now, will be the tool with which to approach this task.

I define the *elementary history* $H_e f^{k+1}$ of a function f^{k+1} to be the function

$$H_e f^{k+1}(\alpha,n) = \Pi < prino(i)^{f^{k+1}(\alpha,i)+1} \mid i \leq n > .$$

Then $H_e f^{k+1}$ determines f^{k+1} by $f^{k+1}(\alpha,n) = exp(n, H_e f^{k+1}(\alpha,n)) \dot{-} 1$. Observe that if one of f^{k+1} and $H_e f^{k+1}$ is elementary, respectively primitive recursive, then so is the other. Because $H_e f^{k+1}$ is the bounded product Πh^{k+1} defined from

$$h(\alpha,i) = prino(i)^{f^{k+1}(\alpha,i)+1} ,$$

i.e. the superposition of the elementary function x^y with $prino \circ p_k^{k+1}$. If a function u^{k+1} is an upper bound for f^{k+1} then $H_e u^{k+1}$ is an upper bound for $H_e f^{k+1}$.

I define that f^{k+1} arises from g^k and r^{k+2} by *course of values recursion* if

(W)
$$f^{k+1}(\alpha,0) = g^k(\alpha)$$
$$f^{k+1}(\alpha, n+1) = r^{k+2}(\alpha, n, H_e f^{k+1}(\alpha,n)) .$$

THEOREM 1 Given g^k and r^{k+2}, there *exists* a unique function f^{k+1} satisfying (W). A primitive recursively closed class **F** is closed under course of values recursion.

Let h^{k+2} be the function

$$h^{k+2}(\alpha,n,x) = x \cdot prino(n+1)^{r^{k+2}(\alpha,n,x)+1} ;$$

it is in **F** [is elementary] if g^k and r^{k+2} are so. I define a function q^{k+1} in **F** as

$$q^{k+1}(\alpha, 0) = 2^{g^k(\alpha)+1}$$
$$q^{k+1}(\alpha, n+1) = q^{k+1}(\alpha,n) \cdot prino(n+1)^{r^{k+2}(\alpha,n,q^{k+1}(\alpha,n)) + 1}$$

by primitive recursion from the functions 2^{g^k} and $h^{k+2}(\alpha,n,x)$. A first induction on n shows

(a) for all c: if $div(prino(c), q^{k+1}(\alpha,n))$ then $c \leq n$

whence

(b) $exp(n+1, q^{k+1}(\alpha, n+1)) = r^{k+2}(\alpha, n, q^{k+1}(\alpha,n)) + 1$,

and a second induction shows

(c) for all c: if $c \leq n$ then $exp(c, q^{k+1}(\alpha,c)) = exp(c, q^{k+1}(\alpha,n))$.

I define f^{k+1} by

$$f^{k+1}(\alpha,n) = exp(n, q^{k+1}(\alpha,n)) \dot{-} 1 .$$

Then trivially $f^{k+1}(\alpha,0) = g^k(\alpha)$, and since (c) implies

(d) for all n: for all $c \leq n : f^{k+1}(\alpha,c) = exp(c, q^{k+1}(\alpha,n)) \dot{-} 1$ for $c \leq n$

there follows $q^{k+1} = H_e f^{k+1}$ from the definition of $H_e f^{k+1}$. Hence (b) implies $f^{k+1}(\alpha,n+1) = r^{k+2}(\alpha,n, H_e f^{k+1}(\alpha,n))$.

COROLLARY 1 A primitive recursively closed class **F** is closed under recursion with regression functions.

Given the functions g^k, v^{k+p+1} and the t_j, I define

$$r^{k+2}(\alpha,n,x) = v^{k+p+1}(\alpha, n, exp(t_0(n), x) \dot{-} 1, \dots, exp(t_{p-1}(n), x) \dot{-} 1) .$$

and define q^{k+1} and f^{k+1} as in the Theorem. Then (b) implies

$$\begin{aligned} f^{k+1}(\alpha, n+1) &= exp(n+1, q^{k+1}(\alpha,n+1)) \dot{-} 1 \\ &= r^{k+2}(\alpha,n, q^{k+1}(\alpha,n)) \\ &= v^{k+p+1}(\alpha, n, exp(t_0(n), q^{k+1}(\alpha,n)) \dot{-} 1, \dots, exp(t_{p-1}(n), q^{k+1}(\alpha,n)) \dot{-} 1) \end{aligned}$$

which by (d) becomes

$$v^{k+p+1}(\alpha, n, f^{k+1}(\alpha,t_0(n)), \dots, f^{k+1}(\alpha,t_{p-1}(n))) =$$

$$\Pi < prino(j)^{r_j^{k+m+1}(\alpha,n,\, exp(f^{k+1}(\alpha,n),0), \dots,\, exp(f^{k+1}(\alpha,n), m-1)} \mid j<m>$$

and then $f_j^{k+1}(\alpha,n) = exp(f^{k+1}(\alpha,n), j)$.

A class **F** of functions is called closed under *bounded (primitive) recursion* if it contains those functions f^{k+1} which are are majorized by some function h^{k+1} from **F** *and* are defined by primitive recursion from functions g^k and r^{k+2} in **F**. A sufficient criterion for this to be the case is the

LEMMA 1 Let **F** be simply closed and let **F** contain the functions *prino* and *exp* as well as the exponentiation x^y. Then **F** is closed under bounded recursion.

Let f^{k+1} be defined from g^k and r^{k+2}. If $a = f^{k+1}(\alpha,n)$ and $c = H_e f^{k+1}(\alpha,n)$ then $<\alpha,n,a,c>$ belongs to the relation R^{k+3} defined by

$$exp(n,c) = a \text{ and } exp(0,c) = g^k(\alpha)$$
$$\text{and for all } i<n: \; exp(i+1,c) = r^{k+2}(\alpha,n, exp(i,c)) .$$

Conversely, if $<\alpha,n,a,c>$ is in R^{k+3} then $a = f^{k+1}(\alpha,n)$ (and c is a multiple of $H_e f^{k+1}(\alpha,n)$). Consider now the relations

R^{k+4} : the set of all $<\alpha,n,a,c,i>$ such that $exp(n,c) = a$ and $exp(0,c) = g^k(\alpha)$ and $exp(i+1,c) = r^{k+2}(\alpha,n, exp(i,c))$

$R^{k+3} = \forall^{k+1}_{<} R^{k+4}$

$S^{k+3} = \exists_{\leq} R^{k+3}$: the set of all $<\alpha,n,a,d>$ for which there exists c such that $c \leq d$ and $<\alpha,n,a,c>$ in R^{k+3}.

They belong to **R(F)** since **F** is simply closed and contains *exp*. If now f^{k+1} is majorized by h^{k+1} in **F** then $H_e f^{k+1}(\alpha,n) \leq H_e h^{k+1}(\alpha,n)$. Given α, n, a, then $H_e h^{k+1}(\alpha,n)$ is a number d such that $<\alpha,n,a,d> \epsilon S^{k+3}$. But

$$H_e h^{k+1}(\alpha,n) = \Pi < prino(i)^{h^{k+1}(\alpha,i)+1} \mid i \leq n > \leq prino(n)^{1+max\, h^{k+1}(\alpha,n)},$$

and it follows from my assumptions that the majorizing function $q^{k+1}(\alpha,n)$, standing on the right, belongs to **F**. Thus also $q^{k+1}(\alpha,n)$ is such a number d. Hence **R(F)** and **F** also contain

S^{k+2} : the set of all $<\alpha,n,a>$ such that $<\alpha,n,a, H_e h^{k+1}(\alpha,n)> \epsilon S^{k+3}$

$\mu_{\leq} S^{k+2}$: the function assigning to $<\alpha,n,b>$ the smallest a (provided it exists) such that $a \leq b$ and $<\alpha,n,a> \epsilon S^{k+2}$

$$e^{k+1} = (\mu_{\leq} S^{k+2}) \circ < p_0^{k+1}, \dots, p_k^{k+1}, h^{k+1}> .$$

I complete the proof by showing $e^{k+1} = f^{k+1}$. Because $e^{k+1}(\alpha,n)$ is the smallest (if that exists) a such that $a \leq h^{k+1}(\alpha,n)$ and such that there exists c with $c \leq H_e h^{k+1}(\alpha,n)$ for which $<\alpha,n,a,c> \epsilon R^{k+3}$. But $f^{k+1}(\alpha,n) \leq h^{k+1}(\alpha,n)$ implies $H_e f^{k+1}(\alpha,n) \leq H_e h^{k+1}(\alpha,n)$, and so $f^{k+1}(\alpha,n)$ actually is such an a which for $c = H_e f^{k+1}(\alpha,n)$ satisfies my condition. And the remark made at the beginning of the proof shows that $f^{k+1}(\alpha,n)$ is the *only* such a which satisfies $<\alpha,n,a,c> \epsilon R^{k+3}$ for *some* c.

THEOREM 2 An elementarily closed class **F** is closed under bounded recursion, bounded simultaneous recursion, bounded course of values recursion and bounded recursion with regression functions.

Closure under bounded recursion follows from the preceding Lemma. If f^{k+1} is majorized by u^{k+1} then $H_e f^{k+1}$ is majorized by $H_e u^{k+1}$. If f^{k+1} now is defined by course of values recursion then the proof of Theorem 1 shows that $H_e f^{k+1}$ is a function q^{k+1} defined by primitive recursion from functions belonging to **F**; hence $H_e f^{k+1}$ arises by bounded recursion and, therefore, belongs to **F**. But together with $H_e f^{k+1}$ also f^{k+1} is in **F**. The case of simultaneous recursion can be treated analogously since the function cau^m is monotonic in each of its arguments.

A frequent application of recursion with regression function occurs in proofs that the characteristic function χ_M of a set M is elementary or primitive recursive. In that case there usually is an elementary regression function fxp satisfying $fxp(j,n) < n$ for all j and n, and the set M is defined in the form

$n \epsilon M$ if, and only if, $n \epsilon A$ or
($n \epsilon B$ and for all j: if $j \leq L(n)$ then $fxp(j,n) \epsilon M$)

with auxiliary sets A, B and a function L. This is equivalent to

$$(M_0) \qquad \chi_M(n) \;=\; sg\,(\,\chi_A(n) + \chi_B(n)\cdot \Pi < \chi_M(\mathrm{fxp}(j,n)) \mid j \le L(n) >)$$

for $n > 0$. Thus χ_M is defined by the course of values recursion

$$(W_M) \qquad \begin{aligned} &\chi_M(0) = 0 \\ &\chi_M(n+1) = r^2(n,\, H_e\,\chi_M(n)) \end{aligned}$$

where

$$(M_1) \qquad r^2(n,z) = \\ sg\,(\,\chi_A(n+1) + \chi_B(n+1)\cdot \Pi < exp(\mathrm{fxp}(j,\, n+1), z) \dot- 1 \mid j \le L(n+1) >).$$

It follows from the definition of $H_e\,\chi_M$ that $\chi_M(c) = exp(c,\, H_e\,\chi_M(n+1)) \dot- 1$ for every $c \le n+1$; choosing c as $\mathrm{fxp}(j,n)$, it follows from (M_1) that the second equation in (W_M) is indeed (M_0) with $n+1$ in place of n.

If now A, B, L are primitive recursive then so is M. But χ_M is majorized by the elementary function c_2^1. Hence if A, B, L are elementary then so is M .

LEMMA 2 Let f^{k+1} be recursively defined from polynomials g^k, r^{k+1}. Then f^{k+1} is elementary.

If a is the maximum of the coefficients of the polynomial

$$p(x) = a_0 x^v + a_1 x^{v-1} + \dots$$

then $p(x) \le (v+1)\cdot a \cdot x^v$; thus for every $p(x)$ there are constants v and d_1 such that $p(x) \le d_1 \cdot x^v$. Analogously, if k, m, r are the maximal exponents with which x, y, z occur in

$$q(x,y,z) = a x^k y^n z^s + \dots + b x^i y^m z^t + \dots + c x^j y^p z^r + \dots$$

then $q(x,y,z) \le d_3 \cdot x^k y^m z^r$ where d_3 is a multiple of the maximum of the coefficients. Assume now that $f = f^2$ is recursively defined from p and q; then $f(a,0) \le d_1 \cdot a^v$. Abbreviate $e(b,x,n) = x^{bn}$ and $g(b,x,n) = b\cdot(x^n + x^{n-1} + \dots + 1)$; then $e(b,x,n)\cdot x = e(b,x,n+1)$ and $b + g(b,x,n)\cdot x = g(b,x,n+1)$. So if

$$\text{(a)} \quad f(a,n) \le d_3^{\,g(1,r,n-1)} \cdot d_1^{\,e(1,r,n)} \cdot n^{g(m,r,n-2)} \cdot a^{g(k,r,n-1) + e(v,r,n)}$$

is used as inductive hypothesis, then $(n+1)^m \cdot n^{g(m,r,n-2)\cdot r} \le (n+1)^{m + g(m,r,n-2)\cdot r} = (n+1)^{g(m,r,n-1)}$ implies

$$\begin{aligned} f(a,n+1) &= q(a,n,\, f(a,n)) \le d_3 \cdot a^k \cdot (n+1)^m \cdot f(a,n)^r \\ &\le d_3^{\,g(1,r,n)} \cdot d_1^{\,e(1,r,n+1)} \cdot (n+1)^{g(m,r,n-1)} \cdot a^{g(k,r,n) + e(v,r,n+1)}. \end{aligned}$$

This completes the inductive proof (a), and since the majorizing functions on the right are elementary in a and n, also f is elementary.

COROLLARY 2 The class **FEF** of elementary functions is the smallest class containing s, c_0^1, p_1^k, $\max(x,y)$, 2^x which is closed under superposition and bounded primitive recursion.

That **FEF** does have these properties follows from the Theorem above. Let **F** be some other class having these properties. Then **F** contains together with $b_0 = p_0^1$ and 2^x all the elementary scaling functions $b_{n+1} = 2^x \circ b_n$. Together with the 2-ary maximum function also the k-ary maximum function is in **F**, hence for every k also the k-ary function E_n^k defined by $E_n^k(\alpha) = b_n(\max(\alpha))$. From the primitive recursive initial functions s, c_0^1 there arise c_1^1, *cs*, $\dot{-}$, $+$ and $\cdot$ by primitive recursion, and they can be majorized by functions E_n^k for suitable n. Being majorized by functions from **F**, they belong to **F** again, and so **F** contains the elementary initial functions. It remains to be seen that for every f^{k+1} from **FEF**, which is in **F**, also Σf^{k+1} and Πf^{k+1} are in **F**. But this is the case because these functions arise by primitive recursion, and by BERECZKI's result about **FEF**, mentioned earlier, they are majorized by suitable E_n^k since f^{k+1} is in **FEF**.

6. Simple Arithmetization

The developments of the last section depended on the coding of finite sequences of numbers - and in particular of values of functions - based on the unique prime factor decomposition. The basic coding functions *exp* were elementary, and the elementary history of elementary function was elementary again.

During the discussion of Full Arithmetic in Chapter 1.16, I used a coding of finite sequences invented by GÖDEL which relayed on the two arithmetical facts

(A1) Given a sequence of k pairwise coprime numbers g_i, $i<k$. For every sequence σ of k numbers there exists a number a such that for every $i<k$: $mod(g_i, a) = \sigma(i)$,

(A2) For $k>0$ let b be such that $b>0$ and for every i : if $0<i<k$ then i divides b . Then the k numbers $g_i = b\cdot(i+1)+1$ are pairwise coprime ,

which were proven there in detail. I shall now describe this coding process more systematically and shall use it in order to prove some more theorems about classes **R** for which **F(R)** is closed under primitive recursion.

I define the four simple functions

$$\beta_0 = mod \circ \langle s \circ \cdot \circ \langle p_0^3, s \circ p_2^3 \rangle, p_1^3 \rangle ,$$

$$\beta = \beta_0 \circ \langle cro_0 \circ p_0^2, cro_1 \circ p_0^2, p_1^2 \rangle ,$$

$$L = \beta \circ \langle p_0^1, c_0^1 \rangle \ ,$$

$$K = \beta \circ \langle p_0^2, s \circ p_1^2 \rangle \ ,$$

i.e.

$$\beta_0(d,c,i) = mod\,(d \cdot (i+1)+1,\ c) \quad , \quad \beta(b,i) = \beta_0(cro_0(b),\ cro_1(b),\ i) \ ,$$

$$L(b) = \beta(b,0) \quad , \quad K(b,i) = \beta(b,\ i+1) \ .$$

β_0 and β are called the 3-ary and the 2-ary GÖDEL function respectively. Observe that $mod\,(x,y) < y$ for $y > 0$ whence $\beta_0(d,c,i) < c$ for $c > 0$. It follows from $cro_1(b) \leq b$ that

$$\text{if } 0 < b \text{ then } \beta(b,i) < b \ .$$

I next shall show

(0) for every k, $k > 0$, and for every ξ in ω^k there exists a number b such that for every $i < k$: $\beta(b,i) = \xi(i)$.

Given k and ξ in ω^k, I choose d in (A2). Applying (A1) I find a number c such that für all $i < k$

$$\beta_0(d,c,i) = \xi(i) \text{ and } \beta_0(d,c,i) < c \ .$$

Thus for $b = cau\,(d,c)$ there follows $\beta(b,i) = \xi(i)$. Applying (0) to $\langle k, \xi \rangle$, I find

(1) for every k, $k > 0$, and for every ξ in ω^k there exists a number b such that $L(b) = k$ and, for every $i < k$, also $K(b,i) = \xi(i)$,

i.e. L(b) is the length of ξ and K(b,i) is the i-th component of ξ.

Making use of β, I can obtain coding maps Γ from ω^∞ into ω by choosing, for α in ω^k, the value $\Gamma(\alpha)$ to be *some* number b determined by (1). Whichever way this choice may have been made, Γ will be injective, but it should be noticed that k and α do *not* determine the value b uniquely : while (1) holds, the values K(b,i) for $i \geq L(b)$ remain undetermined.

For instance, define Γ_e by choosing d in (A2) as the smallest common multiple of 2,..., k−1; in the case $k = 4$ e.g. this will give $d = 6$. Since c in (A1) is uniquely determined modulo the product of the numbers d_i, choose it as the smallest number, determined by the algorithms contained in the proofs of (a6) and (a7). For the sequence

$$\alpha = \langle 1,2,3 \rangle$$

the four congruences $3 \equiv c \pmod 7$, $1 \equiv c \pmod{13}$, $2 \equiv c \pmod{19}$, $3 \equiv c \pmod{25}$ have the solution 25728. Thus $\Gamma_e(\alpha) = cau(d,c) = cau(6,\ 25728) =$ 331 132 251 will be *one* possible value $\Gamma(\alpha)$.

I define the *special* coding map Γ_0 by choosing $\Gamma_0(\alpha)$ to be the *smallest* b satisfying (1). But to find this b requires a step by step search − which here stops at 20 914 282. In this case, therefore, the value $\Gamma_0(\alpha)$ is smaller by about one power of 10 than the value $\Gamma(\alpha)$ computed from the Chinese Remainder theorem (A1). The elements b of $\mathrm{im}(\Gamma_0)$ are characterized by

(2a) for all a: if $L(a) = L(b)$ and, for all $i < L(b)$, also $K(b,i) = K(a,i)$, then $b \leq a$

or

(2b) for all a: if $a < b$ then ($L(b) \neq L(a)$ *or* there exists i such that $i < L(b)$ and $K(b,i) \neq K(a,i)$) .

Actually, I shall not need Γ_0 itself but only $\mathrm{im}(\Gamma_0)$. Thus I shall use (2a) or (2b) as the *definition* of the set *sic* of *simple codes*. Rewriting (2b), I first form the set

$$S^3 = \exists_<((-E^2)[K\circ<p^3_1, p^3_2>, K\circ<p^3_0, p^3_2>])$$

of all $<a,b,c>$ such that there exists i with $i < c$ and $K(b,i) \neq K(a,i)$, and then obtain

$$sic = \forall^1_<(((-E^2)[L\circ p^2_1, L\circ p^2_0]) \cup S^3[p^2_0, p^2_1, L\circ p^2_1]).$$

Since L and K are simple, there follows $sic \in \mathbf{R}(\mathbf{FSF})$.

For every function f^{k+1} let H^{k+2} be the set of all $<\delta,n,b>$ such that

$b \in sic$,
$L(b) = n+1$,
for all $i < n+1$: $K(b,i) = f(\delta,i)$.

It then follows from (2b) that if $<\delta,n,b> \in H^{k+2}$ then b is the *only* b' such that $<\delta,n,b'> \in H^{k+2}$; hence H^{k+2} is the graph of a total function $H_s f^{k+1}$ which I may call the *simple history* of f^{k+1}. $H_s f^{k+1}$ determines f^{k+1} as

(3) $K\circ<H_s f^{k+1}, p^{k+1}_k>$, i.e. $f^{k+1}(\delta,n) = K(H_s f^{k+1}(\delta,n), n)$.

Since H^{k+2} is the intersection of the three relations

$sic\,[p^{k+2)}_{k+1}]$

$E^2[s\circ p^{k+2}_k, L\circ p^{k+2}_{k+1}]$

$\forall^k_<(E^2[K\circ<p^{k+2}_{k+1}, p^{k+2}_k>, f^{k+1}\circ<p^{k+2}_0, \ldots, p^{k+2}_{k-1}, p^{k+2}_k>>])$,

it follows that for a simple function f^{k+1} the relation H^{k+2} is in $\mathbf{R}(\mathbf{FSF})$ whence $H_s f^{k+1}$ is in $\mathbf{F}(\mathbf{R}(\mathbf{FSF}))$.

Consider now two functions g^k, r^{k+2} and define $H^{k+2} = H^{k+2}(g^k, r^{k+2})$ to be the set consisting of all $<\delta, n, b>$ such that

(4) (0.b) $b \in sic$,
(1.b) $L(b) = n+1$,
(2.b) $K(b,0) = g^k(\delta)$,
(3.b) for all $i < n$: $K(b, i+1) = r^{k+2}(\delta, i, K(b,i))$;

the identity $K(b, i+1) = r^{k+2}(\delta, i, K(b,i))$ in (3.b) I abbreviate as [b.i] .

LEMMA 3 H^{k+2} is the graph of the function $H_s f^{k+1}$ for the function f^{k+1} defined by primitive recursion from g^k, r^{k+2}.

Observe first

(5) If $<\delta, n, b>$ and $<\delta, n, c>$ are in H^{k+2} then $b = c$.

Here b,c are in *sic* by (0.b), (0.c). Thus (2a) will imply $b = c$ if I can show that $L(b) = L(c)$ - which follows from (1.b), (1.c) - and if $K(b,i) = K(c,i)$ for all $i < L(b)$. For $i = 0$ this follows from (2.b), (2.c), and (3.b), (3.c) permit to conclude from $K(b,i) = K(c,i)$ upon $K(b,i+1) = r^{k+2}(\delta, i, K(b,i)) = r^{k+2}(\delta, i, K(c,i)) = K(c, i+1)$ for $i < n+1$. - I next prove by induction on i :

(6) for all n,m,b,c : if $i \lesssim n$ and $i \lesssim m$ and $<\delta, n, b> \in H^{k+2}$ and $<\delta, m, c> \in H^{k+2}$ then $K(b,i) = K(c,i)$.

For $i = 0$ this follows from $K(b,0) = K(c,0) = g^k(\delta)$ by (2.b) and (2.c). Assume (6) to hold for i and let $<\delta,n,b>$ and $<\delta,m,c>$ be in H^{k+2} where $i+1 \lesssim n$ and $i+1 \lesssim m$. Then $K(b,i) = K(c,i)$ implies $K(b, i+1) = r^{k+2}(\delta, i, K(b,i)) = r^{k+2}(\delta, i, K(c,i)) = K(c, i+1)$ by [b.i] and [c.i]. - Now I prove by induction on n

(7) for every $<\delta,n>$ there exists a number b such that $<\delta,n,b> \in H^{k+2}$.

For $n = 0$ I obtain from (1) a minimal b such that $L(b) = 1$ and $K(b,1) = g^k(\delta)$; it belongs to *sic* since it is minimal and $L(b) = 1$ and $K(b,1)$ are given. Assume now that for $<\delta,n>$ a number b satisfying (7) has been found; then (1) produces a number b_0 which is minimal for the properties

(7.1) $L(b_0) = n+2$,
(7.2) for all $i \lesssim n$: $K(b_0,i) = K(b,i)$,
(7.3) $K(b_0, n+1) = r^{k+2}(\delta, n, K(b,n))$.

Being minimal, b_0 belongs to *sic* since $L(b_0) = n+2$ and since the first n+1 values $K(b,i)$ are given by (7.2), (7.3). Condition $(1.b_0)$ for $<\delta, n+1, b_0>$ in H^{k+2} is obvious. Condition $(2.b_0)$ follows from (2.b) and from $K(b_0,0) = K(b,0)$ by (7.2). Conditions $[b_0.i]$, i.e. $K(b_0, i+1) = r^{k+2}(\delta, i, K(b_0,i))$ for $i \lesssim n$, hold for $i < n$ since then $K(b_0, i+1) = K(b, i+1) = r^{k+2}(\delta, i, K(b,i)) = r^{k+2}(\delta, i, K(b_0,i))$ by [b.i] and by (7.2). The condition $[b_0.n]$ holds since $K(b_0, n+1) = r^{k+2}(\delta, n, K(b,n)) = r^{k+2}(\delta, n, K(b_0,n))$ by (7.3), (7.2). This concludes the induction proof of (7).

It follows from (5) and (7) that H^{k+2} is the graph of a function h^{k+1} defined on ω^{k+1}. Define f^{k+1} by (3), i.e. $f^{k+1}(\delta,n) = K(h^{k+1}(\delta,n), n)$. Then $f^{k+1}(\delta,0) = K(h^{k+1}(\delta,0),0) = g^k(\delta)$ follows from $(2.h^{k+1}(\delta,0))$. Applying (6) to $m = n+1$ and $i = n$, I obtain $K(h^{k+1}(\delta, n+1), n) = K(h^{k+1}(\delta,n), n)$, hence

$$\begin{aligned} f^{k+1}(\delta, n+1) &= K(h^{k+1}(\delta,n+1), n+1) && \text{by definition of } f^{k+1} \\ &= r^{k+2}(\delta,n, K(h^{k+1}(\delta,n+1), n)) && \text{by } [(h^{k+1}(\delta,n+1),n] \\ &= r^{k+2}(\delta,n, K(h^{k+1}(\delta,n), n)) && \text{applying (6)} \\ &= r^{k+2}(\delta,n, f^{k+1}(\delta,n)) && \text{by definition of } f^{k+1} . \end{aligned}$$

Thus f^{k+1} is defined by primitive recursion from g^k and r^{k+2}, and $h^{k+1} = H_s f^{k+1}$, $H^{k+2} = G(H_s f^{k+1})$.

[The construction of f^{k+1} from H^{k+2} is nothing but a reformulation of DEDEKIND's *set theoretical* proof for the existence of recursively defined functions. In that situation, if g^k, r^{k+2} are given, and if δ is a fixed sequence of parameters, then an n-*germ* is defined to be a (partially defined) function k_n, defined for all i with $i<n$ and satisfying the recursion equations for all i (in place of n) such that $i<n$. Induction on i shows that any two germs k_n, k_m with $i<n$, $i<m$ attribute the same value to i (which corresponds to (4)). Applying induction on n, there follows for every n that there exists *one*, hence *exactly one* n-germ k_n (which corresponds to (5)). This having been established, f^{k+1} is defined explicitly by $f^{k+1}(\delta, n) = k_n(n)$.

It was an idea of J. VON NEUMANNs that the set theoretical proof can be arithmetized with help of a map Γ which codes finite sequences. Because a germ k_n is nothing but the n-member sequence of its values, beginning with $g^k(\delta)$ and continued according to the recursion equations. Making use of the coding map Γ_0, the set theoretical proof translated into the arithmetical one. The direct use (2), made in the proof of (7), may be replaced by applications of Γ_0: The number b for $<\delta,0,b> \epsilon H^{k+2}$ I define explicitly as $\Gamma_0(<1, g^k(\delta)>)$, and from the b with $<\delta, n, b> \epsilon H^{k+2}$ I define the b' for $<\delta, n+1, b'> \epsilon H^{k+2}$ as

$$\Gamma_0(<n+2, <K(b,i) \mid i \leq n>, r^{k+2}(\delta, n, K(b,n))> .$$

It then is evident that the arithmetized germs $<\delta,n,b>$ will all be such that b is in $im(\Gamma_0)$.]

LEMMA 4 Let f^{k+1} be defined by primitive recursion from functions g^k, r^{k+2} from a class **D**. Assume that

(RD0) **D** is closed under superposition and contains the functions p_i^k, c_0^1, s, + , · , *mod*, cro_0, cro_1 ,
(RD1) **R** contains E^2 and $-E^2$,
(RD2) **R** is positively-Boolean closed ,
(RD3) **R** is closed under superposition with functions from **D**,
(RD4) **R** is closed under $\forall_<$ and $\exists_<$.

Then $H_s f^{k+1}$ is in **F**(**R**). If, moreover,

(RD5) **F**(**R**) is closed under superpositions of functions in **D** with functions from **F**(**R**) .

then f^{k+1} is in **F**(**R**).

It follows from (RD0) that **D** contains the functions β_0, β, L and K. It follows from (RD1), (RD3), (RD4), (RD2) that **R** contains the relation S^3

used to approach *sic* and then the set *sic* itself. The set $H^{k+2} = G(H_s f^{k+1})$ is, by its definition in (4), the intersection of the four relations

$$sic[p^{k+2}_{k+1}]$$

$$E^2[s \circ p^{k+2}_k, L \circ p^{k+2}_{k+1}]$$

$$E^2[K \circ \langle p^{k+2}_{k+1}, c^1_0 \circ p^{k+2}_k \rangle, g^k \circ \langle p^{k+2}_0, \dots, p^{k+2}_{k-1} \rangle]$$

$$\forall^k_< (E^2[K \circ \langle p^{k+2}_{k+1}, s \circ p^{k+2}_k \rangle,$$
$$r^{k+2} \circ \langle p^{k+2}_0, \dots, p^{k+2}_{k-1}, p^{k+2}_k, K \circ \langle p^{k+2}_{k+1}, p^{k+2}_k \rangle \rangle])$$

all of which now also belong to **R**. Thus $G(H_s f^{k+1})$ is in **R**, hence $H_s f^{k+1}$ is in **F**(**R**). Since $G(p^{k+1}_k) = E^2[p^{k+2}_k, p^{k+2}_{k+1}]$, also p^{k+1}_k is in **F**(**R**), and only now I use (RD5) and conclude that $f^{k+1} = K \circ \langle h^{k+1}, p^{k+1}_k \rangle$ is in **F**(**R**).

THEOREM 3 If a class **R** has the properties (RD0)–(RD4) for **D** = **F**(**R**) then **F**(**R**) is closed under primitive recursion.

This follows from the Lemma since by (D3) the condition (RD5) is a consequence of (RD3).

COROLLARY 3 If **F** is simply closed and if **F**(**R**(**F**)) is closed under superposition then **F** is closed under bounded primitive recursion.

I apply the Lemma to **D** = **F** as **R** = **R**(**F**). There holds (RD0) by (FS3), (FS12), (FS15), and (RD1), (RD2) hold by (FS9). (RD3) follows from (FS1) and (RD4) from (FS10). (RD5) follows from $\mathbf{F} \subseteq \mathbf{F}(\mathbf{R}(\mathbf{F}))$ and the assumption about **F**(**R**(**F**)). So if f^{k+1} is defined by primitive recursion from functions g^k and r^{k+2} in **F** then it is in **F**(**R**(**F**)), i.e. $G(f^{k+1})$ is in **R**(**F**). And if f^{k+1} is also majorized by a function in **F** then it belongs to **F** by (FS4).

The function of $H_s f^{k+1}$ was defined with the special coding map Γ_0 on which, therefore, all the results so far depend. For an arbitrary map Γ the set $\mathrm{im}(\Gamma)$ cannot be described as easily as *sic* could. Adding in Lemma 4 the hypothesis

(RD6) If R^{k+1} is in **R** and is full then so is νR^{k+1}

(with νR^{k+1} as defined in (CC7)), I still can define a function h^{k+1} which in Lemma 4 can take the place of $H_s f^{k+1}$, and then Theorem 3 and its Corollary remain in effect. To that end, I change the definition (4) of H^{k+2} by omitting the clause (0.b); under the hypotheses of Lemma 4 (where $-E^2$ in (RD2) now is not needed) *this* H^{k+2} still will in **R** . In the proof of Lemma 3 then (5) cannot be proved anymore, but the proofs of (6) and (7) remain in effect. So H^{k+2} still will be a *full* relation in **R**, and then (RD6) implies that νH^{k+2} is the graph of a function h^{k+1} which, therefore, is in $\mathbf{F}(\mathbf{R})$. Now h^{k+1} is *another* history of f^{k+1} which it again determines as $K(h^{k+1}(\delta, n), n)$.

Assuming that I know how to compute the coding map Γ by (1) (e.g. by defining $\Gamma_e(\alpha)$ as above), the induction proofs of (6) and (7) in Lemma 3 produce an algorithm by which to compute $h^{k+1} = \nu H^{k+2}$. But if $\Gamma(\alpha)$ is chosen more arbitrarily (e.g. with random values as $K(b, n+2)$ for the b coding a sequences of length n), the relation H^{k+2} still remains full, and (RD6) does *not* ask by which methods the fullness of its relation R^{k+1} has been established. I here have the situation that, in order to secure insights about the *formal structure* of H^{k+2} (its belonging e.g. to $\mathbf{R}(\mathbf{FSF})$), I may make use of (RD6) whose hypothesis, the fullness of R^{k+1}, may have been secured in a perfectly inconstructive manner: That R^{k+1} is full must be a *true* arithmetical fact, but not necessarily proved by controlled means. For (A1) and (A2), of course, the means by which to define Γ_e were controlled in the arithmetical proofs of Chapter **1.16** .

The presentation chosen for this and the following Chapter is based on selected parts of FELSCHER 93 . While being recursively enumerable, the textbooks on recursive functions are many in number; the ones by MAL'T-SEV 66 and TOURLAKIS 84 listed below may be particularly worth the reader's attention.

References

W. Ackermann: Zum Hilbertschen Aufbau der reellen Zahlen. Math.Ann 99 (1928) 118–133

I. Bereczki: Nem-elemi rekurziv függvény létezése. C.R.Premier Congrès Math. Hongrois 1950 409–417

A.L. Cauchy: Cours d'Analyse de l'Ecole Royale Polytechnique, 1e partie: Analyse Algébrique. Paris 1821

W. Felscher: Berechenbarkeit. Rekursive und Programmierbare Funktionen. Berlin 1993

A.I. Mal'tsev: Algoritmy i rekursivnye funkcii. Moskva 1966 . English translation: Algorithms and Recursive Functions. Groningen 1970

I. Niven, H.S. Zuckerman: An Introduction to the Theory of Numbers. New York 1960

R. Peter: Konstruktion nichtrekursiver Funktionen. Math.Ann. 111 (1935) 42–60

G.J. Tourlakis: Computability. Reading 1984

Chapter 6. Recursive Relations and Recursive Functions

The study of primitive recursive functions has lead, in Theorem 5.3, to a description of classes **R** of relations for which the class **F(R)** of functions is closed under primitive recursion. What is important about this description is that (in the conditions (RD0)-(RD4)) *no* reference to recursion had to mentioned in its hypotheses, and only *such* descriptions can be susceptible to a formulation by formulas of my (1st order) languages as required to arrive at results about representability. What is unsatisfactory about these results is their hypothesis that **R** be closed under superposition with functions from **F(R)**, a situation for which I have not given a single example, such that in particular it is not clear whether there *are* any classes **R** to which my insights may be applied. It is the purpose of the present Chapter to provide such examples, and to do so in a perspicuous manner.

I shall beginn this Chapter by presenting the class **RAB** of arithmetically bounded relations which consists, essentially, of what can be obtained from the graphs of + and · by Boolean operations and bounded quantifications. Obviously, that cannot be much, and **RAB** will still not be closed under superposition with functions from **F(RAB)**. An operation on relations R^{k+1}, closure with respect to which suffices to fulfill this desire, is the formation of *projections* $\exists R^{k+1}$, defined as the k-ary relation of all α for which there exists b such that $<\alpha,b>\epsilon R^{k+1}$. A first important result, (AB20), then will be that the hypotheses of Theorem 5.3 are satisfied by the class **E(RAB)** of projections of relations in **RAB**; hence **E(RAB)** will be closed under superposition with functions from **F(E(RAB))**, and that class will be closed under primitive recursion.

While this would suffice to arithmetize the situation given in the model theoretic theorems which refer to semantical *definability,* the syntactical theorems make use of *representability* of relations R, and that is a property which involves both R and its *complement* −R. This leads to the study of the class **QE(RAB)**, consisting of the R such that both R and −R are in **E(RAB)**, which again will satisfy the hypotheses of Theorem 5.3; this **QE(RAB)** then is defined to be the class **RFR** of *recursive relations.* The class **FRF** = **F(RFR)**) is called the class *recursive functions*, and in harmony there also holds **RFR** = **R(FRF)**; the Chapter then ends with several other characterizations both of **RFR** and **FRF** which are useful for later applications.

For further applications, one more notion is required, viz. that of *recursively enumerable* relations. Within the present setting, they can be defined as the relations in **E(RAB)**, and it will turn out that this class coincides with each of **E(R(FSF)))**, **E(R(FEF)))**, **E(R(FPF)))** and **E(R(FRF)))**.

1. Bounded Arithmetical Relations

A class **R** of relations is called AB-*closed* if

(RAB0) the graphs of the operations + and · are in **R** ,

(RAB1) **R** is trivially closed,

(RAB2) **R** is Boolean closed,

(RAB3) **R** is closed under the bounded quantification $\forall_{\leq}$.

It follows from (RAB2) and **5**.(C3) that **R** is closed under the bounded quantification $\exists_{\leq}$, and as noticed in **5**.(CC5) it then is also closed internally bounded quantifications $\forall^{i}_{\leq}$, $\exists^{i}_{\leq}$.

The elements of the *smallest* AB-closed class **RAB** I call the *bounded arithmetical relations*.

If **F** is simply closed then **R**(**F**) is AB-closed; in particular **RAB** $\subseteq$ **R**(**FSF**). This follows from **5**.(FS9–10).

Given a function $f = f^1$ such that $G(f) \epsilon \mathbf{R}$ and given a relation $R = R^2$ in **R**, I may substitute f^1 into the first place of R and obtain the relation S of all $<x,y>$ such that

there exists z such that $<x,z> \epsilon G(f)$ and $<z,y> \epsilon R$.

As long as z here cannot be bound by x or by y, I do not know whether S again is in **R**. In special cases, though, such bounds can be found; the following relations and functions belong to **RAB** or to **F**(**RAB**):

(AB0) The 1-ary relation $\{0\}$.

Because $G(+)[p^1_0,p^1_0,p^1_0]$ is the set of all x such that $x+x = x$, and 0 is the only such x.

(AB1) The 1-ary relation $\{1\}$.

Because 1 is the only x such that $x \cdot x = x$ *and* $x \epsilon -\{0\}$, and together with $\{0\}$ also $-\{0\}$ is in **R** .

(AB2) The successor function s .

It follows from (AB1) that together with G(+) the class **R** also contains the set R^3 of all $<x,z,y>$ satisfying

$x+y = z$ and $y \epsilon \{1\}$.

Here y is bounded by z and $G(s)$ becomes $\exists^1{}_{\leq} R^3$.

(AB3) The relations $<$, $\leq$, E^2.

Because $\leq$ is $\exists^1{}_{\leq} R^3$ for the set R^3 of all $<x,z,y>$ such that $<x,y,z>$ is in $G(+)$. Further, $<$ is the complement of the relation $\geq$ of all $<x,y>$ such that $y \leq x$. And E^2 is the intersection of $\leq$ and $\geq$.

It now follows from **5**.(C5)

$$\forall_< R^{k+1} = \exists^k{}_{\leq} (G(s)[p^{k+2}_{k+1}, p^{k+2}_k] \cap (\forall_{\leq} R^{k+1})[p^{k+2}_0, \ldots, p^{k+2}_{k-1}, p^{k+2}_{k+1}])$$

that every AB-closed **R** is also closed under bounded quantifications $\forall_<$. by **5**.(C5), and for $\exists_<$ then from $\exists_< R^{k+1} = -E^2[p^{k+1}_k, c^{k+1}_0] \cap -\forall_< -R^{k+1}$ by **5**.(C2).

(AB4) For every n die 1-ary relation $\{n\}$.

Because if this is known for n then $\{n+1\} = \exists^0{}_{\leq} R^2$ where R^2 contains the $<x,y>$ such that $y \in \{n\}$ and $s(y) = x$, i.e. $R^2 = \{n\}[p^2_1] \cap G(s)[p^2_1, p^2_0]$.

(AB5) Every finite subset of ω^k.

Because if $\alpha \in \omega^k$ then $\{\alpha\}$ is the intersection of the k sets $\{\alpha(i)\}[p^k_i]$, $i < k$, which are in **R** by (AB4). So every finite set is in **R** by (RAB2). (Cf. also **5**.(C2) and the following.)

(AB6) The constant functions c^k_n and the projections p^k_i .

Because $G(c^k_n) = \{n\}[p^{k+1}_k]$ and $G(p^k_i) = E^2[p^{k+1}_i, p^{k+1}_k]$.

(AB7) The relation *div*.

Because *div* is $\exists^1{}_{\leq} R^3$ where R^3 is the set of all $<x,z,y>$ such that $x \cdot y = z$, i.e. $<x,y,z> \in G(\cdot)$.

(AB8) The functions *qu* and *mod*.

I begin by rewriting $y = q \cdot x + z$ as

$$y = u + z \text{ and } u = q \cdot x \text{ .}$$

The set R^5 of these $<x,y,z,q,u>$ is in **R**, and there holds $u \leq y$. Thus **R** also contains the set $R^4 = \exists^1{}_{\leq} R^5$ of all $<x,y,z,q>$ such that $y = q \cdot x + z$. As $\{0\}$ is in **R**, also the set S^4 of all $<x,y,z,q>$ such that

$$y = q \cdot x + z \text{ and } (z < x \text{ or } x = 0 \text{ and } z = 0 \text{ and } q = y)$$

is in **R**. Hence **R** contains the set $S^3 = \exists^1{}_{\leq} S^4$ of all $<x,y,z>$ satisfying this condition with $q \leq y$, and that is $G(mod)$. The set of all $<x,y,q>$ satisfying the condition with $z \leq x$ is $G(qu)$.

(AB9) The functions *cau*, cro_0, cro_1 .

Here I rewrite $cau(x,y) = z$ as

$$z = qu(2, (x+y)(x+y+1)) + x$$

and then this as

$$z = t+x \text{ and } ((t = (x+y)\cdot qu(2,x+y+1) \text{ and } div(2,x+y+1)) \text{ or } (t = (x+y+1)\cdot qu(2,x+y) \text{ and } div(2,x+y))).$$

If $<x,y>$ is not $<0,0>$ or $<0,1>$ then $x+y+1 \leq z$ and $2 \leq z$. Hence the above becomes

$(z = 0$ and $x = 0$ and $y = 0)$ or $(z = 1$ and $x = 0$ and $y = 1)$
or $(z = t+y$ and $u = 2$ and
$($ $(t = v\cdot r$ and $v = x+y$ and $r = qu(u,w)$ and $w = v+1$ and $div(u,w)$ $)$
or $(t = s\cdot w$ and $v = x+y$ and $w = v+1$ and $s = qu(u,v)$ and $div(u,v)$ $)$ $)$.

By (AB3) the set of all u such that $u = 2$ is in **R**. Thus here I have a relation R^9, belonging to **R**, containing certain $<x,y,z,u,v,w,r,s,t>$ in which u,v,w,r,s,t are bounded by z. Hence $\exists^2_{\leq}\exists^2_{\leq}\exists^2_{\leq}\exists^2_{\leq}\exists^2_{\leq}\exists^2_{\leq}R^9$ is $G(cau)$. From the set R^3 of all $<z,x,y>$ such that $cau(x,y) = z$, I obtain $G(cro_0)$ as $\exists^0_{\leq}R^3$; from the set S^3 of all those $<z,y,x>$ I obtain cro_1 as $\exists^0_{\leq}S^3$.

(AB10) GÖDELs 3-ary function $\beta_0(b,c,i) = mod(b\cdot(i+1) + 1, c)$.

Observe that $mod(x,y) = z$ implies $z \leq y$. I rewrite $z = mod(b\cdot(i+1)+1,c)$ as

$$z = mod(u,c) \text{ and } u = v+1 \text{ and } v = b\cdot w \text{ and } w = i+1$$

where u, v and w are bounded by c. These $<b,c,i,z,u,v,w>$ form a set R^7 with which $G(\beta_0)$ becomes $\exists^1_{<}\exists^1_{<}\exists^1_{<}R^7$.

(AB11) GÖDELs 2-ary function $\beta(a,i) = \beta_0(cro_0(a), cro_1(a), i)$.

Because **R** contains the set R^5 of all $<a,i,z,b,c>$ such that $\beta_0(b,c,i) = z$ and $a = cau(b,c)$. Hence $G(\beta)$ is $\exists^0_{\leq}\exists^0_{\leq}R^5$.

Henceforth I shall call the functions in **F**(**R**) also *functions belonging* to **R**.

(AB12) If **R** is AB-closed then **R** is closed under keeping an argument n constant, and is closed under superpositions with the functions c^1_n .

If $R^{k+1} \in \mathbf{R}$ then the set of all $<\alpha,n>$ in R^{k+1} is $R^{k+1} \cap G(c^k_n)$ and so is in **R** by (RAB2). As for superpositions for the c^1_n, I use an idea of ULRICH MAYER. Let U^k_n be the set of all α such that $<\alpha,n> \in R^{k+1}$, then

$$R^{k+1}[p^{k+1}_0,\dots, p^{k+1}_{k-1}, c^1_n \circ p^{k+1}_k] = U^k_n[p^{k+1}_0,\dots, p^{k+1}_{k-1}] .$$

So it remains to be shown that U^k_n is in **R**. Let $U = U^{k+1}_n$ be the set of all $<\alpha,n>$ in R^{k+1} as above. If $i<k$ then $\exists^i_{\leq}U$ is the set of all α such that $<\alpha,n> \in R^{k+1}$ and $\alpha(i) \geq n$. The union V of the k sets $\exists^i_{\leq}U$ remains in **R**, and U^k_n is the union of V and the set W of all α satisfying $<\alpha,n> \in R^{k+1}$ and $\alpha(i) < n$ for every $i<k$. Being finite, W is in **R** by (AB5), and so also $V \cup W = U^k_n$ is in **R**.

It now follows from $\exists_{<} R^{k+1} = -E^2[p^{k+1}_k, c^{k+1}_0] \cap -\forall_{<}-R^{k+1}$ by **4**.(C2) that every AB-closed **R** is also closed under bounded quantifications $\exists_{<}$.

The class POF of *polynomial functions* shall be the smallest class containing $+$, $\cdot$, the p_i^k and the c_n^k, and closed under superposition. It follows from the associative law for superposition

$$f^k \circ \langle g_0^m \circ \langle h_0^n, \ldots, h_{m-1}^n \rangle, \ldots, g_{k-1}^m \circ \langle h_0^n, \ldots, h_{m-1}^n \rangle \rangle$$
$$= (f^k \circ \langle g_0^m, \ldots, g_{k-1}^m \rangle) \circ \langle h_0^n, \ldots, h_{m-1}^n \rangle$$

that POF is also the smallest class S containing the p_i^k and the c_n^k which is closed under superpositions of $+$ and $\cdot$ with functions from S.

(AB13) If R is AB–closed then R contains the graphs of the polynomial functions.

For the initial functions of S this follows from (AB6). Consider now the case $\cdot \circ \langle f^k, g^k \rangle = f^k \cdot g^k$. Then R contains together with $G(f^k)$, $G(g^k)$, $G(\cdot)$ also the set R^{k+3} of all $\langle \alpha, a, b, c \rangle$ such that

$$\langle \alpha, b \rangle \in G(f^k) \text{ and } \langle \alpha, c \rangle \in G(g^k) \text{ and } \langle b, c, a \rangle \in G(\cdot) \ ,$$

hence also the set V^{k+1} of all $\langle \alpha, a \rangle$ such that there exist $b \lesssim a$, $c \lesssim a$ satisfying $\langle \alpha, a, b, c \rangle \in R^{k+3}$. But $\langle \alpha, 0, b, c \rangle \in R^{k+3}$ if, and only if, $b = 0$ or $c = 0$, i.e. if, and only if, $\langle \alpha, 0 \rangle \in G(f^k)$ or $\langle \alpha, 0 \rangle \in G(g^k)$. Hence

$$G(f^k \cdot g^k) = V^{k+1} \cup G(f^k)_0^{k+1} \cup G(g^k)_0^{k+1}$$

where the new relations are defined as in the proof of (AB12) and so belong to R again.

I now turn to the question which functions besides the c_n^1 may be superposed into the relations of an AB–closed class R.

Let CRO be the smallest class of 1–ary functions containing p_0^1 and all cro_i^k and closed under superposition. A function g^k *descends in the argument* i ($i < k$) if $g^k(\alpha) \lesssim \alpha(i)$ for all α. The functions u from CRO descend in their (only) argument; the function β is descends in its 0th argument since *mod* descends in its argument 1 and since $\beta(a,i) = \beta_0(cro_0(a), cro_1(a), i)$ implies $\beta(a,i) \lesssim cro_1(a) \lesssim a$.

(AB14) If $u = u^1$ and g belong to R, and if g^k descends in an argument i, then also $u \circ g^k$ belongs to R.

Because R contains the set R of all $\langle \alpha, z, y \rangle$ such that $u(y) = z$ and $g^k(\alpha) = y$ by (RAB1), (RAB2), and since $y \lesssim \alpha(i)$ the graph of $u \circ g^k$ is $\exists^i_{\lesssim} R^3$.

(AB15) If g^k belongs to and R^n is in R, and if g^k descends in an argument i, then R contains the superposition

$$T^{n+k-1} = R^n[p_0^{n+k-1}, \ldots, p_{j-1}^{n+k-1}, g^k(p_j^{n+k-1}, \ldots, p_{j+k-1}^{n+k-1}), p_{j+k}^{n+k-1}, \ldots, p_{n+k-2}^{n+k-1}]$$

of R^n with g^k at an argument j.

Because R contains by (RAB2) the set S^{n+k} of all

$$\langle x_0, \ldots, x_{j-1}, y_0, \ldots, y_{k-1}, x_{j+1}, \ldots, x_{n-1}, x_j \rangle$$

such that $<x_0,\dots,x_{n-1}> \epsilon R^n$ and $g^k(y_0,\dots,y_{k-1}) = x_j$. But $\exists^{j+i}{}_{\leq}S^{n+k}$ is the set T^{n+k-1} of all $<x_0,\dots,x_{j-1},y_0,\dots,y_{k-1},x_{j+1},\dots,x_{n-1}>$ such that

$$<x_0,\dots,x_{i-1},g^k(y_0,\dots,y_{k-1}),x_{i+1},\dots,x_{k-1}> \ \epsilon\ R^n .$$

(AB16) The functions from **CRO** belong to **R**, and **R** is closed under superposition with functions from **CRO**.

A much more general result was obtained 1992 by ULRICH MAYER:

THEOREM 0 The class **RAB** is closed under superposition with polynomial functions.

The proof requires too much space to be presented here.

As observed earlier, **R(F)** is AB-closed if **F** is simple. A limited converse result is:

(AB17) If **R** is AB-closed and closed under superposition with functions from **F(R)** then **F(R)** is simply closed.

The initial functions c_1^1, p_i^k, $+$, $\cdot$, $\chi_<$, $\chi_=$ are in **F(R)** by (AB6),(AB2),(RAB0),(AB3). Together with $\chi_<$ and c_0^1 also $\chi_< \circ <c_0^1, p_0^1>$ is in **F(R)**, and this is the function *sg*. This proves (FSF0), and (FSF1) follows from 5.(D3). Now 5.(D6) implies $\mathbf{R}(\mathbf{F}(\mathbf{R})) \subseteq \mathbf{R}$; hence if $R^{k+1} \epsilon \mathbf{R}(\mathbf{F}(\mathbf{R}))$ then $R^{k+1} \epsilon \mathbf{R}$, and as **R** is Boolean closed, it also contains the graph

$$G(\mu_< R^{k+1}) = \nu R^{k+1} = R^{k+1} \cap (\forall_< - R^{k+1} \cup E^2[p_k^{k+1}, c_0^{k+1}])$$

by 5.(C9). Thus $R^{k+1} \epsilon \mathbf{R}(\mathbf{F}(\mathbf{R}))$ implies $\mu_< R^{k+1} \epsilon \mathbf{R}(\mathbf{F})$, proving (FSF2).

(AB18) Let **R** be the smallest AB-closed class containing the graphs of a class **A** of functions. Then the characteristic functions of the relations in **R** belong to the smallest elementarily closed class **F** containing **A**. In particular, the characteristic functions of relations in **R** are elementary.

If $S = R^k[p^k_{\pi(0)},\dots,p^k_{\pi(k-1)}]$ then $\chi_S = \chi_R \circ <p^k_{\pi(0)},\dots,p^k_{\pi(k-1)}>$; hence $\chi_R \epsilon \mathbf{F}$ implies $\chi_S \epsilon \mathbf{F}$. Further, $\chi_{R \cap S} = \cdot \circ <\chi_R, \chi_S>$ and $\chi_{-R} = \mathrm{CSG} \circ \chi_R$. Finally, if $S = \forall_{\leq} R$ then $\chi_S(\alpha,a) = \Pi <\chi_R(\alpha,b) \mid b \leq a>$ whence $\chi_S = \Pi\, \chi_R$.

2. Projections

For a relation R^{k+1}, $k>0$, the *projection* $\exists R^{k+1}$ is the **k**-ary relation of all α such that there exists some b satisfying $<\alpha,b> \epsilon R^{k+1}$; more precisely, this is the projection *along the* (k+1)-*th axis*, and projections along other axis' are defined analogously..

In this section, I restrict myself to classes **R** satisfying

(ER0) **R** is trivially closed and closed under superpositions with cro_0, cro_1.

(ER1) **R** is positively–Boolean closed.

I define **E(R)** to be the set of all projections of relations in **R**.

(RE0) $\mathbf{R} \subseteq \mathbf{E(R)}$ and **E(R)** is trivially closed.

Every R^k in **R** may be viewed as projection of a relation R^{k+1} obtained by adding a dummy argument. Consider next $\exists R^{k+1}$ with R^{k+1} in **R** and form, for a map π from k into m, $S^m = (\exists R^{k+1})[p^m_{\pi(0)}, \ldots, p^m_{\pi(k-1)}]$. Then S consists of the $\alpha \epsilon \omega^m$ with $\alpha\pi \epsilon \exists R^{k+1}$, i.e. those for which there exists a such that $<\alpha\pi, a> \epsilon R^{k+1}$. I extend π to a map σ from k+1 into m+1 by sending k to m; I then form the relation $T^{m+1} = R^{k+1}[p^{m+1}{}_{\sigma(0)}, \ldots, p^{m+1}{}_{\sigma(k)}]$ in **R**. It consists of the β in ω^{m+1} such that $\beta\sigma \epsilon R^{k+1}$. But every β in ω^{m+1} can be written as $<\alpha, a>$ with $\alpha \epsilon \omega^m$; hence T^{m+1} consists of all $<\alpha, a>$ in ω^{m+1} such that $<\alpha, a> \cdot \sigma = \; <\alpha\pi, a>$ is in R^{k+1}. Thus S^m is $\exists T^{m+1}$.

(RE1) **E(R)** is positively–Boolean closed.

If R, S are projections of V, W along the same axis then $R \cup S$ is projection of $V \cup W$ along this axis. Also $R \cap S$ is projection of a relation in **R** because $\alpha \epsilon R \cap S$ holds if, and only if, there exists a such that $<\alpha, cro_0(a)> \epsilon V$ *and* $<\alpha, cro_1(a)> \epsilon W$; by (ER0) the relations to be intersected here belong to **R**.

(RE2) **E(R)** is closed under projections.

Assume $V^k = \exists S^{k+1}$ and $S^{k+1} = \exists R^{k+2}$ with R^{k+2} in **R**. Thus α is in V^k if, and only if, there exist a and b such that $<\alpha, a, b> \epsilon R^{k+2}$, and this is equivalent to the existence of c such that $<\alpha, cro_0(c), cro_1(c)> \epsilon R^{k+2}$.

(RE3) If **R** is closed under $\exists_<$ then so is **E(R)** .

Assume $R^{k+1} = \exists S^{k+2}$ for S^{k+2} in **R** and set $T^{k+2} = S^{k+2}[p^{k+2}_0, \ldots, p^{k+2}_{k-1}, p^{k+2}_{k+1}, p^{k+2}_k]$. It is $<\alpha, a>$ in $\exists_< R^{k+1}$ if, and only if, there exists i such that $i < a$ and $<\alpha, i> \epsilon R^{k+1}$, i.e. if, and only if,

there exists i and there exists y such that $i < a$ and $<\alpha, i, y> \epsilon S^{k+2}$

or

there exists y and there exists i such that $i < a$ and $<\alpha, y, i> \epsilon T^{k+2}$.

Thus $\exists_< R^{k+1} = \exists(\exists_< T^{k+2})$, and $\exists_< T^{k+2}$ was in **R**.

(RE4) Let **R** contains the graphs of + , · and that of GÖDELs function β_0. If **R** is closed under $\forall_<$ and under $\exists_<$ then **E(R)** is closed under $\forall_<$.

Assume $R^{k+1} = \exists S^{k+2}$ for S^{k+2} in **R**. Thus $<\alpha, a>$ is in $\forall_< R^{k+1}$ if, and only if, $a \neq 0$ and if $i < a$ implies $<\alpha, i> \epsilon R^{k+1}$, i.e. if, and only if

(x) $a\neq 0$ and for all i: if $i<a$ then there exists y such that $<\alpha,i,y>\in S^{k+2}$.

Thus there exists a sequence $<y_i \mid i<a>$ such that $<\alpha,i,y_i>\in S^{k+2}$ for all $i<a$. Making use of the properties of β_0, I find numbers b and c such that $\beta_0(b,c,i)=y_i$ for all $i<k$, and conversely every pair b,c determines such values y_i. Thus $<\alpha,a>$ is in $\forall_< R^{k+1}$ if, and only if,

$a\neq 0$ and there exist b, c: for all i :
if $i<a$ then there exists y with $\beta_0(b,c,i)=y$ and $<\alpha,i,y>\in S^{k+2}$.

But $i<a$ implies $1\leq a$, and $\beta_0(b,c,i)=\mathrm{MOD}(b\cdot(i+1)+1,\ c)=y$ then implies $y<b\cdot(i+1)+1\leq\ b\cdot a+1\leq b\cdot a+a$. Hence $<\alpha,a>$ is in $\forall_< R^{k+1}$ if, and only if

$a\neq 0$ and there exist b, c: for all i :
if $i<a$ then there exists $y<b\cdot a+b$ with $\beta_0(b,c,i)=y$ and $<\alpha,i,y>\in S^{k+2}$,

i.e. if, and only if,

(y) $a\neq 0$ and there exist b, c, z_0, z_1: $z_0=b\cdot a$ and $z_1=z_0+a$ and for all i:
if $i<a$ then there exists $y<z_1$ with $\beta_0(b,c,i)=y$ and $<\alpha,i,y>\in S^{k+2}$.

Making use of the hypotheses about **R**, I find in **R**

the set V^{k+4} of all $<\alpha,b,c,z_1,i>$ such that there exists $y<z_1$
such that $\beta_0(b,c,i)=y$ and $<\alpha,i,y>\in S^{k+2}$
the set W^{k+4} of all $<\alpha,a,b,c,z_1>$ with $a\neq 0$ and for all $i<a$:
$<\alpha,a,b,c,z_1,i>\in V^{k+4}$
the set U^{k+5} of all $<\alpha,a,b,c,z_0,z_1>$ such that $z_0=b\cdot a$ and $z_1=z_0+a$
and $<\alpha,a,b,c,z_1>\in W^{k+4}$.

Applying (RE2), I find $\exists\exists\exists\exists U^{k+5}$ in **E(R)**, and that is $\forall_< R^{k+1}$ because of (y). - I may mention that in the special case of **R** = **R(F)** for an elementarily closed **F** the simple coding with help of β_0 can be replaced by an elementary coding with prime power exponents. Because setting $y(i)=y_i$ and

$$d=\Pi<prino(i)^{y(i)+1} \mid i<a> ,$$

condition (x) becomes equivalent to

(z) $a\neq 0$ and there exists d such that for all i :
if $i<a$ then $<\alpha,i,exp(i,d)\dot{-}1>\in R^{k+2}$.

Here the existentially quantified expression defined a relation which belongs to **R(F)** because **R(F)** is closed under bounded quantification. Thus the relation defined by (z) still belongs to **ER(F)**.

(RE5) Let **R** be trivially closed, positively-Boolean closed, and closed under projections. Then **R** is closed under superposition with functions from **F(R)**.

Let R^k be in **R** and let $S^m=R^k[g_i^m \mid i<k]$ be a superposition with functions g_i^m whose graphs $G(g_i^m)$ are in **R**. As functions have unique values, α from ω^m will be in S^m if, and only if,

there exists β in ω^k such that $\beta \epsilon R^k$ and

$$<\alpha,\beta(0)> \epsilon G(g_0^m), \ldots, <\alpha,\beta(k-1)> \epsilon G(g_{k-1}^m) \ .$$

Adding suitable dummy arguments, I obtain (m+k)-ary relations $G_i^{m+k} = G(g_i^m)[p_0^{m+k}, \ldots, p_{m-1}^{m+k}, p_{m+i}^{m+k}]$ and $R^{m+k} = R^k[p_m^{m+k}, \ldots, p_{m+k-1}^{m+k}]$ which are in **R** by (ER0). Hence the above may written as

there exists β in ω^k such that $<\alpha,\beta> \epsilon R^{m+k}$

$$\text{and} \quad <\alpha,\beta> \epsilon G_0^{m+k}, \ldots, <\alpha,\beta> \epsilon G_k^{m+k} \ .$$

Thus $<\alpha,\beta>$ belongs to a relation which is in **R** by (ER1), and after k projections I arrive at S^m in **R**.

I now turn to the classes **E**(**R**) for AB-closed **R**. Such **R** satisfy (ER0), (ER1) as follows from (RAB1), (RAB2) and (AB16). Further, **R** is closed under all bounded quantifications. Since the hypotheses of (RE4) are satisfied by (AB10), it follows from (RE3), (RE4) that **E**(**R**) is closed under quantifications $\exists_<$ and $\forall_<$. Since by **5**.(C3)

$$\forall_{\leq} R^{k+1} = (\forall_< R^{k+1} \cup E^2[p_k^{k+1}, c_0^{k+1}]) \cap R^{k+1} \text{ and}$$

$$\exists_{\leq} R^{k+1} = \exists_< R^{k+1} \cup R^{k+1} \ ,$$

E(**R**) is also closed under quantifications $\forall_{\leq}$ and $\exists_{\leq}$. Since **E**(**R**) satisfies the hypotheses of (RE5), I obtain

(AB19) If **R** is AB-closed then **E**(**R**) is positively-Boolean closed, closed under projections, closed under bounded quantifications, and closed under superposition with functions from **F**(**E**(**R**)).

(AB20) If **R** is AB-closed then **F**(**E**(**R**)) is primitive recursively closed, and there holds **FPF** $\subseteq$ **F**(**E**(**R**)) for the class **FPF** of primitive recursive functions.

Here I only need to verify the hypotheses (RD0-4) of Theorem 5.3. Since (RE0) implies **F**(**R**) $\subseteq$ **F**(**E**(**R**)), I conclude (RD0) from (AB6),(AB2), (RAB0),(AB8),(AB9). I conclude (RD1) from (RG0) and (AB3), (RAB2). Finally, (AB19) implies (RD2)-(RD4).

The relations in **E**(**RAB**) may be called the *arithmetically definable* relations.

3. P-closed Classes and their Core

In this section S shall be a class of relations such that

(PP0) E^2, $-E^2$ and the graphs of c_0^1, c_1^1 are in S ,

(PP1) S is trivially closed ,

(PP2) S is positively-Boolean closed ,

(PP3) If R^{k+1} is in S then so is $\forall_{<}R^{k+1}$,

(PP4) If R^{k+1} is in S then so is the projection $\exists R^{k+1}$;

I then say that S is *P–closed*. It follows from (AB19) and (AB3) that

If **R** is AB–closed then **E**(**R**) is P–closed.

It follows from **5**.(D3–4) and (RE5) that

If S is P–closed then it is closed under superposition with functions from **F**(S), and **F**(S) contains the p_i^k and is closed under superposition.

For a P–closed class S I define the POST *core* **Q**(S) as the subclass of all those R in S for which also the complement –R is in S.

(PQ0) **Q**(S) is Boolean closed, and E^2 is in **Q**(S) .

(PQ1) If f^k is in **F**(S) then $G(f^k)$ is in **Q**(S) .

$G(f^k)$ is in S by hypothesis; is remains to show that also $-G(f^k)$ is in S. Set $g^{k+1} = f^k \circ <p_0^{k+1},\dots, p_{k-1}^{k+1}>$ whence $g^{k+1}(\alpha,a) = f^k(\alpha)$. The $<\alpha,a>$ in $G(f^k)$ are precisely those for which $g^{k+1}(\alpha,a) = p_k^{k+1}(\alpha,a)$ holds. Thus $-G(f^k)$ is the superposition $(-E^2)[g^{k+1},p_k^{k+1}]$.

(PQ2) **F**(S) = **F**(**Q**(S)) .

Q(S) $\subseteq$ S implies **F**(**Q**(S)) $\subseteq$ **F**(S), and (PQ1) implies **F**(S) $\subseteq$ **F**(**Q**(S)).

(PQ3) **Q**(S) is closed under superposition with functions from **F**(S) .

Let R^k be in **Q**(S) and let $S^m = R^k[g_i^m \mid i<k]$ be a superposition with functions g_i^m whose graphs $G(g_i^m)$ are in S. Then S^m in is S by (RE5), and it remains to show that also $-S^m$ is in S. As the functions g_i^m are defined for every α and as they unique values, the α from ω^m which are in $-S^m$, i.e. those with $<g_0^m(\alpha),\dots, g_{k-1}^m(\alpha)> \epsilon -R^k$, can be characterized by

there exists β in ω^k such that $\beta \epsilon -R^k$ and
$$<\alpha,\beta(0)> \epsilon G(g_0^m),\dots, <\alpha,\beta(k-1)> \epsilon G(g_{k-1}^m) .$$

Adding suitable dummy arguments leads to (m+k)–ary relations

$$G_i^{m+k} = G(g_i^m)[p_0^{m+k},\dots, p_{m-1}^{m+k}, p_{m+i}^{m+k}]$$
$$\text{and} \quad R^{m+k} = (-R^k)[p_m^{m+k},\dots, p_{m+k-1}^{m+k}]$$

in S with which the above becomes

there exists β in ω^k such that and $<\alpha,\beta> \epsilon R^{m+k}$ and
$$<\alpha,\beta> \epsilon G_0^{m+k},\dots, <\alpha,\beta> \epsilon G_k^{m+k} .$$

Thus $<\alpha,\beta>$ belongs to a relation which by (PP2) is in S. Applying k projections I arrive at $-S^m$ in S.

The following observation will be most useful:

(PQ4) $\mathbf{Q}(S) = \mathbf{R}(\mathbf{F}(S)) = \mathbf{R}(\mathbf{F}(\mathbf{Q}(S)))$.

$\mathbf{Q}(S)$ is closed under superposition with c_0^1, c_1^1 by (PQ3),(PP0). Then $\mathbf{Q}(S) = \mathbf{R}(\mathbf{F}(\mathbf{Q}(S)))$ follows from 5.(D7) since (PQ0) holds. $\mathbf{R}(\mathbf{F}(S)) = \mathbf{R}(\mathbf{F}(\mathbf{Q}(S)))$ follows from (PQ2).

My aim now is to prove (PQ10) below, and the difficulty in doing that consists in the verification of (RAB3) for $\mathbf{Q}(\mathbf{E}(\mathbf{R}))$. In order to approach this task, I shall make use of the notions connected with general minimization as introduced in 5.(CC7). Observe first that $R^{k+1} \epsilon \mathbf{Q}(S)$ implies that the relation is νR^{k+1} in S. Because (PP2), (PP4) permit to conclude this from the representation 5.(C9):

$$\nu R^{k+1} = R^{k+1} \cap (\forall_< -R^{k+1} \cup E^2[p_k^{k+1}, c_0^{k+1}]) .$$

In (PQ8) I then shall show that νR^{k+1} actually is in $\mathbf{Q}(S)$. Already from (PQ2) there follows

(PQ5) If $R^{k+1} \epsilon \mathbf{Q}(S)$ and if R^{k+1} is full then the function μR^{k+1} is in $\mathbf{F}(\mathbf{Q}(S))$.

(PQ6) If $R^{k+1} \epsilon \mathbf{Q}(S)$ then $\mu_< R^{k+1}$ is in $\mathbf{F}(\mathbf{Q}(S))$; if S is closed under superposition with s then $\mu_\leq R^{k+1}$ is in $\mathbf{F}(\mathbf{Q}(S))$.

Because it follows from 5.(C11) that the graph of $\mu_< R^{k+1}$ is νS^{k+2} for the full relation

$$S^{k+2} = R^{k+1}[p_0^{k+2}, \dots, p_{k-1}^{k+2}, p_{k+1}^{k+2}] \cup E^2[p_k^{k+2}, p_{k+1}^{k+2}] .$$

Hence $\mu_< R^{k+1} = \mu S^{k+2}$, Since S^{k+2} is in $\mathbf{Q}(S)$ by (PP1), (PQ0), it follows that νS^{k+2} is in S. Hence the function μS^{k+2} with this graph is in $\mathbf{F}(S) = \mathbf{F}(\mathbf{Q}(S))$. Making use of 5.(C12), the statement about $\mu_\leq R^{k+1}$ follows analogously.

(PQ7) $\mathbf{Q}(S)$ is closed under bounded quantification.

$R^{k+1} \epsilon \mathbf{Q}(S)$ implies by (PQ6) that $G(\mu_< R^{k+1})$ is in $\mathbf{Q}(S)$, hence $-G(\mu_< R^{k+1})$ is in $\mathbf{Q}(S)$. Now the representation 5.(C7)

$$\exists_< R^{k+1} = (-G(\mu_< R^{k+1}))[p_0^{k+1}, \dots, p_k^{k+1}, p_k^{k+1}]$$

implies by (PQ3) that $\exists_< R^{k+1}$ is in $\mathbf{Q}(S)$. Analogously, $-R^{k+1} \epsilon \mathbf{Q}(S)$ implies that $-\exists_< -R^{k+1}$ is in $\mathbf{Q}(S)$ whence by 5.(C2) also $\forall_< R^{k+1} \epsilon \mathbf{Q}(S)$. Since $\exists_\leq R^{k+1} = \exists_< R^{k+1} \cup R^{k+1}$ by 5.(C3) also $\exists_\leq R^{k+1}$ and, by 5.(C1), $\forall_\leq R^{k+1}$ are in $\mathbf{Q}(S)$.

(PQ8) If $R^{k+1} \epsilon \mathbf{Q}(S)$ then νR^{k+1} is in $\mathbf{Q}(S)$.

This follows by (PQ7), (PQ3) from the representation 5.(C9) used above.

(PQ9) Let S contain the graphs of s , + , · , *mod*, cro_0, cro_1. Then $\mathbf{F}(S) = \mathbf{F}(\mathbf{Q}(S))$ is primitive recursively closed.

I only need to verify that **Q**(**S**) satisfies the hypotheses of Theorem 5.3. (RD0) follows from the stated assumptions in view of (PQ2). (RD1) and (RD2) follow from (PQ0). (RD3) follows from (PQ3), and (RD4) follows from (PQ7).

Now (AB20) is a special case of (PQ9) since **S** = **E**(**R**) is P-closed if **R** is AB-closed, and since **F**(**S**) = **F**(**Q**(**S**)) .

(PQ10) If **R** is AB-closed then so is **Q**(**E**(**R**)), and there holds **R** $\subseteq$ **Q**(**E**(**R**)) .

(RAB0) is inherited by classes containing **R**, (RAB1) follows from (PQ3), (RAB2) follows from (PQ0), and (RAB3) follows from (PQ7). And (RAB2) and **R** $\subseteq$ **E**(**R**) imply **R** $\subseteq$ **Q**(**E**(**R**)).

4. Recursively Enumerable Relations

I shall write **ER**(G) instead of **E**(**R**(G)).

If G is simply closed then G $\subseteq$ **F**(**ER**(G)) since fϵG implies G(f)ϵ**R**(G) and since **R**(G) $\subseteq$ **ER**(G). As **R**(G) is AB-closed, **ER**(G) will be P-closed, and **F**(**ER**(G)) = **F**(**Q**(**ER**(G))) is primitive recursively closed by (PQ9). – Recall that **FSF**, **FEF**, **FPF** are the smallest classes which are simply, elementarily and primitive recursively closed respectively.

THEOREM 1 (a) **ER**(**FSF**) = **ER**(**FEF**) = **ER**(**FPF**)

(aa) **E**(**RAB**) = **ER**(**FPF**)

I observed earlier that **R**(**F**) is AB-closed if **F** is simply closed. **FSF** *is* simply closed, and thus **RAB** $\subseteq$ **R**(**FSF**). But **FSF** $\subseteq$ **FEF** $\subseteq$ **FPF** whence also **RAB** $\subseteq$ **R**(**FSF**) $\subseteq$ **R**(**FEF**) $\subseteq$ **R**(**FPF**), and the inclusions remain preserved when projections are taken. It so remains to show the inclusion **ER**(**FPF**) $\subseteq$ **E**(**RAB**). As **F**(**E**(**RAB**)) is primitive recursively closed, there holds

FPF $\subseteq$ **F**(**E**(**RAB**))

whence

R(**FPF**) $\subseteq$ **R**(**F**(**E**(**RAB**))) ,

and as **E**(**RAB**) is P-closed, (PQ4) implies

R(**F**(**E**(**RAB**))) = **Q**(**E**(**RAB**)) .

Hence

R(**FPF**) $\subseteq$ **Q**(**E**(**RAB**))

and so

ER(**FPF**) = **E**(**R**(**FPF**)) $\subseteq$ **E**(**Q**(**E**(**RAB**))) $\subseteq$ **E**(**RAB**) since $\mathbf{Q}(\mathbf{E}(S)) \subseteq \mathbf{E}(S)$ for any S, hence $\mathbf{E}(\mathbf{Q}(\mathbf{E}(S)) \subseteq \mathbf{E}(\mathbf{E}(S)) = \mathbf{E}(S)$ if $\mathbf{E}(S)$ is closed under projections.

The class characterized in the Theorem is called the class **RER** of *recursively enumerable relations.* This name actually refers to another characterization. Although no use of it will be made here, I shall state it for the sake of completeness.

If G is a simply closed class then I call G-*enumerable* the projections of relations in $\mathbf{R}(G)$. There hold:

> A non-empty set M of numbers is G-enumerable if, and only if, there exists a function f^1 in G which maps ω onto M .
>
> A non-empty relation R^k, $k>1$, is G-enumerable if, and only if, there exists a function f^1 in G which maps ω onto the image of R^k under cau^k.

Consider first a function f^1 from G. Then $G(f^1)$ is in $\mathbf{R}(G)$, hence also $R^2 = G(f^1)[p_1^2, p_0^2]$ is in $\mathbf{R}(G)$, and $\exists R^2$ consists of the a for which there exists b such that $<a,b> \epsilon\ G(f^1)[p_1^2, p_0^2]$, i.e. $<b,a> \epsilon G(f^1)$, $f^1(b) = a$; thus $\exists R^2 = \mathrm{im}(f^1)$. Conversely, let M be non-empty and projection of R^2 in $\mathbf{R}(G)$. I choose t in M and define f^1 by cases as

$$\begin{array}{lll} f^1(z) = t & \text{if not} & <cro_0(z), cro_1(z)> \epsilon R^2 \\ f^1(z) = cro_0(z) & \text{if} & <cro_0(z), cro_1(z)> \epsilon R^2 \ . \end{array}$$

Then f^1 is in G by 5.(FS2) and (FS15), (FS0), and also in the second case of the definition of f^1 the values $f^1(z)$ are in $\exists R^2 = M$. On the other hand, for every a in M there is some b with $<a,b> \epsilon R^2$; hence for $z = cau(a,b)$ there holds $cro_0(z) = a$, $cro_1(z) = b$ whence $f^1(z) = a$.

Assume now that a relation R^k, $k>1$, is mapped by cau^k onto the image M of ω under f^1 from G. Then $\alpha \epsilon R^k$ is equivalent to the existence of some b such that $cau^k(\alpha) = f^1(b)$. This is a relation R^{k+1} with $R^k = \exists R^{k+1}$ and with the characteristic function $\chi = \circ < cau^k \circ < p_0^{k+1}, \dots, p_{k-1}^{k+1} >, f^1 \circ p_k^{k+1} >$, and it is in $\mathbf{R}(G)$ since f^1 is in G. Conversely, given an non-empty set $\exists R^{k+1}$, I choose one of its elements ϑ and define f^1 by

$$\begin{array}{lll} f^1(z) = cau^k(\vartheta) & \text{if not} & uac^{k+1}(z) \epsilon R^{k+1} \\ f^1(z) = cro_0(z) & \text{if} & uac^{k+1}(z) \epsilon R^{k+1} \ . \end{array}$$

If $R^{k+1} \epsilon \mathbf{R}(G)$ then $f^1 \epsilon G$, and $\mathrm{im}(f^1)$ is contained in the image of $\exists R^{k+1}$ under cau^k because $uac^{k+1}(z) = < uac^k(cro_0(z)), cro_1(z) >$, $cau^k(uac^k(cro_0(z))) = cro_0(z)$. But for every c in this image I find α, b such that $<\alpha, b> \epsilon R^{k+1}$ and $c = cau^k(\alpha)$. Thus for $z = cau^{k+1}(\alpha, b)$ there holds $<\alpha, b> = uac^{k+1}(z) = < uac^k(cro_0(z)), cro_1(z) >$ and $c = cau^k(uac^k(cro_0(z))) = cro_0(z) = f^1(z)$.

In this sense, then, the relations in **RER** are the ones enumerable by simple, by elementary, and by primitive recursive functions. As as variation on this theme, I add

(AB21) Let $\mathbf{R}$ be AB-closed. If $M = \exists R^2$ is not empty and R^2 in $\mathbf{R}$ then there is a function f^1 in $\mathbf{F}(\mathbf{R})$ such that $M = im(f^1)$; if $M^k = \exists R^{k+1}$, $k>1$, is not empty and R^{k+1} in $\mathbf{R}$ then there is a function f^1 in $\mathbf{F}(\mathbf{R})$ which maps ω onto the image of M^k under cau^k.

The function f^1 is defined as above, and in the case of M it is, in the first case of its definition, the 1-ary function c_t. As c_t and cro_0 are in $\mathbf{F}(\mathbf{R})$, it follows from (AB16) that $R^2[cro_0, cro_1]$ is in $\mathbf{R}$; hence f^1 is in $\mathbf{F}(\mathbf{R})$ by 5.(D5). In the case of M^k with $uac_{k+1} = \langle cro_i^{k+1}(z) \mid i < k+1 \rangle$ I rewrite the definition of f^1 as

$$f^1(z) = cau^k(\vartheta) \quad \text{if not} \quad z \in R^{k+1}[cro_0^{k+1}, \ldots, cro_k^{k+1}]$$
$$f^1(z) = cro_0(z) \quad \text{if} \quad z \in R^{k+1}[cro_0^{k+1}, \ldots, cro_k^{k+1}] .$$

Now $R^{k+1}[cro_0^{k+1}, \ldots, cro_k^{k+1}]$ is in $\mathbf{R}$ by (AB16) since $cro_i^{k+1} = cro_i^k \circ cro_0$ and $cro_k^{k+1} = cro_1$. That cau^k with $cau^k(\alpha, a) = cau(cau^{k-1}(\alpha), a)$ is in $\mathbf{F}(\mathbf{R})$ follows from (AB9) by an obvious induction.

There are both proper subclasses $\mathbf{S}$ of **RAB** and proper subclasses $\mathbf{G}$ of **FSF** which generate the recursively enumerable relations in the sense that $\mathbf{E}(\mathbf{S}) = \mathbf{ER}(\mathbf{G}) = \mathbf{RER}$; for a discussion of one of them cf. FELSCHER 93.

A slightly different generation of **RER** has been obtained from the class DIO of all *Diophantine* relations, defined as the relations $E^2[f^k, g^k]$ where f^k and g^k are polynomial functions in the sense used in (AB13). Due to its very simplicity, the class **E(DIO)** of projections of relations in DIO will itself not be closed under projections, and I will have to form the classes $\mathbf{E}_n(\mathrm{DIO})$ with $\mathbf{E}_0(\mathrm{DIO}) = \mathrm{DIO}$, $\mathbf{E}_{n+1}(\mathrm{DIO}) = \mathbf{E}(\mathbf{E}_n(\mathrm{DIO}))$ to obtain the union $\mathbf{E}^*(\mathrm{DIO})$ of these $\mathbf{E}_n(\mathrm{DIO})$ as a class closed under projections. Obviously, $\mathbf{E}^*(\mathrm{DIO}) \subseteq \mathbf{RER}$, and it is not hard to see that $\mathbf{E}^*(\mathrm{DIO})$ is positively Boolean closed. The proof that $\mathbf{E}^*(\mathrm{DIO})$ is closed under bounded quantifications is much harder, and this having been secured, one arrives at the theorem of MATIYASEVIC that $\mathbf{E}^*(\mathrm{DIO}) = \mathbf{RER}$. A very perspicuous discussion of this and related matters can be found in

Y.V. Matiyasevich: Hilbert's Tenth Problem. Cambridge Mass. 1993.

5. Recursive Relations and Recursive Functions

I shall write **QE(S)** instead of **Q(E(S))** and **QER(G)** instead of **QE(R(G))**.

Q(RER) is defined to be the class **RFR** of *recursive relations*, **F(Q(RER))** is defined to be the class **FRF** of *recursive functions*.

THEOREM 2 (a) **F(RFR) = FRF** and **R(FRF) = RFR** .

(b) **FRF** is primitive recursively closed, and there holds
$$\mathbf{FPF} \subseteq \mathbf{FRF}\ .$$

(c) $\mathbf{RFR} = \mathbf{QE(RAB)} = \mathbf{QER(FSF)} = \mathbf{QER(FEF)} = \mathbf{QER(FPF)}\ .$

(d) $\mathbf{FRF} = \mathbf{F(E(RAB))} = \mathbf{F(ER(FSF))} = \mathbf{F(ER(FEF))} = \mathbf{F(ER(FPF))}\ .$

(e) $\mathbf{RER} = \mathbf{E(RFR)}\ .$

Theorem 1 implies the statements (c), and they then, together with (PQ2), imply the statements (d). In (a) $\mathbf{F(RFR)} = \mathbf{FRF}$ holds by definition; making use of (d) there follows $\mathbf{R(FRF)} = \mathbf{R(F(QE(RAB)))} = \mathbf{QE(RAB)}$ by (PQ4). That $\mathbf{FRF} = \mathbf{F(E(RAB))}$ is primitive recursively closed follows from (PQ9) since $\mathbf{E(RAB)}$ is P–closed. Thus also $\mathbf{FPF} \subseteq \mathbf{FRF}$. In order to see (e), I use $\mathbf{E(RFR)} = \mathbf{E(QE(RAB))} \subseteq \mathbf{E(E(RAB))} \subseteq \mathbf{E(RAB)}$ whence $\mathbf{FPF} \subseteq \mathbf{FRF}$ implies $\mathbf{E(RAB)} = \mathbf{ER(FPF)} \subseteq \mathbf{ER(FRF)} = \mathbf{E(RFR)}$.

Every characterization of the class **RER** gives raise to a characterization of $\mathbf{FRF} = \mathbf{F(RER)}$. Every characterization of **FRF** gives raise to a characterization of $\mathbf{RER} = \mathbf{ER(FRF)}$.

Recursive relations are recursively enumerable since $\mathbf{Q(RER)} \subseteq \mathbf{RER}$ follows from the definitions. An example for a recursively enumerable, but not recursive relation will appear only in Theorem 8.3 .

I now wish to define *recursively closed* classes S of relations and G of functions. I consider the families

$\mathscr{S}$ of all AB–closed classes S ,
$\mathscr{G}$ of all classes G such that $G \subseteq \mathbf{F(R(}G\mathbf{))}$ and $\mathbf{R(}G\mathbf{)}$ is AB–closed ;

these classes G I sometimes call *generic*. Clearly, every simply closed G is generic. I define an operator Φ for arguments from $\mathscr{S}$ (only !) and an operator Ψ for arguments from $\mathscr{G}$ (only !) by

$$\Phi S = \mathbf{QE(}S\mathbf{)} \quad \text{and} \quad \Psi G = \mathbf{F(QER(}G\mathbf{))} = \mathbf{F(}\Phi\,\mathbf{R(}G\mathbf{))}\ .$$

I define that

S is *recursively closed* if S is AB–closed and $\Phi S \subseteq S$,

G is *recursively closed* if $G = \mathbf{F(R(}G\mathbf{))}$ and $\mathbf{R(}G\mathbf{)}$ is recursively closed.

It follows from the definitions that S then is in $\mathscr{S}$ and that G is in $\mathscr{G}$.

THEOREM 3 (a) ΦS is the smallest recursively closed class containing S .
ΨG is the smallest recursively closed class containing G .

(b) If S is recursively closed then $S = \Phi S$.
If G is recursively closed then $G = \Psi G$.

(c) S is recursively closed if, and only if, $\mathbf{S} = \mathbf{R}(\mathbf{G})$ for some recursively closed G .

G is recursively closed if, and only if, $\mathbf{G} = \mathbf{F}(\mathbf{S})$ for some recursively closed S .

(d) **RFR** is the smallest recursively closed class of relations. **FRF** is the smallest recursively closed class of functions.

(e) Every recursively closed class G is primitive recursively closed.

Clearly, (b) is a consequence of (a). As the operators making up Φ and Ψ preserve inclusions, there holds

(0) If $\mathbf{S} \subseteq \mathbf{T}$ then $\Phi\mathbf{S} \subseteq \Phi\mathbf{T}$, if $\mathbf{G} \subseteq \mathbf{H}$ then $\Psi\mathbf{G} \subseteq \Psi\mathbf{H}$.

If S is in $\mathscr{S}$ then $\mathbf{S} \subseteq \Phi\mathbf{S}$. Because $\mathbf{S} \subseteq \mathbf{E}(\mathbf{S})$ by (RE0), and this implies $\mathbf{S} \subseteq \Phi\mathbf{S}$ as S is closed under complements. Also, if S is in $\mathscr{S}$ then $\mathbf{E}(\mathbf{S})$, being P-closed, is closed under projections. Thus $\mathbf{E}(\mathbf{QE}(\mathbf{S})) \subseteq \mathbf{E}(\mathbf{S})$ whence $\mathbf{QE}(\mathbf{QE}(\mathbf{S})) \subseteq \mathbf{QE}(\mathbf{S})$. Thus $\Phi\mathbf{S}$ is recursively closed. Finally, $\mathbf{S} \subseteq \mathbf{T}$ implies $\Phi\mathbf{S} \subseteq \Phi\mathbf{T}$, and if T is recursively closed then $\Phi\mathbf{T} \subseteq \mathbf{T}$ implies $\Phi\mathbf{S} \subseteq \mathbf{T}$.

If G is in $\mathscr{G}$ then $\mathbf{G} \subseteq \Psi\mathbf{G}$. Because then $\mathbf{R}(\mathbf{G})$ is AB-closed, hence $\mathbf{ER}(\mathbf{G})$ is P-closed whence $\mathbf{G} \subseteq \mathbf{F}(\mathbf{R}(\mathbf{G})) \subseteq \mathbf{F}(\mathbf{ER}(\mathbf{G})) = \mathbf{F}(\mathbf{QER}(\mathbf{G})) = \Psi\mathbf{G}$ by (PQ2). Observe now that (PQ4) has the consequence

(1) If S is in $\mathscr{S}$ then $\Phi\mathbf{S} = \mathbf{QE}(\mathbf{S}) = \mathbf{R}(\mathbf{F}(\mathbf{QE}(\mathbf{S}))) = \mathbf{R}(\mathbf{F}(\Phi\mathbf{S})$ and
$$\mathbf{F}(\Phi\mathbf{S}) = \mathbf{F}(\mathbf{R}(\mathbf{F}(\Phi\mathbf{S})))$$

whence

(2) If $\mathbf{R}(\mathbf{G})$ is in $\mathscr{S}$ then $\Phi\mathbf{R}(\mathbf{G}) = \mathbf{R}(\mathbf{F}(\Phi\mathbf{R}(\mathbf{G})))$ and
$$\Psi\mathbf{G} = \mathbf{F}(\Phi\mathbf{R}(\mathbf{G})) = \mathbf{F}(\mathbf{R}(\mathbf{F}(\Phi\mathbf{R}(\mathbf{G})))) = \mathbf{F}(\mathbf{R}(\Psi\mathbf{G})).$$

So if G is in g then (2) implies $\Psi\mathbf{G} = \mathbf{F}(\mathbf{R}(\Psi\mathbf{G}))$, and so $\Psi\mathbf{G}$ is recursively closed. Finally, $\mathbf{G} \subseteq \mathbf{H}$ implies $\Phi\mathbf{R}(\mathbf{G}) \subseteq \Phi\mathbf{R}(\mathbf{H}))$, and if H is recursively closed then $\Phi\mathbf{R}(\mathbf{H}) \subseteq \mathbf{R}(\mathbf{H})$ implies $\Phi\mathbf{R}(\mathbf{G}) \subseteq \mathbf{R}(\mathbf{H}))$. Thus $\Psi\mathbf{G} = \mathbf{F}(\Phi\mathbf{R}(\mathbf{G}))$ $\subseteq \mathbf{F}(\mathbf{R}(\mathbf{H}))$ and $\mathbf{F}(\mathbf{R}(\mathbf{H})) \subseteq \mathbf{H}$ again if H is recursively closed. This concludes the proof of (a).

As for (c), if G is recursively closed then so is $\mathbf{R}(\mathbf{G})$ by definition. Conversely, if S is recursively closed then $\mathbf{S} = \Phi\mathbf{S}$ implies $\mathbf{R}(\mathbf{F}(\mathbf{S})) = \mathbf{R}(\mathbf{F}(\Phi\mathbf{S})) = \Phi\mathbf{S} = \mathbf{S}$ by (1). Concerning $\mathbf{F}(\mathbf{S})$, this implies that $\mathbf{R}(\mathbf{F}(\mathbf{S}))$ is AB-closed. But (1) also implies $\mathbf{F}(\Phi\mathbf{S}) = \mathbf{F}(\mathbf{R}(\mathbf{F}(\Phi\mathbf{S})))$, and with $\mathbf{S} = \Phi\mathbf{S}$ this gives $\mathbf{F}(\mathbf{S}) = \mathbf{F}(\mathbf{R}(\mathbf{F}(\mathbf{S})))$. Thus $\mathbf{F}(\mathbf{S})$ is recursively closed. This proves the forward implication in the first part of (c) as well as the backward implication in the second part. Finally, if G is recursively closed then $\mathbf{G} = \Psi\mathbf{G} = \mathbf{F}(\Phi\mathbf{R}(\mathbf{G}))$ where $\Phi\mathbf{R}(\mathbf{G})$ is recursively closed.

Turning to (d), a recursively closed S is AB-closed, hence $\mathbf{RAB} \subseteq \mathbf{S}$ and so $\mathbf{RFR} = \Phi(\mathbf{RAB}) \subseteq \Phi\mathbf{S} = \mathbf{S}$. For a recursively closed G this implies $\Phi\mathbf{RAB}) \subseteq$

R(G) whence $\mathbf{F}(\Phi\ \mathbf{RAB})) \subsetneq \mathbf{F}(\mathbf{R}(G)) = G$. As for (e), since R(G) is AB-closed, (AB20) implies that F(E(R(G))) is primitive recursively closed. But that is $G = \Psi G = \mathbf{F}(\mathbf{QER}(G))$ by (PQ2). This concludes the proof of the Theorem.

If F(S) is recursively closed then S may not be so. For instance, let G be recursively closed and let S the set of graphs of functions in G; then F(S) = G, but S is a proper subset of R(G). If R(G) is recursively closed then G may be not so. For instance, let S be recursively closed and let G be the set of all 0–1–functions in F(S); then R(G) = S, but G is a proper subset of F(S).

6. Recursivity and Minimization

A class G of functions is called *closed under total minimization* if there holds

if R^{k+1} is in R(G) and if R^{k+1} is full then the function μR^{k+1} is in G.

If G contains *sg* and is closed under superposition then the use of R(G) here can be avoided by defining, for every f^{k+1} a k–ary *partial* function $\mu^* f^{k+1}$ such that $\alpha \in \mathrm{def}(\mu^* f^{k+1})$ and $\mu^* f^{k+1}(\alpha) = b$ if

$$f^{k+1}(\alpha,b) \neq 0 \text{ and for all } b<a : f^{k+1}(\alpha,a) = 0 .$$

If $R = R^{k+1}$ is in R(G) then χ_R is in G and $\mu R^{k+1} = \mu^*(\chi_R)$. Conversely, if f^{k+1} is in G then so is $sg \circ f^{k+1}$ which is the characteristic function of the set R^{k+1} of all $<\alpha,b>$ such that $f^{k+1}(\alpha,b) \neq 0$; hence $\mu^* f^{k+1} = \mu R^{k+1}$. Thus

If G contains *sg* and is closed under superposition then G is closed under total minimization if, and only if, $f^{k+1} \in G$ implies $\mu^* f^{k+1} \in G$ provided the function $\mu^* f^{k+1}$ is total.

In place of $\mu^* f^{k+1}$ one usually considers the k–ary partial function μf^{k+1} such that $\alpha \in \mathrm{def}(\mu f^{k+1})$ and $\mu f^{k+1}(\alpha) = b$ if

$$f^{k+1}(\alpha,b) = 0 \text{ and for all } a<b : f^{k+1}(\alpha,a) \neq 0 .$$

Clearly, $f^{k+1}(\alpha,b) \neq 0$ if, and only if, $csg \circ f^{k+1}(\alpha,b) = 0$. Hence $\mathrm{def}(\mu^* f^{k+1}) = \mathrm{def}(\mu(csg \circ f^{k+1}))$ whence $\mu^* f^{k+1} = \mu(csg \circ f^{k+1})$, and $f^{k+1} = csg \circ csg \circ f^{k+1}$ then implies $\mu f^{k+1} = \mu^*(csg \circ f^{k+1})$. Thus

If G contains *csg* and is closed under superposition then G is closed under total minimization if, and only if, $f^{k+1} \in G$ implies $\mu f^{k+1} \in G$ provided the function $\mu^* f^{k+1}$ is total.

A class G of functions is called *μ–recursively closed* if it is primitive recursively closed and closed under total minimization.

It is easily seen that recursively closed classes are μ-recursively closed. As shown in the last Corollary, they are primitive recursively closed. And since $\mathbf{R}(G) = \mathbf{QER}(G)$, it follows from (PQ5) that for any full R^{k+1} from $\mathbf{R}(G)$ also μR^{k+1} is in $\mathbf{F}(\mathbf{QER}(G)) = G$.

THEOREM 4 A class G is recursively closed if, and only if, $\mathbf{R}(G)$ is AB-closed and G contains the projections p^k_i and is closed under superposition and under total minimization.

The conditions stated hold for recursively closed classes. Conversely, assume them to hold for G; I shall first show that $G = \mathbf{F}(\mathbf{R}(G))$. There always holds $G(f^k) = E^2[f^k \circ \langle p_0^{k+1}, \ldots, p_{k-1}^{k+1} \rangle, p_k^{k+1} \rangle]$, and if f^k is in G then so are the functions imposed here on E^2. But E^2 is in the AB-closed class $\mathbf{R}(G)$, and together with G also $\mathbf{R}(G)$ is closed under superposition with functions from G. Hence $G(f^k)$ is in $\mathbf{R}(G)$ if f^k was in G, i.e. $G \subseteq \mathbf{F}(\mathbf{R}(G))$. On the other hand, if f^k is in $\mathbf{F}(\mathbf{R}(G))$ then $G(f^k)$ is a full relation in $\mathbf{R}(G)$; hence $f^k = \mu G(f^k)$ is in G, and this proves $\mathbf{F}(\mathbf{R}(G)) \subseteq G$.

It remains to be shown that $S = \mathbf{R}(G)$ is core-closed, i.e. that $\mathbf{QE}(S) \subseteq S$. Let R^k be in $\mathbf{QE}(S)$. There then exist A^{k+1} and B^{k+1} in S such that $R^k = \exists A^{k+1}$ and $-R^k = \exists B^{k+1}$. Since S is (positively) Boolean closed, also $C^{k+1} = A^{k+1} \cup B^{k+1}$ is in S. Every α in ω^k is either in R^k or in $-R^k$; in every case, therefore, there is an a such that $\langle \alpha, a \rangle \in C^{k+1}$. Hence C^{k+1} is full, and thus μC^{k+1} belongs to G. Further, $D^k = A^{k+1}[p_0^k, \ldots, p_{k-1}^k, \mu C^{k+1}]$ is in Ssince $S = \mathbf{R}(G)$ is closed under superposition. But this is the set of all α such that $\langle \alpha, \mu C^{k+1}(\alpha) \rangle \in A^{k+1}$. I now conclude the proof by showing $D^k = R^k$. My last description of D^k gives $D^k \subseteq \exists A^{k+1} = R^k$. On the other hand, $\alpha \in R^k$ implies that *not* α in $-R^k$. Hence $\langle \alpha, \mu C^{k+1}(\alpha) \rangle$ must belong to A^{k+1}, and therefore α is in D^k.

COROLLARY 4 A class G is recursively closed if, and only if, it is closed under total minimization and is one of

primitive recursively closed , or
elementarily closed , or
simply closed .

Because in all these cases $\mathbf{R}(G)$ is AB-closed. - It follows in particular that the recursively closed classes are precisely those which are *μ-recursively* closed.

It is an interesting feature of the characterizations, employing simple or elementary closedness, that closure under primitive recursion is enforced by closure under minimization, while only relatively weak additional assumptions have to be made. Still, simple closure includes the request of closure under bounded minimization. But **5**.(C11) provides me with the representation of the graph $G(\mu_< R^{k+1})$ as νS^{k+2} for the full relation

$$S^{k+2} = R^{k+1}[p_0^{k+2}, \dots, p_{k-1}^{k+2}, p_{k+1}^{k+2}] \cup E^2[p_k^{k+2}, p_{k+1}^{k+2}] \, ,$$

and this permits a reduction of bounded to total minimization. I so find the

COROLLARY 5 A class G is recursively closed if, and only if, it is closed under superposition and total minimization and if it contains the functions from one of the two following lists:

$$c_1^1, p_i^k, +, \cdot, sg, \chi_<, \chi_= \quad \text{or} \quad p_i^k, +, \cdot, csg, \chi_= \, .$$

The first list consists of the initial simple functions. In order to see that G is simply closed, it remains to verify **5**.(FSF2), i.e. that $R^{k+1} \in \mathbf{R}(G)$ implies that $\mu_< R^{k+1} \in G$. But $\chi_{R \cup S} = sg \circ + \circ \langle \chi_R, \chi_S \rangle$ shows that **R**(G) is closed under unions. Thus the relation S^{k+2} from the above representation is in **R**(G) and its total minimization μS^{k+2} will be in G. As E^2 is in **R**(G), the graph $G(f^{k+1}) = E^2[f^{k+1} \circ \langle p_0^{k+2}, \dots, p_k^{k+2} \rangle, p_{k+1}^{k+2}]$ of every function f^{k+1} from G is in **R**(G), hence also the graph νS^{k+2} of μS^{k+2}. Consequently $G(\mu_< R^{k+1})$ is in **R**(G) whence $\mu_< R^{k+1} = \mu G(\mu_< R^{k+1})$ will be in G.

For the second list, observe first that $sg = csg \circ csg$ is in G. Next, the above verification of **5**.(FSF2) remains in effect without change. Further, **R**(G) now is Boolean closed since $\chi_{-R} = csg \circ \chi_R$. The representation **5**.(C6),

$$\exists_< R^{k+1} = (-E^2)[\mu_< R^{k+1}, p_k^{k+1}]$$

now shows that **R**(G) is closed under $\exists_<$. The relation $<$ is $\exists_< E^2$ since $\langle a,b \rangle \in \exists_< E^2$ if, and only if, there is a c such that $c<b$ and $a=c$. Consequently, $\chi_<$ is in G. There remains the function c_1^1, and it will be in G if c_0^1 is so since $c_1^1 = csg \circ c_0^1$. As observed before, **R**(G) contains the graphs of the functions from G, hence $R^2 = G(sg) \cup G(csg)$ is in **R**(G). This relation is full whence μR^2 will be in G. But $\langle 0,0 \rangle \in G(sg)$ and $\langle n,0 \rangle \in G(csg)$ for $n>0$ such that $c_0^1 = \mu R^2$.

I shall say that a class S of relations is *closed under total minimization*

if R^{k+1} is in S and if R^{k+1} is full then νR^{k+1} is in S.

COROLLARY 6 A class S is recursively closed if, and only if, it contains E^2 and the graphs of $+$ and $\cdot$ and is closed

trivially, and
Boolean, and
under superposition with functions from **F**(S), and
under total minimization.

The verification that S is core-closed, i.e. $\mathbf{QE}(S) \subseteq S$, can be taken from the proof of Theorem 4 with the following changes, writing S in place of $\mathbf{R}(G)$. For the full relation C^{k+1} now νC^{k+1} is in S, hence the function μC^{k+1} in $\mathbf{F}(S)$. But now S is closed under superposition with such functions; hence $D^k = A^{k+1}[p_0^k, \dots, p_{k-1}^k, \mu C^{k+1}]$ is in S again.

It remains to be shown that S is AB-closed, and for that I only have to verify (RAB3), i.e. closure under $\forall_\leq$. Again making use of **5**.(C11) and of the closure under total minimization, I see that for R^{k+1} from S the graph $G(\mu_< R^{k+1})$ is in S, and since S is closed under complements, it contains also $-G(\mu_< R^{k+1})$. The representation **5**.(C7)

$$\exists_< R^{k+1} = (-G(\mu_< R^{k+1}))[p_0^{k+1}, \dots, p_k^{k+1}, p_k^{k+1}]$$

then shows that $\exists_< R^{k+1}$ is in S. Hence also $\exists_\leq R^{k+1} = \exists_< R^{k+1} \cup R^{k+1}$ and $\forall_\leq R^{k+1} = -\exists_\leq - R^{k+1}$ belong to S. This concludes the proof of Corollary 6.

The description of recursive relations in Corollary 6 will be used in the proof of Theorem **8**.3 in order to verify that they are representable with respect to axiom systems of arithmetic. A reader wishing to avoid the structure theory of recursive relations, as it was developed here, may of course use the characterization in the Corollary as a *definition* of recursively closed classes.

In that case, it will be useful to observe that from this new definition a direct proof, *not* employing the machinery of AB-closed classes, can be given for the fact that $\mathbf{F}(S)$ is primitive recursively closed. To this end, Theorem **5**.3 will be applied, and so only its hypotheses (RD0)-(RD4) have to be verified. Here (RD1)-(RD3) follow directly from the assumed properties of S, and (RD4) I derived in the proof above. It remains to verify the presence of the functions in (RG0), of which $+$ and $\cdot$ are there by hypothesis.

Since S is trivially closed, the presence of the p_i^k follows again from $G(p_i^k) = E^2[p_i^{k+1}, p_k^{k+1}]$. Since E^2 is in S, also $\omega^2 = E^2 \cup -E^2$ will be there, and as ω^2 is full, $\nu\omega^2$ is in S; but $\nu\omega^2 = G(c_0^1)$. Also $-\nu\omega^2$ is in S and still is full; hence also $\nu(-\nu\omega^2)$ is in S; but $\nu(-\nu\omega^2) = G(c_1^1)$. As $\mathbf{F}(S)$ is closed under superposition by (D3), now $s = + \circ \langle p_1^1, c_1^1 \rangle$ is in $\mathbf{F}(S)$. It remains to show that also *mod*, cro_0, cro_1 are in $\mathbf{F}(\mathbf{RFR})$, and this I shall do by repeating the arguments used in **5**.(FS12), (FS15), once I have acquired the tools they require.

As in the above proof for closure under bounded quantifications, **5**.(C11) shows that $R^{k+1} \in S$ implies $\mu_< R^{k+1} \in \mathbf{F}(S)$. Now I find $\dot{-}$ in $\mathbf{F}(S)$ by repeating the argument from **5**.(FS6). Further, definition by cases in **4**.(FS2) remains in effect, and $c_0^2 = c_0^1 \circ \langle p_0^2, p_1^2 \rangle$ is in $\mathbf{F}(S)$, hence $A = \omega^2 \cap E^2[c_0^2, p_1^2]$ in S. Thus I can repeat the argument from **5**.(FS12) and obtain that *qu* and *mod* are in $\mathbf{F}(S)$. Repeating the argument from **5**.(FS15), I finally find cro_0, cro_1 in $\mathbf{F}(S)$. (I might add that at this point I have shown that $\mathbf{F}(S)$ has all the properties of a simply closed class, with exception of the presence of *sg*.

But *sg* can be defined by cases from c_0^1, c_1^1 since $\{0\} = E^2[p_0^1, c_0^1]$ and $\omega = E^2[p_0^1, p_0^1]$ are in **S**.)

All the characterizations, stated in Theorem 4 and its Corollaries, may be read as *definitions* of recursively closed classes of functions and relations. There is, however, a methodological difference between *such* definitions and the ones chosen here.

The definition of **RFR** as **QE(RAB)** (and analogously in a case such as **QE(R(FSF)))** describes a process of how to generate the relations of **RFR**. In order to see this in detail, I begin with a generation of **RAB**. For any (finite) set **M** of relations I define

$\Theta(n, \mathbf{M})$ as the set of all relations S^m such that

there exist π and R^k, π a map from k into m, R^k in **M** and $m \leq n+3$ and $S^m = R^k[p^m_{\pi(0)}, \ldots, p^m_{\pi(k-1)}]$, or

there exist P^m, R^m in **M** such that $S^m = P^m \cup R^m$, or

there exist P^m, R^m in **M** such that $S^m = P^m \cap R^m$, or

there exists R^m in **M** such that $S^m = -R^m$, or

there exists R^{m+1} in **M** such that $S^m = \forall_{\leq} R^{m+1}$.

I then define

$\mathbf{A}(0)$ contains the two graphs $G(+)$ and $G(\cdot)$

$\mathbf{A}(n+1) = \Theta(n, \mathbf{A}(n))$.

Now $\mathbf{A}(0)$ contains only 3–ary relations, and if I know that $\mathbf{A}(n)$ contains only relations of arities m with $m \leq n+3$, then $R^m = R^m[p_0^m, \ldots, p_{m-1}^m]$ implies $\mathbf{A}(n) \subseteq \Theta(n, \mathbf{A}(n)) = \mathbf{A}(n+1)$. It thus follows that **RAB** is the union of the sets $\mathbf{A}(n)$. Inspecting the constructions in the proofs of (AB8) and (AB9), the reader may verify the the graphs of *quo* and *mod* occur at least in $\mathbf{A}(13)$ and those of cro_0, cro_1 at least in $\mathbf{A}(27)$.

The functions + and · provide me with algorithms by which to decide whether a tuplet $\langle x,y,z \rangle$ belongs, or does not belong, to $G(+)$ or to $G(\cdot)$; such algorithms, for instance, may be taken as programs computing the characteristic functions. If, for each relation R in **M**, I have an algorithm by which to decide whether a tuplet α belongs to R or not, then I can form such decision algorithms for the relations in $\Theta(n, \mathbf{M})$. In this sense, the formation of $\Theta(n, \mathbf{M})$ from **M** is a *constructive* operation.

On the other hand, given a set of relations **M**, decidable in this sense, I do *not* know how to decide whether some α in ω^k belongs to $\exists R^{k+1}$ for a given R^{k+1} in **M**: I do not know for how many numbers a I will have to test whether $\langle \alpha, a \rangle$ is in R^{k+1}. In this sense, the formation of the set

$\Psi\mathbf{M}$ of all relations $\exists R$ with R in **M** is an *inconstructive* operation. Also, the operation assigning to any two (finite) sets **M**, **N** the set

$\Omega(\mathbf{M}, \mathbf{N})$ of all relations in **M** which have their complement in **N**

must be viewed as inconstructive, because, in general, I do not know whether two programs compute the same (or compute complementary) relations. Observe that always $\Omega(\mathbf{K}, \Omega(\mathbf{M}, \mathbf{N})) \subseteq \mathbf{N}$.

I generate **QE**(**RAB**) as the union of the sequence **Q**(n) defined by

$$\mathbf{D}(0) \quad \text{is empty}$$

$$\mathbf{L}(0) = \mathbf{A}(0)$$

$$\mathbf{L}(n+1) = \mathbf{A}(n+1) - \mathbf{A}(n)$$

$$\mathbf{D}(n+1) = \mathbf{D}(n) \cup \Psi\mathbf{L}(n)$$

$$\mathbf{Q}(n) = \mathbf{A}(n) \cup \Omega(\Psi\mathbf{L}(n), \mathbf{D}(n)) .$$

If R is in **QE**(**RAB**) and not already in **RAB** then $R = \exists S$ and $-R = \exists T$ with S, T in **RAB**. Hence $S \epsilon \mathbf{A}(n)$ and $T \epsilon \mathbf{A}(m)$ where I choose n, m to be minimal and may assume that $m \leq n$. Thus S is in **L**(n) and R is in $\Psi\mathbf{L}(n)$. Also, $-R$ is in $\Psi\mathbf{L}(m)$ and thus in **D**(m). But $m \leq n$ implies $\mathbf{D}(m) \subseteq \mathbf{D}(n)$. Hence R is in $\Omega(\Psi\mathbf{L}(n), \mathbf{D}(n))$ and so in **Q**(n). Thus **QE**(**RAB**) is contained in the union of the sets **Q**(n).

Conversely, if R is in **Q**(n) and not already in **A**(n) then it is in $\Omega(\Psi\mathbf{L}(n), \mathbf{D}(n))$. Thus R is in $\Psi\mathbf{L}(n)$ and therefore in **E**(**RAB**). But also $-R$ is in **D**(n) and thus in $\Psi\mathbf{L}(m)$ for some $m < n$. Hence also $-R$ is in **E**(**RAB**), and thus R is in **QE**(**RAB**).

Inconstructivity enters here in the formation of $\Omega(\Psi\mathbf{L}(n), \mathbf{D}(n))$, first in the inconstructive formation of $\Psi\mathbf{L}(n)$ and then in the comparition of the relations from $\Psi\mathbf{L}(n)$ with the (possibly also inconstructively obtained) relations from **D**(n). Observe, however, that I do *not* use inconstructively obtained relations in order to *form new* inconstructively obtained relations: there is *no nesting* of inconstructive formations of relations.

In the same manner now, the definition of the class **RFR** contained in Corollary 6 leads to a process generating this class with help of the operation assigning to every **M** the set

$\nu\mathbf{M}$ of all relations νR for which R is in **M** and is full .

This again is an inconstructive operation, because I have no algorithm by which to decide whether a relation R is full. And during the generation of **RFR** this operation will have to be *nested arbitrarily often*.

An analogous situation presents itself in the case of the definition of **FRF** by the properties of Corollary 4. Given a function f^{k+1}, it is clear how to test whether $\mu^* f^{k+1}(\alpha) = b$, but it is, in general, not possible to decide in finitely many steps whether $\mu^* f^{k+1}$ has any value at all for the argument α .

Quite generally, there is good reason to also call the recursive relations the *decidable* ones. Because R^k is in **RFR** = **QE**(**FSF**) if, and only if, there are two simple 1-ary functions f and g such that cau^k maps R^k onto im(f) and $-R^k$ onto im(g). In order to decide whether α is in R^k, I then start two simultaneous computations of successive values of f and of g and test whether one of these values is $cau^k(\alpha)$. Would there only be f (i.e. would R^k be only recursively enumerable), then I could not decide in finitely many steps that $cau^k(\alpha)$ is not a value under f, but if there also is g then one of the two computations will after finitely many steps decide that $cau^k(\alpha)$ is a value of f or of g, i.e. is in R^k or is in $-R^k$.

Chapter 7. The Arithmetization of Syntax

The study of undefinability and incompleteness in Chapter 4 did depend on assumptions about functions g_T, g_F, g_S which coded the objects of a language L. I now shall show that these functions can be defined in many situations arising in mathematical practise, and I shall also show that such definitions then have the consequence that the sets, relations and functions, which correspond to these objects, become recursive, if not elementary - in particular the substitution function *sub* and the deduction relation *ded*.

Already in Chapter 1.5, I described how term algebras can be *realized* by natural numbers if there are at most countably many variables and operation symbols. It thus remains to be shown that, under suitable assumptions, these realizations produce objects, such as the set of formulas or the the function *sub*, which are recursive. This should be no surprise if I recall that those objects, if not already defined explicitly, arose by course of values recursion.

I shall, therefore, construct *arithmetically realized* languages L_A such that their terms and formulas are natural numbers themselves and are their own codes, i.e. the coding maps g_T and g_F become the identity. As term algebras are naturally isomorphic if they have the same signature and equipotent sets of generators, any abstractly given language L with countably many variables then has natural isomorphisms, g_T of its term algebra and g_F of its algebra of formulas, to the terms and formulas of an arithmetically realized language L_A, and these already will be its coding functions. The function g_S, coding sequences of L-formulas, I obtain as $g_A \circ g_F$ when g_A is a coding function for sequences of arithmetically realized formulas.

Let Δ be a first order signature such that in $\Delta_a = <n_i \mid i \epsilon I_a>$ and $\Delta_b = <m_i \mid i \epsilon I_b>$ the index sets I_a and I_b are at most countably infinite. So I assume right away that I_a and I_b are sets of integers i with $i > 1$ whose characteristic functions *ina* and *inb* shall have one of the properties $\mathcal{R}$

recursive, *primitive recursive* or *elementary*,

and I further assume that Δ_a and Δ_b are restrictions to I_a and I_b of functions with the property $\mathcal{R}$. I shall say then that Δ is $\mathcal{R}$*-presented* (and in the concrete case that it be $\mathcal{R}$ly-presented where $\mathcal{R}$ly is the adverbial form of the adjective $\mathcal{R}$).

The particular arithmetization to be developed in this Chapter will be a slight refinement of the Realization 1 of terms by numbers from Chapter 1.5. Finite sequences α of positive numbers will be coded by numbers x which have the members $\alpha(i)$ as exponents $exp(x,i)$ in their prime factor decomposition; in view of $\alpha(i) > 0$ there is no need to rise these exponents by 1. I shall write

$\varepsilon_i(x)$ in place of $exp(x,i)$,
$\delta_i(x)$ in place of $\varepsilon_i(\varepsilon_0(x))$,
$|x|$ in place of $len(x)$

where $len(x)$ is the elementary function from 5.(FE6) whose value is the smallest prime number p such that no prime number q with $q \geq p$ divides x. All objects of my language shall be of the form

$$2^c \cdot \Pi < prino(k+1)^{\alpha(k)} \mid k < j(c) >$$

where c is characteristic for the kind of object and the $\alpha(j)$ are arguments of that object. As for Δ_a and Δ_b, this arrangement permits to define them as the restrictions to I_a and I_b of the elementary functions

$$art_a(x) = \varepsilon_0(\delta_1(x)) \qquad art_b(x) = \varepsilon_0(\delta_2(x)) + 1$$

which lets the $\mathcal{R}$-presentability depend only on the characteristic functions *ina* and *inb* of I_a and I_b.

Let *va* be the elementary function $va(n) = 2^{2^{n+2}}$; the set *var* of variables x_n shall consist of its values, i.e. of all m such that there is an n with $n<m$ and $va(n) = m$, and so it is elementary.

I define the operations F_i belonging to the i in I_a by

$$F_i(\alpha) = 2^{3^i} \cdot \Pi < prino(k+1)^{\alpha(k)} \mid k < art_a(i) > \ .$$

Under these operations, the variables generate a term algebra T(L) whose underlying set I call *term*. [So the numbers from I_a in the sequences

$2n+1$	3, 5, 7, ...
$2\cdot(2n+1)$	2, 6, 10, 14, ...
$2^2\cdot(2n+1)$	4, 12, 20, 28, ...

are the indices of 0-ary, 1-ary, 2-ary operations respectively.]

In particular, the set *con* of constants consists of all m such that there exists an n with $n<m$ and $ina(2n+1) = 1$ and $m = 2^c$ with $c = 3^{2n+1}$. [So if I wish to construct a language containing an ω-sequence of constants κ_n, then I choose these as the values of the elementary function $\kappa(n) = 2^{c(n)}$ with $c(n) = 3^{2n+1}$.] The set *term* of terms consists of all m such that

$m \in var$ or $m \in const$
or $\delta_0(m) = 0$ and $ina(\delta_1(m)) = 1$
and $|\varepsilon_0(m)| = 2$
and $art_a(m) > 0$
and $|m| = 1 + art_a(m)$
and for all k: if $k < art_a(m)$ then $\varepsilon(k+1, m) \in term$.

This is a recursion defining the characteristic function of *term* with the regression function ε. Therefore *term*, depending on I_a, is an $\mathcal{R}$-set.

I define the set *atom* of atomic formulas with predicate symbols p_i, $i \in I_b$, to consist of the values of the functions

$$P_i(\lambda) = 2^{5^i} \cdot \Pi < \mathit{prino}(k+1)^{\lambda(k)} \mid k < \mathit{art}_b(i) >$$

for λ in term^{m_i}. [So the numbers from I_b in the sequences above now are the indices of 1-ary, 2-ary, 3-ary predicate symbols respectively.] The set atom can be characterized to consist of the numbers m such that

$$\begin{aligned} \delta_0(m) = \delta_1(m) = 0 \text{ and } & \mathit{inb}(\delta_2(m)) = 1 \\ & \text{and } |\varepsilon_0(m)| = 3 \\ & \text{and } |m| = 1 + \mathit{art}_b(m) \\ & \text{and for all } k\text{: if } k < \mathit{art}_b(m) \text{ then } \varepsilon(k+1, m) \in \mathit{term}\,. \end{aligned}$$

This is a recursion defining the characteristic function of *atom* with the regression function ε. Therefore *atom*, depending on I_a and I_b, is an $\mathcal{R}$-set.

The results of the propositional connectives $\wedge, \vee, \rightarrow, \neg$ I define with the functions *et*, *vel*, *imp*, *ne* defined as assigning to m and n the values

$$2^7 \cdot 3^m \cdot 5^n \quad , \quad 2^{7^2} \cdot 3^m \cdot 5^n \quad , \quad 2^{7^3} \cdot 3^m \cdot 5^n \quad , \quad 2^{7^4} \cdot 3^m$$

respectively. The quantification with $\forall x_m$ and $\exists x_m$ of a formula n I define as values of 2-ary functions *quniv*, *qexi* assigning to m and n the values

$$2^{11^{2^{2^{m+2}}}} \cdot 3^n \text{ and } 2^{13^{2^{2^{m+2}}}} \cdot 3^n \text{ respectively.}$$

The algebra Fm(L) generated by *atom* under these operations is a term algebra whose underlying set is the set *form* of formulas. It can be characterized as the set of all m such that

$$\begin{aligned} & m \in \mathit{atom} \\ & \text{or } \delta_0(m) = \delta_1(m) = \delta_2(m) = 0 \\ & \quad \text{and } ((\delta_3(m) = 1 \text{ or } \delta_3(m) = 2 \text{ or } \delta_3(m) = 3\ ) \text{ and } |\varepsilon_0(m)| = 4 \text{ and } |m| = 3 \\ & \qquad\qquad\qquad\qquad\qquad\qquad \text{and } \varepsilon_1(m) \in \mathit{form}, \\ & \qquad\qquad\qquad\qquad\qquad\qquad \text{and } \varepsilon_2(m) \in \mathit{form}\ ) \\ & \qquad \text{or } (\delta_3(m) = 4 \text{ and } |\varepsilon_0(m)| = 4 \text{ and } |m| = 2 \text{ and } \varepsilon_1(m) \in \mathit{form}\ ) \\ & \qquad \text{or } (\delta_3(m) = 0 \text{ and } |\varepsilon_0(m)| = 5 \text{ and } |m| = 2 \text{ and } \delta_4(m) \in \mathit{var} \\ & \qquad\qquad\qquad\qquad\qquad\qquad \text{and } \varepsilon_1(m) \in \mathit{form}\ ) \\ & \qquad \text{or } (\delta_3(m) = \delta_4(m) = 0 \text{ and } |\varepsilon_0(m)| = 6 \text{ and } |m| = 2 \text{ and } \delta_5(m) \in \mathit{var} \\ & \qquad\qquad\qquad\qquad\qquad\qquad \text{and } \varepsilon_1(m) \in \mathit{form}\ ). \end{aligned}$$

This is a recursion defining the characteristic function of *form* with the regression function ε. Therefore *form* is an $\mathcal{R}$-set.

Formulas with $\delta_4(m) \neq 0$ or $\delta_5(m) \neq 0$ are ***quantified***, and $2^{\delta_4(m)}$ or $2^{\delta_5(m)}$ then is their ***quantifying variable***.

I now obtain various syntactical relations as $\mathcal{R}$. To begin with, the 2-ary relation *occt*, saying that a variable m occurs in a term n,

$m \in \mathit{var}$ and $n \in \mathit{term}$
and $m = n$ or ($|n| > 1$
and there exists $k < |n| : 0 < k$ and $\langle m, \varepsilon_k(n) \rangle \in \mathit{occt}$),

has a characteristic function defined by recursion with regression functions. The relation *occf*, that a variable m occurs in a formula n, I define by

$m \in \mathit{var}$ and $n \in \mathit{form}$
and($n \in \mathit{atom}$ and there exists $k < |n| : 0 < k$ and $\langle m, \varepsilon_k(n) \rangle \in \mathit{occt}$)
or (not $n \in \mathit{atom}$
and there exists $k < |n| : 0 < k$ and $\langle m, \varepsilon_k(n) \rangle \in \mathit{occf}$).

The relation *occfree*, saying that a variable m occurs freely in a formula n, I define by

$m \in \mathit{var}$ and $n \in \mathit{form}$
and $\langle m, n \rangle \in \mathit{occf}$
and $n \in \mathit{atom}$
or (not $n \in \mathit{atom}$ and $\delta_4(n) \neq m$
and $\delta_5(m) \neq m$
and there exists $k < |n| : 0 < k$
and $\langle m, \varepsilon_k(n) \rangle \in \mathit{occfree}$).

The relation *freefor*, of all $\langle b, a, m \rangle$ expressing that $\zeta(x, t)$ is free for v, I define by employing the characterization from Lemma 1.13.1 :

$b \in \mathit{var}$ and $a \in \mathit{term}$
and $m \in \mathit{form}$
and $m \in \mathit{atom}$
or not $m \in \mathit{atom}$
and (($\delta_3(m) = 1$ or $\delta_3(m) = 2$ or $\delta_3(m) = 3$)
and $\langle b, a, \varepsilon_1(m) \rangle \in \mathit{freefor}$
and $\langle b, a, \varepsilon_2(m) \rangle \in \mathit{freefor}$)
or $\delta_3(m) = 4$ and $\langle b, a, \varepsilon_1(m) \rangle \in \mathit{freefor}$
or $\delta_4(m) \neq 0$ and $b = \delta_4(m)$
or $\langle b, a, \varepsilon_1(m) \rangle \in \mathit{freefor}$
and not $\langle b, \varepsilon_1(m) \rangle \in \mathit{occfree}$
or not $\langle b, a \rangle \in \mathit{occt}$
or $\delta_5(m) \neq 0$ and $b = \delta_5(m)$
or $\langle b, a, \varepsilon_1(m) \rangle \in \mathit{freefor}$
and not $\langle b, \varepsilon_1(m) \rangle \in \mathit{occfree}$
or not $\langle b, a \rangle \in \mathit{occt}$.

It is an immediate consequence of the construction of terms and formulas that for a formula m, say, subformulas of m, or terms occurring in m, are numbers smaller than m. This has the effect that often only bounded quantifications need to be used (leading e.g. from elementary sets to elementary ones again). For instance, the set *sentence* of all sentences consists of all m in *form* such that, for all $n < m$, if $n \in \mathit{var}$ then not $\langle n, m \rangle \in \mathit{occfree}$.

Replacement I need only for maps η which change *at most one* variable. I begin by defining the 3-ary function *rept* which acts as homomorphism on terms n. I make use of the operation F_i of T(L) and set *rept*(b,a,n) to be

a	if b ϵ *var* and n = b
$F_i(\textit{rept}(b,a,-)\cdot\alpha)$	if $n = F_i(\alpha)$
n	in all other cases .

This is a recursion with regression functions; hence *rept* is primitive recursive again. In the Lemma below I shall show that *rept* can be majorized by an elementary function; hence if $\mathscr{R}$ = elementary then *rept* will be elementary.

I define an auxiliary function *d* by $d(0) = d(1) = 0$, $d(n) = 1$ if n is prime, and

$$d(n) = 1 + \max < d(\varepsilon_i(n)) \mid i < |n| > .$$

otherwise. I claim that $d(n) < n$ for $0 < n$, and this is obvious if, say, $n \leq 3$. Assume that it holds for all $k < n$. If $\varepsilon_i(n) \leq 1$ for all i then $d(n) = 1 < n$. Otherwise $1 < \varepsilon_i(n)$ and $0 < d(\varepsilon_i(n))$ for at least one $i < |x|$, and $d(n) = 1 + d(\varepsilon_k(n))$ for some k of them. But for $1 < e$ and $1 < p$ always $e+1 < p^e$. Thus $d(\varepsilon_k(n)) < \varepsilon_k(n)$ implies

$$d(n) = 1 + d(\varepsilon_k(n)) < 1 + \varepsilon_k(n) < \textit{prino}(k)^{\varepsilon_k(n)} \leq n .$$

Possessing an elementary majorant, *d* is itself elementary. I observe that if n contains a square then $2 \leq d(n)$.

I define a function *h* of arguments a,n,j by

$$h(a,n,j) = n \quad \text{if } n \leq 1 \text{ or } j = 0 ,$$
$$h(a,n,1) = \Pi < \textit{prino}(i)^{\varepsilon(n,i)+a} \mid i < |n| > ,$$
$$h(a,n,j+1) = \Pi < \textit{prino}(i)^{sg(\varepsilon(n,i))\cdot h(a,n,j)} \mid i < |n| > ,$$

and a function *g* by $g(a,n) = h(a,n,\ d(n) \dot{-} 1)$. Now the functions $h(a,n,0)$ and $h(a,n,1)$ are elementary; hence induction shows that also $h(a,n,j+1)$ is elementary. Consequently, also *g* is elementary. Observe that

if $n > 1$, $a > 0$ then $n < h(a,n,1)$
if $n > 1$ then $h(a,n,j) < h(a,n,j+1)$

whence $n \leq h(a,n,j)$ for $n > 1$ and every j.

LEMMA 1 If $a > 0$ then for every n in *term* : $\textit{rept}(b,a,n) < g(n,a)$.

The proof is by induction on terms. Observe first that every term n contains a square since either $n = 2^c \cdot q$ with $c = 2^{m+2}$ or $c = 3^i$ where $i \epsilon I_a$ implies $i > 1$. Thus $2 \leq d(n)$, $1 \leq d(n) \dot{-} 1$ and $h(a,\ n,1) \leq h(a,\ n, d(n) \dot{-} 1) = g(a,n)$. If n is a variable 2^c then *rept*(b,a,n) is either n or a whence

$$\mathit{rept}(b,a,n) \leq \max(a, 2^c) < 2^{c+a} = h(a, 2^c, 1) \leq g(a,n) .$$

The case of $c = 3^i$ is analogous. Assume now that $n = f_i(\alpha)$, $0 < n_i$, and that for every $k < n_i$

$$\mathit{rept}(b, a, \alpha(k)) < h(a, \alpha(k), d(\alpha(k)) - 1) = g(a, \alpha(k)) .$$

Since

$$(x) \quad n = 2^c \cdot \Pi < \mathit{prino}(k+1)^{\alpha(k)} \mid k < n_i \mid >$$

there holds $d(n) > d(\alpha(k))$. Since the $\alpha(k)$ are terms there holds $2 \leq d(\alpha(k))$, $0 < d(\alpha(k)) - 1 \leq d(n) - 2$. Hence

$$\mathit{rept}(b, a, \alpha(k)) < h(a, \alpha(k), d(n)-2) .$$

Since $1 < c$ also implies $c < h(a, c, d(n)-2)$, there follows

$$\begin{aligned} \mathit{rept}(b,a,n) &= 2^c \cdot \Pi < \mathit{prino}(k+1)^{\mathit{rept}(b,a,\alpha(k))} \mid k < n_i > \\ &\leq 2^{h(a,c,\, d(n)-2)} \cdot \Pi < \mathit{prino}(k+1)^{h(a,\, \alpha(k),\, d(n)-2)} \mid k < n_i > . \end{aligned}$$

But (x) implies that this is $h(a, n, d(n)-1) = g(a,n)$, and this concludes the proof of the Lemma.

I extend *rept* to a function *repf* acting on formulas by:

$$\mathit{repf}(b,a,m) = 2^{\varepsilon_0(m)} \cdot \Pi < \mathit{prino}(k+1)^{\mathit{rept}(b,a,\varepsilon_k(m))} \mid k+1 < |m| >$$

if $m \in \mathit{atom}$,

$$\mathit{repf}(b,a,m) = 2^{\varepsilon_0(m)} \cdot \Pi < \mathit{prino}(k+1)^{\mathit{repf}(b,a,\varepsilon_k(m))} \mid k+1 < |m| >$$

if $m \in \mathit{form}$ and $b \neq \delta_4(m)$ and $b \neq \delta_5(m)$

$\mathit{repf}(b,a,m) = m$ in all other cases .

It follows from Lemma 1 that $\mathit{repf}(b,a,-)$ has an elementary majorant if its arguments are restricted to *atom*, and then this Lemma immediately translates into one which produces an elementary majorant $g'(a,m)$ of $\mathit{repf}(b,a,m)$. Thus $\mathit{repf}(b,a,m)$ is $\mathscr{R}$ again.

The functions F_i forming terms, the functions P_i forming atomic formulas, and the functions *et*, *vel*, *imp*, *ne*, *quniv*, *qexi* : they all are strictly monotonic and, therefore, injective in each of their arguments. Hence $\mathit{rept}(b,a,n)$ and $\mathit{repf}(b,a,m)$ are injective in the argument a.

While *repf* arithmetizes the replacement map rep, substitution requires the arithmetization of the renaming map tot. The recursive clause in the definition of tot(j , t | Qx w) was of the form

$\mathsf{Q}x\,w$	if $j = x$ and $fr(v) \cap occ(t) = 0$
$\mathsf{Q}x\,tot(j,t \mid w)$	if $j \neq x$ and $fr(v) \cap occ(t) = 0$
$\mathsf{Q}y\,rep(x,y \mid tot(j,t \mid w))$	if $j \neq x$ and $fr(v) \cap occ(t) \neq 0$
$\mathsf{Q}y\,rep(x,y \mid w)$	if $j = x$ and $fr(v) \cap occ(t) \neq 0$

where y was the first variable (with respect to a given enumeration β of variables which here is $\beta(m) = 2^{m+2}$) satisfying certain (elementarily expressible) properties (a)-(d). Of them, the property (b) was critical in that it referred to variables in $rep(x,y \mid tot(j,t \mid w))$ and so caused a series of dependent choices. But then I did prove in Lemma **1.13**.4a that another choice of y is possible which depends on v and t alone: let $i_0 = i_0(v)$ be the minimal number such that the variable $\beta(i_0)$ does not belong to var(v), bd(v), $\bigcup < occ(\eta(z)) \mid z \in var(v)$; then $y = \beta(i_0 + |v| - 1)$ will do where $|v|$ is any function on formulas such that $|w| < |v|$ for subformulas w of v.

If now m is in *form* then the variables occurring in m are smaller than m and, therefore, smaller than the variable 2^c, $c = 2^{m+2}$, and the same holds for the variables occurring bound in m. Thus the function *t* is $\mathcal{R}$ for which $t(a,m)$ is the first variable not occurring in the term a nor in the formula m or as one of its quantified variables. I therefore define the function *tot* by primitive recursion with help of *t* and with $|m|$ in place of $|v|$. For future use, I also define tot_k with the additional stipulation that its values shall contain no bound variables $\beta(n) = 2^c$, $c = 2^{n+2}$ with $n \leq k$. With *repf* and *tot* being available, I define *subf* by

$$subf(j,t,m) = repf(j,t,\ tot(j,t,m)) \ .$$

Lemma **1.13**.4a then ensures that Lemma **1.13**.4 remains in effect for the redefined function *tot*, and it follows again that $subf(b,a,m) = repf(b,a,m)$ holds if $<j,t,m> \in freefor$. In particular, this is the case if t is *constant*.

The functions *tot* and, therefore, *subf* then are primitive recursive. If $\mathcal{R}$ means "elementary" then again an elementary majorant of *tot* can be found, showing them to be elementary. To this end, I use the second part of Lemma **1.13**.4a which states that the new $tot(j,t \mid \mathsf{Q}x\,w)$ contains only variables $\beta(k)$ with $k \leq i_0 + |v| - 1$. Thus $g'(t(a,m), m)$ is the desired majorant.

The terms and formulas, of the language constructed here, are numbers and, as such, they code themselves. Deductions, however, in calculi operating on formulas of L, are *not* numbers and *must* be coded. I shall code sequences α of formulas as

$$g_A(a) = \Pi < prino(k)^{\alpha(k)} \mid k < def(\alpha) > \ .$$

These numbers n are different from all codes of terms and formulas. Because $\varepsilon_0(n)$ being code of a formula, neither $\varepsilon_0(\varepsilon_0(n))$ nor $\varepsilon_1(\varepsilon_0(n))$ is 0; on the other hand, variables n have $\varepsilon_1(\varepsilon_0(n)) = 0$ and other terms and formulas n have $\varepsilon_0(\varepsilon_0(n)) = 0$.

As for the calculi, it will be convenient to work with a modus ponens calculus, because deductions then are sequences of formulas (while the arithmetic would become slightly more complex would I have to code trees and trees carrying sequents of formulas). Also, matters will be simplified by considering deductions from axiom systems, because I then only have to describe a calculus ccqt without subsidiary deductions. My choice shall be the calculus $ccqt_1$ studied in Chapter **2.9**.

I begin by defining the subset *axiom* of *form*. For the propositional axioms, I can use one of the sets of finitely many schemata defining the MP-calculus cc discussed in Chapter **2.7**. If (s0),..., s(h−1) are the schemata of such a set, containing, for instance, as (s0), (s5) the schemata

$$v \to (w \to v) \quad , \quad (u \to v) \to ((u \to w) \to (u \to v \wedge w))$$

then the clauses defining the set *axiomp* of propositional axioms m contain

(S0) there exist $a \leq m$, $b \leq m$, $c \leq m$ such that m = ***imp***(a, ***imp***(b,c))

(S5) there exist $a \leq m$, $b \leq m$, $c \leq m$ such that
m = ***imp***(***imp***(a,b), ***imp***(***imp***(a,c), ***imp***(a, ***et***(b,c))))

where I have used the elementary functions ***imp*** and ***et*** arithmetizing → and ∧. The translations (Si) of the remaining schemata (si) are obtained analogously. Enlarging ***axiomp*** by the translations of the four schemata

$$\begin{array}{ll} \forall x\, w \to rep(x,t \mid w) & \text{where } \zeta(x,t) \text{ is free for } w \\ rep(x,t \mid w) \to \exists x\, w & \text{where } \zeta(x,t) \text{ is free for } w \\ \forall x (u \to w) \to (u \to \forall x\, w) & \text{if } x \text{ is not in } fr(u) \\ \forall x (w \to u) \to (\exists x\, w \to u) & \text{if } x \text{ is not in } fr(u) \quad , \end{array}$$

I arrive at the $\mathcal{R}$-set ***axiom***. The first of these translations is

there exist p<m and q<m and r<m and b<m and a<m
and p ,q ,r in ***form*** and b in *var* and a in *term* and m = ***imp***(p,q) and
p = ***quniv***(b,q) and r = ***repf***(b,a,q) and <b,a,q> in ***freefor***,

and the others are formulated analogously. It should be noticed that in r = ***repf***(b,a,q) the number a is indeed uniquely determined by r since the function ***repf***(b,a,q) is injective in a. In particular, a is also the smallest a' such that r = ***repf***(b,a',q).

Given a subset G of ***form***, I define the set ***dedu(G)*** of arithmetizations of deductions from G, i.e. sequences of formulas, each of which is either in G or in ***axiom*** or arises from previous ones by ***modus ponens*** or by the rule

$$w \vdash \forall x\, w \quad .$$

Hence m is in ***dedu(G)*** if, and only if,

for every k: if $k < |m|$ then $\varepsilon_k(m) \epsilon$ *form*
and ($\varepsilon_k(m) \epsilon G$ or $\varepsilon_k(m) \epsilon$ *axiom*)
(S) or there exist i,j : $i<k$, $j<k$ and $\varepsilon_j(m) = imp(\varepsilon_i(m), \varepsilon_k(m))$
or there exist i,b: $i<k$, $b<m$ and $\varepsilon_k(m) = quniv(b, \varepsilon_i(m))$.

Obviously, the set *dedu(G)* will be $\mathcal{R}$ if G is so. If $G$is only recursively enumerable, i.e. in **RER** = **E**(**RAB**), then so is *dedu(G)*, because the constructions used in (S) are unions, intersections and bounded quantifications, applied to recursive or recursively enumerable relations, and **E**(**RAB**) is closed under these constructions by **6**.(AB14).

Finally, I define the relation *ded(G)* to be the set of the pairs $<n,m>$ such that $m \epsilon$ *dedu(G)* and $n = \varepsilon(|m|-1, m)$. Together with *dedu(G)* also *ded(G)* is $\mathcal{R}$ or recursively enumerable. This concludes the proof of Theorem 1.

Let me observe that the constructions of *dedu(G)* and *ded(G)* are *uniform* in G: the characterization (S) is a *schema* which, for any G, produces *dedu(G)* in a uniform manner.

I collect the facts established now in the

THEOREM 1 Let Δ be an $\mathcal{R}$-presented signature. Then a language L_Δ of signature Δ can be constructed which has the following properties:

The set of variables *var* of L and the sets *term* = u(T(L)) and *form* = u(Fm(L)) are $\mathcal{R}$-subsets of ω . The operations both of T(L) and of Fm(L) are restrictions to *term* and *form* of $\mathcal{R}$-functions which are strictly monotonic in all of their arguments.

There are 3-ary $\mathcal{R}$-functions *repf* and *sub* such that, for $b \epsilon$ *var*, $a \epsilon$ *term* and $m \epsilon$ *form*, there holds *repf*(b,a,m) = rep(b,a | m) and *sub*(b,a,m) = sub(b,a | m) .

There is an injection g_Δ, mapping finite sequences of formulas into ω, whose image is disjoint both to *term* and to *form*. Given one of the modus ponens calculi for quantifier logic, for every subset G of the set *form* :

if G is $\mathcal{R}$, or is recursively enumerable, then so is *dedu(G)*, the image under g_Δ of the set of deductions from the axioms G , and so is also the relation *ded(G)* which consists of all $<n, g_\Delta(\alpha)>$ such that $n \epsilon$ *form* and α is a deduction of n from the axioms G .

A language with the properties stated in the Theorem I shall call an $\mathcal{R}$-*arithmetized language*.

Other constructions of $\mathcal{R}$-arithmetized languages can be performed using the coding techniques employed e.g. in

R.S. Smullyan: Gödel's Incompleteness Theorems. Oxford 1992

R.S. Smullyan: Recursion Theory for Metamathematics. Oxford 1993

Y.V. Matiyasevich: Hilbert's Tenth Problem. Cambridge Mass. 1993.

Chapter 8. Consequences of Arithmetization

The properties of codings for arithmetized languages having been established, I now shall express the effect which recursiveness or recursive enumerability of an axiom system G have upon decidability and incompleteness. My approach still remains *abstract* in the sense the axiom systems G are only specified with respect to certain syntactical properties they can prove.

In Section 1, Theorem 1 will reduce the question of incompleteness to that of undecidability. Theorem 2 is of a more technical character and expresses which arithmetical properties - e.g. recursiveness - of relations follow from their syntactical definability properties. Section 2 contains the main result, namely Theorem 3, saying that a recursively enumerable axiom system G has an undecidable theory provided it makes every relation formally definable. Then G will also be incomplete, and this result now will be compared with what can be obtained by applying the abstract incompleteness theorems from Chapter 4. Section 3 is devoted to the conditions ensuring *radical* undecidability, i.e. the undecidability of an axiom system G_1 in which a given undecidable axiom system G_0 can be interpreted.

While so far notions about recursive relations are employed to prove logical properties of axiom systems, in Section 4 the direction is turned around, and properties of arithmetized deducibility are used in order to obtain insights about recursive relations and functions, namely the *uniformization* (or *normal form)* theorems of KLEENE 36 , obtained here as Theorem 5 and Corollaries 5-7. Slightly more laborious is the proof of Theorem 6 which secures the SMN-property for this uniformization; it is required in order to conclude from ROGERS' 58 isomorphism theorem that the uniformization obtained here is isomorpic to the uniformizations constructed by KLEENE or by the tools of computability theory. Finally, an immediate application is KLEENE's (2nd) recursion theorem, showing e.g. that the ACKERMANN-PETER function is recursive.

References

W. Craig: On axiomatizability within a system. J.Symb.Logic 18 (1953) 30-32

S.C. Kleene: General recursive functions of natural numbers. Math.Ann. 112 (1936) 727-742

S.C. Kleene: Introduction to Metamathematics. Amsterdam 1952

H. Rogers: Gödel numberings of partial recursive functions. J.Symb.Logic 23 (1958) 331-341

H. Rogers: Recursive Functions and Effective Computability. New York 1967

1. Recursively Enumerable Axiom Systems

Let L be an $\mathcal{R}$-arithmetized language L as defined in the last Chapter. Its decisive property is the connection between sets G of formulas and the relations $ded(G)$, providing a deducibility relation which, together with G, is $\mathcal{R}$ or recursively enumerable. From $ded(G)$, I immediately obtain the sets

$$prov(G) = \exists\, ded(G) \quad \text{and} \quad provs(G) = prov(G) \cap sentence$$

of all n which are formulas or sentences provable from G. Hence if G is recursively enumerable then so are $prov(G)$ and $provs(G)$.

The following observation was made, essentially, by CRAIG 53 :

LEMMA 1 If G is (arbitrary, but) such that $prov(G)$ is recursively enumerable, then I can find an elementary subset G' of *form* such that $prov(G') = prov(G)$. If G consists of sentences then also G' can be chosen to consist of such.

The proof uses the results from Chapter 6 that **RER** = **E(FEF)** and that for any set M from **E(FEF)** there is an elementary function with im(f) = M. So let f be such a function with im(f) = $prov(G)$, i.e.

$$n \in prov(G) \text{ if, and only if, there exists k such that } n = f(k)\ .$$

From f, I construct an elementary function g, mapping ω into *form*, such that

for every k : $k \leq g(k)$, and the formulas f(k) and g(k) are provably equivalent.

For instance, I may choose g(k) to be $et(f(k), j(k))$ where, in the previous Chapter's coding, j(k) is the tautology

$$imp(2^{k+2}, 2^{k+2}) = 7^3 \cdot 19^{2^{k+2}} \cdot 23^{2^{k+2}}\ .$$

Since j is elementary in k, also g is elementary. Setting $G' = \text{im}(g)$, it follows that $prov(G) = prov(G')$ since f(k) is provably equivalent to g(k). But since $k \leq g(k)$ for every k, I may characterize the set $G' = \text{im}(g)$ by

$$n \in G' \text{ if, and only if, there exists } k \leq n \text{ such that } n = g(k)\ .$$

The quantification on k being bounded, together with g also G' will be elementary.

Let Δ be an $\mathcal{R}$-presented signature; let L be an arbitrary and let L_A be the $\mathcal{R}$-arithmetized language of this signature. If the term algebra T(L) is generated by variables y_m then the bijection sending y_m to 2^c with $c = 2^{m+2}$

extends to an isomorphism g_T from T(L) onto T(L_A). The bijection sending an atomic L-formula $p_i(\alpha)$ to

$$5^i \cdot \Pi < prino(k+7)^{g_T(\alpha(i))} \mid k < m_i >$$

in *atom* extends to an isomorphism g_F from Fm(L) onto Fm(L_A). As I did already in Chapter 4, I shall mostly write ⌜t⌝ and ⌜v⌝ in place of $g_T(t)$ and $g_F(v)$. The map g_F establishes a bijection between deductions β using L-formulas and deductions α using L_A-formulas; prolonging it with the coding map s for sequences of L_A-formulas, I obtain an arithmetization map $g_S = g_A \circ g_F$ for sequences of L-formulas.

Throughout the following, I consider an arbitrary set G of L-formulas, and G shall always denote the subset g_F(G) of *form* = u(Fm(L_A)). I shall say that G is $\mathscr{R}$ or is recursively enumerable if G is so. In that case both *prov*(G) and *provs*(G) are recursively enumerable.

Employing the notion that recursive relations may also be called decidable, I shall say that *the theory of* G *is decidable,* respectively *undecidable,* if *provs*(G) is recursive respectively not recursive.

Recall that G is called complete if, for every sentence s, either $\vdash_G$ s or $\vdash_G \neg$s. Recall that G is called consistent if there is no formula v such that both $\vdash$ v and $\vdash \neg$v. If G is consistent and complete then a sentence s is *not* provable if, and only if, ¬s is provable. It follows from the isomorphism of the languages L and L_A that G is complete, or consistent, in L if and only if G is complete, or consistent, in L_A.

THEOREM 1 Let G be recursively enumerable and complete. Then *provs*(G) is recursive.

If G is recursively enumerable and if the theory of G is undecidable then G is incomplete.

Obviously, the second statement is but a reformulation of the first one. If G is inconsistent then *provs*(G) is the recursive set *sentence*; assume now that G is consistent. Thus *provs*(G) is recursively enumerable. If G, hence G, is consistent and complete then a number n is in *sentence* − *provs*(G) if, and only if, *ne*(s) is in *provs*(G). The function *ne* being recursive, it follows that together with *provs*(G) also *sentence* − *provs*(G) will be recursively enumerable, hence recursive. Therefore, together with *sentence*, also −*provs*(G) = (*sentence* − *provs*(G)) ∪ −*sentence* will be recursive.

I now define a notion which will basic for the following developments. A triplet L, L_A, κ shall be *in arithmetical situation* if

L is a language with equality and with a recursively presented signature

L_A is a recursively arithmetized copy of L determining the maps g_T, g_F, g_S

κ is an injection of ω into the set of constant terms of L (terms without variables) such that the function $\delta = g_T \cdot \kappa$ is recursive.

In such situation I shall call κ a *recursive sequence of arithmetical constants*. I write the value $\ulcorner \kappa_n \urcorner$ of n under δ as δ_n when its rôle as a term, and as $\delta(n)$ when its rôle as a number is to be emphasized.

I shall also say that L, κ is in arithmetical situation if the rôle of L_A is clear from the context, and occasionally I shall also omit to mention κ here. Given a set G of L-sentences, its image under g_F will be written as G, and provability in L_A from G will be denoted by $\vdash_{GA}$.

Consider now an arithmetical situation with the recursive function δ from ω into the constant terms of L_A. The twofold rôle of L_A-formulas $u(\psi)$ as *formulas* and as *numbers* is decisively used in the following *Uniformization Lemma* :

LEMMA 2 Let $\psi = \langle y_0, \dots, y_{k-1} \rangle$ be a given sequence of L_A-variables. Then the (k+1)-ary function ϑ_k, defined for all L_A-formulas $u = u(\psi)$ and for all λ in ω^k as

$$\vartheta_k(u, \lambda) = u(\delta \cdot \lambda) \qquad \text{(and elsewhere as the identity)}$$

is recursive. If $\delta = g_T \cdot \kappa$ is primitive recursive then so is ϑ_k.

By definition, $u(\delta \cdot \lambda)$ is $sub(\eta \mid u)$ where η maps the variable y_i to $\delta_{\lambda(i)}$. The $\delta_{\lambda(i)}$ being without variables, $sub(\eta \mid u)$ is $rep(\eta \mid u)$, and the result of this simultaneous replacement is the same as that of k singular replacements:

$$u(\delta \cdot \lambda) = rep(\eta \mid u(\psi)) = rep(y_{k-1}, \delta_{\lambda(k-1)} \mid \cdots$$
$$\cdots \mid rep(y_1, \delta_{\lambda(1)} \mid rep(y_0, \delta_{\lambda(0)} \mid u(\psi)) \dots).$$

Now L_A is recursively arithmetized and $u(\delta \cdot \lambda)$ and u are numbers. Thus the last line may be written as

$$u(\delta \cdot \lambda) = \mathit{repf}(y_{k-1}, \delta_{\lambda(k-1)}, \dots, \mathit{repf}(y_1, \delta_{\lambda(1)}, \mathit{repf}(y_0, \delta_{\lambda(0)}, u) \dots) .$$

Here *repf* is a 3-ary primitive recursive function, and also δ is recursive. Thus $\mathit{repf}(y, \delta(z), u)$ is recursive. Hence also the (2k+1)-ary function

$$\vartheta^*_k(\psi, u, \lambda) = \mathit{repf}(y_{k-1}, \delta_{\lambda(k-1)} \mid \dots \mid \mathit{repf}(y_1, \delta_{\lambda(1)} \mid \mathit{repf}(y_0, \delta_{\lambda(0)} \mid u)) \dots)$$

will be recursive, as well as the function $\vartheta_k(u, \lambda) = \vartheta^*_k(\psi, u, \lambda)$ where ψ is fixed.

LEMMA 3 If G is recursively enumerable then so is, for every u, the set of all λ such that $\vdash_{GA} u(\delta\cdot\lambda)$.

$\vdash_{GA} u(\delta\cdot\lambda)$ says the same as $\vartheta_k(u,\lambda)\in prov(G)$ since $u(\delta\cdot\lambda)$ is $\vartheta_k(u,\lambda)$. Together with G also $prov(G)$ is recursively enumerable. If u is kept fixed, then together with ϑ_k also the function $\vartheta_k(u,-)$ is primitive recursive. Hence the set characterized by the right side of the equivalence will be recursively enumerable as well.

THEOREM 2 Let L be in arithmetical situation, and let G be recursively enumerable. Let R^k be a relation on ω .

If R^k is G-formally defined by a formula $v(\xi)$ then R^k is recursively enumerable.

If R^k is G-represented by a formula $v(\xi)$ then R^k is recursive .

A function f^k with a G-representable graph is recursive .

Recall that R^k is G-formally defined by $v(\xi)$ if, and only if, $\lambda\in R^k$ is equivalent to $\vdash_G v(\kappa\cdot\lambda)$, where $\vdash_G$ indicates provability from G in the language L.

Fm(L) and Fm(L_A) being isomorphic, provability from L is equivalent to provability from G in L_A. As this isomorphism depends on that between T(L) and T(L_A), the sequence ξ of L-variables is mapped to a sequence $\psi = g_T\cdot\xi$ of L_A-variables, and the formula $v(\xi)$ is mapped to a formula $u(\psi)$. Also, with $\delta = g_T\cdot\kappa$ the constant terms κ_n are mapped to constant terms $\delta_n = g_T(\kappa_n)$ of L_A. Hence $v(\kappa\cdot\lambda)$ is mapped to $u(\delta\cdot\lambda)$, and $\vdash_G v(\kappa\cdot\lambda)$ becomes equivalent to $\vdash_{GA} u(\delta\cdot\lambda)$, the provability from G. Consequently, $\lambda\in R^k$ is equivalent to $\vdash_{GA} u(\delta\cdot\lambda)$.

Since δ is recursive, I conclude from Lemma 3 that R^k is recursively enumerable. If, moreover, R^k is G-representable by $v(\xi)$, then the consistency of G implies that R is G-formally defined by $v(\xi)$. Also, $-R^k$ then is G-representable by $\neg v(\xi)$. Consequently, both R^k and $-R^k$ are formally G-defined, and they both are recursively enumerable. Thus R^k is recursive. Finally, if $G(f^k)$ is recursive then f^k is recursive since **FRF** = **F**(**RFR**) by Theorem 6.4 .

2. Undefinability and Incompleteness

Let L be in arithmetical situation as above. If L_A is the arithmetized copy of L, then the isomorphisms g_T, g_F map an L-formula v to $\ulcorner v \urcorner$ and map an

L-formula $\mathrm{rep}(x_0,\kappa_r\,|\,v)$ to the L_A-formula $\mathrm{rep}(\ulcorner x_0\urcorner,\delta_r\,|\,\ulcorner v\urcorner)$ which, as a number, is $repf(\ulcorner x_0\urcorner,\delta(r),\ulcorner v\urcorner)$. Keeping x_0 fixed, together with *repf* and δ also the function $rep(r,k) = repf(\ulcorner x_0\urcorner,\delta(r),k)$ is recursive, and for $k = \ulcorner v\urcorner$, $v = [\![k]\!]$ there holds

$$rep(r,k) = repf(\ulcorner x_0\urcorner,\delta(r),k) = rep(\ulcorner x_0\urcorner,\delta_r\,|\,\ulcorner v\urcorner) = \ulcorner \mathrm{rep}(x_0,\kappa_r\,|\,v)\urcorner$$
$$= \ulcorner \mathrm{rep}(x_0,\kappa_r\,|\,[\![k]\!])\urcorner .$$

Thus the function *rep* is precisely the one denoted so in Chapter 4.

If L is in arithmetical situation, and if the set G of L-sentences is such that

(B_0) Every recursive relation is formally definable

then I shall say that L, G are *in arithmetical situation with* (B_0). Analogous expressions I shall use with respect to conditions to be introduced below.

LEMMA 4 Let L, G be in arithmetical situation with (B_0). Then *provs*(*G*) and *prov*(*G*) are not recursive, and thus G is consistent.

It suffices to consider *provs*(*G*). I cannot recur to the Fixpoint Lemma 4.6, because its proof requires functional representability of *rep*. I shall, however, imitate its argument and derive a contradiction from the assumption that the 1-ary relation $R = provs(G)$ is recursive.

If R were recursive then also $-R$ and $S = (-R)[rep\circ\langle p_0^1,p_0^1\rangle]$ would be so. Now n is in S if, and only if, $rep(n,n)$ is not in *provs*(*G*), and for $n = \ulcorner w\urcorner$ this means that $[\![rep(n,n)]\!] = \mathrm{rep}(x_0,\kappa_n\,|\,w) = w(\kappa_n)$ is not provable, i.e. that *not* $\vdash w(\kappa_n)$.

By (B_0), the recursive relation S is formally definable through a formula w such that $\vdash w(\kappa_n)$ if, and only if, $n\in S$. In particular, for $n = \ulcorner w\urcorner$ then

$$\vdash w(\gamma_w) \text{ if and only if } \ulcorner w\urcorner \in S \text{ if and only if } not \vdash w(\gamma_w) .$$

From Lemma 4 and Theorem 1 there now follows immediately the

THEOREM 3 Let L, G be in arithmetical situation with (B_0), and let G be recursively enumerable. Then

G is incomplete,

the theory of G is undecidable, and

prov(*G*) and *provs*(*G*) are examples of recursively enumerable sets which are not recursive.

For the remainder of this Section, an L-structure A shall always be such that $\equiv$ is interpreted by the identity, ω is a subset of u(A), and every L-term κ_n is interpreted in A by n .

Let Th(A) and Ths(A) be the sets of formulas and sentences true in A, and let *true*(A) and *trues*(A) be their images under g_F. Clearly Th(A) is both complete and consistent, and there hold *true*(A) = *prov*(*true*(A)) , *trues*(A) = *provs*(*true*(A)). If G is such that *provs*(*G*) = *trues*(A) then G is consistent and complete. Hence by Theorem 1, G cannot be recursively enumerable.

Moreover, for any G of which A is a model, also *provs*(*G*) $\subsetneq$ *trues*(A). Thus if G should be recursively enumerable then this inclusion must be proper. Consequently, I have the

COROLLARY 1 There is no recursively enumerable set G which axiomatizes *trues*(A) in the sense that *provs*(*G*) = *trues*(A). If G is recursively enumerable and A is model of G then there is a sentence true in A which is not G-provable .

Let me now compare these results with the consequences which can be drawn from the theorems of Chapter 4. Since the Fixpoint Lemma 4.6 requires the functional representability of the recursive function *rep*, I have to assume, for L and G, one of the the properties

(B_1) Every recursive relation is representable, and every recursive function is functionally representable.

(C) Every recursive relation is representable, and ω is provably initial: L contains a 2-ary predicate symbol $\leq$, and G proves, for every n in ω, the formulas

(MINn) $\vdash y \leq \kappa_n \rightarrow (y \equiv \kappa_0 \vee y \equiv \kappa_1 \vee \ldots \vee y \equiv \kappa_n)$,

(CMPn) $\vdash \neg\, y \leq \kappa_n \rightarrow \kappa_n \leq y$.

Observe that Theorem 2 implies the

OBSERVATION If L and G are in arithmetical situation with (B_1) and if G is recursively enumerable then the G-representable retions are precisely the recursive relations .

It follows from (B_1) that the function *rep* is representable. This was the hypothesis of the Undefinability Theorem 4.1, saying now:

If (B_1) holds then neither both of *provs*(*G*) and $-$*provs*(*G*), nor both of *prov*(*G*) and $-$*prov*(*G*) can be formally definable.

If G is recursive then so is *ded* = *ded*(*G*), hence representable. Choosing any representing formula DED(x_0,y), it follows that (B_1) secures the hypotheses of the Incompleteness Theorem 4.3, saying now:

> Let (B_1) hold and let G be recursive. If G is consistent then it is ω-incomplete; if G is ω-consistent then it is incomplete.

The condition (C), together with the recursiveness of the function *ne*, secures the hypotheses of ROSSER's Incompleteness Theorem 4.5, saying that if G is recursive and consistent then it is incomplete.

If A is a model of G then formally definable relations and functions (e.g. *rep*) will be ω-semantically definable in A. The further assumption that ω be semantically $\leq$-descending secures the hypotheses of Corollary 4.1, and if G is recursive then so is the relation *ded*(*G*). Thus the hypotheses of Theorem 4.2 are satisfied under the assumption (B_0), and I obtain the

COROLLARY 2 Let L and G are in arithmetical situation with (B_1). Let G be recursive, let A be a model of G, and let ω be semantically $\leq$-descending.

Then neither *true*(A) nor *trues*(A) is semantically definable in A.

If G is recursively enumerable then both *provs*(*G*) and *prov*(*G*) are proper subsets of *true*(A) and *trues*(A) respectively, and the fixpoint constructions of both Theorems 4.2 and 4.4 produce a sentence which is true in A, but is not provable.

Observe that here, in contrast to Corollary 1, an unprovable sentence is actually provided.

Under the hypotheses of this Corollary I can supplement the remark (4), made after the proof of Theorem 4.5, concerning sentences true but not provable. Theorem 4.2 produces a sentence s_0 true in A, but not provable from G = G_0. Then both $G_1 = G_0 \cup \{s_0\}$ and $G_0 \cup \{\neg s_0\}$ are recursive and consistent, hence both incomplete and with undecidable theories. While $G_0 \cup \{\neg s_0\}$ is ω-inconsistent, G_1 still has A as model. Repeating the construction, I obtain, for every $n \in \omega$, recursively enumerable sets G_n and sentences s_n, $G_{n+1} = G_n \cup \{s_n\}$, such that G_n has A as model, is incomplete and has an undecidable theory.

LEMMA 5 The function f with f(n) = $\ulcorner s_n \urcorner$ is recursive.

The sentences s_n were defined as follows. I started from the relation ded_n = *ded*(G_nA) of all <n,m> such that [m] is a deduction of [n] from G_n. I then used a formula $\mathrm{DED}_n(x_0,z)$ semantically defining ded_n and set $\mathrm{BEW}_n(x_0)= \exists z\, \mathrm{DED}_n(x_0,z)$. Then Theorem 4.2 produced s_n as fixpoint to $\neg\mathrm{BEW}_n(x_0)$, and by Lemma 4.7 this means that s_n is

$$\exists z\, (\mathrm{REP}'(\kappa_{g(n)}, \kappa_{g(n)} | z) \wedge \neg\mathrm{BEW}_n(z)) \quad \text{with} \quad g(n) = \ulcorner \exists z\, (\mathrm{REP}'(x_0, x_0 | z) \wedge \neg\mathrm{BEW}_n(z)) \urcorner$$

where $\mathrm{REP}'(x',y',z) = \forall z\, \mathrm{REP}(x',y',z') \rightarrow z \leq z')$ and $\mathrm{REP}(x',y',z)$semantically defines the relation $rep(r,k) = \ulcorner rep(x_0,\kappa_r | [k])\urcorner$.

Since $ded_n = ded(G_nA)$ arises from $dedu_n = dedu(G_nA)$, the formula $\mathrm{DED}_n(x_0,z)$ arises from a formula $\mathrm{DEDU}_n(z)$. Making use of the definition (S) in Chapter 7 of $m \in dedu(G)$, I can write down the formula $\mathrm{DEDU}_n(z)$ semantically defining $dedu_n$; I shall do so below. What matters, however, can be seen without knowing all details of $\mathrm{DEDU}_n(z)$, and it is this. For $n>0$ the deductions from G_n employ, besides logical axioms, in addition to the formulas from G also the sentences s_i with $i<n$. Thus the description of $dedu_n$ will (1) contain clauses that certain $\varepsilon(k,m)$ not only may be in *axiom*, but also may be the $f(i)=\ulcorner s_i \urcorner$ for $i<n$, and will (2) contain no further parts referring to any s_i. Hence $\mathrm{DEDU}_n(z)$ will (1) contain clauses with expressions $y \equiv \kappa_{f(i)}$ with variables y, and will (2) contain no further clauses containing references to the $f(i)$.

To simplify matters, let me now assume that I am working already with the arithmetization L_A isomorphic to L; thus $\ulcorner s_i \urcorner = s_i$, $f(i) = s_i$. Also $\mathrm{DEDU}_n(z)$ now is a number which, by the nature of arithmetization, depends recursively (elementarily) on the expressions from which it is built. Viewed as a number, $\kappa_{f(i)}$ is $\delta(f(i))$ with the recursive function δ, and so the number $\mathrm{DEDU}_n(z)$ is the value

$$\mathrm{DEDU}_n(z) = D(f(0), \ldots, f(n-1)) = D(H_e f(n-1))$$

of a recursive function D, applied to the history $H_e f$ of f. But $\mathrm{DEDU}_n(z)$ determines recursively the numbers $\mathrm{DED}_n(x,z)$, $\mathrm{BEW}_n(x)$ and

$$g(n) = \ulcorner \exists x\, (\mathrm{REP}'(x_0,x_0 | x) \wedge \neg\mathrm{BEW}_n(x)) \urcorner = \exists x\, (\mathrm{REP}'(x_0,x_0 | x) \wedge \neg\mathrm{BEW}_n(x)$$

such that also $g(n) = R_0(H_e f(n-1))$ with a recursive function R_0. Since $f(n)$ is $\exists x\, (\mathrm{REP}'(\kappa_{g(n)}, \kappa_{g(n)} | x) \wedge \neg\mathrm{BEW}_n(x))$, there follows

$$f(n) = rep(x_0, \kappa_{g(n)} | g(n)) = repf(x_0, \delta(g(n)) | g(n)) .$$

Since $repf$ is recursive in each of its three arguments, I obtain the number $repf(x_0, \delta(g(n)) | g(n)) = R_1(g(n))$ with a recursive function R_1. Hence

$$f(n) = R_1(R_0(H_e f(n-1))) = R(H_e f(n-1))$$

with a recursive function R. Thus f is defined by a course-of-values recursion. This concludes the proof.

In order to write down $\mathrm{DEDU}_n(z)$ I use that G and the basic notions for arithmetization are recursive. So there are L-formulas which in A semantically define the following subsets and relations:

GA(x)	for *G*	FORM(x)	for *form*
AXIOM(x)	for *axiom*	IMP(x,y,z)	for *imp*(a, b) = c
EXP(x,y,z)	for $\varepsilon(a,b) = c$	QUNIV(x,y,z)	for *quniv*(a,b) = c
LEN(x,y)	for $\|a\| = b$	DE(x,y,z)	for $\varepsilon(a,\varepsilon(0,b)) = c$

and DE(x,y,z) for *there exists* $d<b$ and $\varepsilon(0,b) = d$ and $\varepsilon(a,d) = c$. Expanding the definition (S) in Chapter 7 of *dedu*(*G*) and translating it, the formula $\mathrm{DEDU}_n(z)$ becomes

$$\begin{aligned}
&\mathrm{DE}(\kappa_0,z,\kappa_0) \wedge \mathrm{DE}(\kappa_1,z,\kappa_0) \wedge \mathrm{DE}(\kappa_2,z,\kappa_0) \wedge \mathrm{DE}(\kappa_3,z,\kappa_0)\\
&\qquad\qquad\qquad\qquad \wedge \mathrm{DE}(\kappa_4,z,\kappa_0) \wedge \mathrm{DE}(\kappa_5,z,\kappa_0))\\
&\wedge \mathrm{DE}(\kappa_6,z,\kappa_1) \wedge \forall u<z\ (\mathrm{EXP}(\kappa_0,z,u) \rightarrow \mathrm{LEN}(u,\kappa_7))\\
&\wedge \forall v<z\ (\mathrm{LEN}(z,v) \rightarrow\\
&\quad \forall w<v\ (\kappa_0<w \rightarrow\\
&\quad\ \forall u<z\ (\mathrm{EXP}(w,z,u) \rightarrow \mathrm{FORM}(u) \wedge\\
&\qquad (\ (\mathit{axiom}(u)\ \vee\ \mathrm{GA}(u)\ \vee\ u\equiv\kappa_{f(0)}\ \vee\ u\equiv\kappa_{f(1)}\ \vee \cdots \vee\ u\equiv\kappa_{f(n)}\ )\\
&\qquad\ \vee \exists x\,\exists y\ (\kappa_0<x<w \wedge \kappa_0<y<w \wedge \exists r<z\ \mathrm{EXP}(x,z,r)\\
&\qquad\qquad\qquad\qquad \wedge \exists s<z\ \mathrm{EXP}(y,z,s)\ \rightarrow\ \mathrm{IMP}(r,u,s)\ )\\
&\qquad\ \vee \exists x\,\exists y\ (\kappa_0<x<w \wedge y<z \wedge \exists r<z\ \mathrm{EXP}(x,z,r)\\
&\qquad\qquad\qquad\qquad \rightarrow\ \mathrm{QUNIV}(y,r,u)))\ )\ )\ ).
\end{aligned}$$

The function f being recursive, the set im(f) of all s_n is recursively enumerable. Thus the union $G_\omega = G_0 \cup \mathrm{im}(f)$ of all sets G_n is recursively enumerable, has A as model and, therefore, is still incomplete and has an undecidable theory. Consequently, my constructions may be repeated as long as there are *recursive* ordinals α, obtaining ever larger incomplete axiom systems G_α with undecidable theories, i.e. a transfinite recursive progression of such.

3. Radical Undecidability

In this section, I shall study how arithmetical situations behave under changes of the underlying language frame. Let (X) be one of the conditions (B_0), (B_1) or (C), and let (Y) be one of the conditions (B_0), (B_1).

(L0) Let G and F be two sets of L-sentences such that every sentence in G is F-provable. If L together with G is an arithmetical situation with (X), then so is L together with F. If F is consistent or incomplete then so is G.

Because if every sentence in G is F-provable then $\vdash_G v(\kappa\cdot\lambda)$, say, implies $\vdash_F v(\kappa\cdot\lambda)$. Thus representability of recursive relations as well as (MINn), (CMPn) are inherited from G to F.

From now on, I assume that

> L_1 is a language which is a sublanguage of a language L_2 (i.e. L_2 may have additional function symbols and predicate symbols),
>
> the signature Δ_2 of L_2 is recursively presented, and the signature Δ_1 of L_1, formed by restricting the signature of L_2, is also recursively presented: if $\Delta_{a1} = <n_i \mid i \epsilon I_{a1}>$, $\Delta_{b1} = <m_i \mid i \epsilon I_{b1}>$ and $\Delta_{a2} = <n_i \mid i \epsilon I_{a2}>$, $\Delta_{b2} = <m_i \mid i \epsilon I_{b2}>$ then I_{a1}, I_{b1} are recursive subsets of I_{a2}, I_{b2} respectively.

Then

> If L_{A2} with g_{T2}, g_{F2}, g_{S2} is a recursively arithmetized copy of L_2, then its sublanguage L_{A1} of signature Δ_1 is a recursively arithmetized copy of L_1 with $g_{T1} = g_{T2} \restriction T(L_1)$, $g_{F1} = g_{F2} \restriction Fm(L_1)$, $g_{S1} = g_{S2} \restriction Fm(L_1)^{\infty}$.

Because the terms and formulas of L_{A1} have the same definitions as those of L_{A2}, only with the additional requirement that i belong to the recursive subsets I_{a1} or in I_{b1} respectively.

I now present four more constructions which then will lead to a definition and two Theorems.

(L1) Let G_1 be a set of L_1-sentences. If L_1, G_1 is in arithmetical situation with (X) then so is L_2, G_1.

Because the recursive sequence of arithmetical constants with respect to L_{A1} remains such a sequence with respect to L_{A2}, and every L_1-derivation from G is also an L_2-derivation such that the provability of $\vdash_G v(\kappa \cdot \lambda)$ in L_1 still implies the same provability in L_2.

(L2) Let H be a set of L_2-sentences consisting of definitions, by terms and formulas of L_1, of the operation symbols and predicate symbols of L_2 which are not in L_1.

Let G_1 be a set of L_1-sentences, and let G_2 be $G_1 \cup H$. Let κ with $\delta = g_{T1} \cdot \kappa$ and τ with $\sigma = g_{T2} \cdot \tau$ be two recursive sequences of arithmetical constants for L_1 and for L_2, such that, for every $n \epsilon \omega$, there holds $\vdash_H \kappa(n) \equiv \tau(n)$.

Let L_2, σ, G_2 be in arithmetical situation with (Y) . Then so is L_1, δ, G_1 .

The definitorial extension through H provides, for any L_2-formula $v_2(\xi)$, an L_1-formula $v_1(\xi)$ with the same free variables such that $\vdash_H v_1(\xi) \longleftrightarrow v_2(\xi)$. By assumption, if $\lambda \epsilon \omega^n$ then for all $i<n$ also $\vdash_H \kappa \cdot \lambda(i) \equiv \tau \cdot \lambda(i)$ whence $\vdash_H v_1(\kappa \cdot \lambda) \longleftrightarrow v_2(\tau \cdot \lambda)$. Hence a provability $\vdash_{G2} v_2(\tau \cdot \lambda)$ implies the provability $\vdash_{G1} v_1(\tau \cdot \lambda)$.

(L3) Let L_1, L_2 and H be given as in (L2).

If $prov_2(G_2)$ or $provs_2(G_2)$ is recursively enumerable, or recursive, then so is $prov_1(G_1)$ respectively $provs_1(G_1)$.

If $prov_2(G_2)$ or $provs_2(G_2)$ is *not* recursive then $prov_1(G_1)$, respectively $provs_1(G_1)$, is *not* recursive.

The first statement will hold if I can show that $prov_1(G_1) = prov_2(G_2) \cap form_1$ and $provs_1(G_1) = provs_2(G_2) \cap sentence_1$. It follows from the properties of definitorial extensions that, for any L_1-formula v, $\vdash_{G2}$ v implies $\vdash_{G1}$ v where this deduction can be performed entirely in L_1 (this is obvious for cut-free sequential calculi and can be shown for modus ponens calculi by eliminating the defined function symbols and predicate symbols). Consequently, $<n,m> \epsilon\ dedu_2(G_2)$ and $n \epsilon\ form_1$ implies the existence of m' such that $<n,m'> \epsilon\ dedu_1(G_2)$ whence also $prov_2(G_2) \cap form_1$ is contained in $prov_1(G_1)$. The reverse inclusion is obvious.

In order to prove the second statement, observe that a map from $Fm(L_2)$ to $Fm(L_1)$, assigning to $v_2(\xi)$ a formula $v_1(\xi)$ with $\vdash_H v_1(\xi) \leftrightarrow v_2(\xi)$, can be defined by a recursive algorithm to which there corresponds a recursive function f sending $form_2$ into $form_1$. Thus $n \epsilon\ prov_2(G_2)$ if, and only if, $f(n) \epsilon\ prov_2(G_2)$, and since $f(n) \epsilon\ form_1$, this is equivalent to $f(n) \epsilon\ prov_1(G_1)$. This implies $\chi_1 \circ f = \chi_2$ for the characteristic functions χ_2, χ_1 of $prov_2(G_2)$, $prov_1(G_1)$, whence together with χ_1 also χ_2 would have to be recursive.

(L4) Let L_1, L_2 and H be given as in (L2), and let τ with $\sigma = g_{T2} \cdot \tau$ be a recursive sequence of arithmetical constants for L_2. Then I find a recursive sequence λ for L_1 with $\vdash_H \lambda(n) \equiv \tau(n)$ for every $n \epsilon \omega$.

Let R^2 be the relation of all $<n,d>$ such that $<n,d> \epsilon\ dedu_2(H)$ and d ends with a formula $\sigma(n) \equiv q_n$, q_n in $term_1$. Now the definitorial extension provides for any term t of L_2 a term t_f of L_1 such that $\vdash_H f(\xi) \equiv t_f(\xi)$, and the same holds for L_{A2} and L_{A1} with $\vdash_{HA}$. Consequently, there is, for every $n \epsilon \omega$, a term q_n in L_{A1} such that $\vdash_{HA} \sigma(n) \equiv q_n$, i.e. the relation R^2 is full. The last member of d being $d(len(d) \dot{-} 1) = m$, a characterization of m is

$$m \epsilon\ atom \text{ and } \varepsilon(2,m) = i_= \text{ and } \varepsilon(1,m) = \sigma(n) \text{ and } \varepsilon(2,m) \epsilon\ term_1$$

where $i_=$ in I_b shall be the index of the equality symbol. Hence R^2 is recursive, and so I may choose the δ for $\delta = g_{T1} \cdot \lambda$ as the recursive function μR^2.

I shall show now that a recursively enumerable axiom system G_0, in arithmetical situation for its language L_0 and some κ, and satisfying (B_0) or (B_1), can serve a a *root* for proofs of undecidability: a second recursively enumerable axiom system G_1 will have an undecidable theory if the axioms G_0 are provable in a definitorial expansion of G_1 - and then G_1 itself will be another root of this kind.

To make this precise, let L_0, L_1 be languages with equality and with recursively presented signatures Δ_0, Δ_1. Then a language L_2 with a recursively presented signature Δ_2 can be constructed such that L_2 contains L_0, L_1 as sublanguages and such that the Δ_{ak}, Δ_{bk}, $k = 0,1$, are restrictions of Δ_{a2}, Δ_{b2} to recursive subsets I_{ak} and I_{bk} of I_{a2} and I_{b2}. Given this situation, I state the

THEOREM 4 Assume that G_0, G_1 are recursively enumerable sets of L_1- and L_2-sentences such that

(a) there is a recursive sequence κ of arithmetical constants for L_0,

(b) there is an extension of G_1 to a set $G_2 = G_1 \cup H$ of L_2-sentences by a recursively enumerable set H of definitions such that the sentences of G_0 are G_2-provable,

(c) (Y) holds for G_0 with κ, where (Y) is one of (B_0), (B_1).

If G_1 is consistent then it is incomplete, its theory is undecidable, and $prov(G_1)$, $provs(G_1)$ are examples of recursively enumerable sets which are not recursive.

Morever, G_1 is in arithmetical situation for its language L_1 and a suitable λ, and it satisfies the same (Y) as does G_0.

It follows from the construction of L_2 that κ remains a recursive sequence of arithmetical constants for L_2. It follows from (L1) that (Y) with κ holds also for G_1 as a set of L_2-formulas, hence it follows from (L0) that (Y) with κ holds for G_2. Together with G_1 also its definitorial extension G_2 is consistent, and together with G_1 and H also G_2 is recursively enumerable. Thus Theorem 3 can be applied to G_2. Since G_2 actually contains G_1, it follows from (L0) that together with G_2 also G_1 is incomplete, and it follows from (L3) that together with $prov_2(G_2)$ also $prov_1(G_2)$ is not recursive.

As (Y) with κ holds for G_2, it follows from (L4) that I can find a recursive sequence λ of arithmetical constants for L_1 with which the hypotheses of (L2) are satisfied. It follows from (L2) that (Y) with λ holds for G_1.

In view of this theorem, a recursively enumerable axiom system G_0, in arithmetical situation with (B_0) or (B_1) for its language L_0 and some κ, shall be called *radically* (often also: *essentially*) *undecidable*. Under the assumptions of Theorem 4, therefore, together with G_0 also G_1 is radically undecidable.

Let L_0 be a sublanguage of L_2, both with recursively presented signatures Δ_0, Δ_2 and such that Δ_{a0}, Δ_{b0} are restrictions of Δ_{a2}, Δ_{b2} to recursive subsets I_{a0}, I_{b0} of I_{a2} and of I_{b2}. In this situation there holds the

COROLLARY 3 Let G_0 be a finite, radically undecidable set of L_0-sentences; let G_2 be a recursively enumerable set of L_2-sentences.

If $G_0 \cup G_2$ is consistent then the theory of G_2 is undecidable.

It follows from (L1), (L0) that $F = G_0 \cup G_2$ in L_2 is in arithmetical situation with (Y). Since F is consistent and recursively enumerable, its theory is undecidable. Since G_0 is finite, I may form the conjunction g_0 of formulas in G_0, such that provability from F is equivalent to provability from $F' = \{g_0\} \cup G_2$. But a sentence s is provable from this F' if, and only if, $g_0 \to s$ is provable from G_2. Making use of the [elementary] recursive function *imp* representing implication, this means that

$$\ulcorner s \urcorner \in provs_2(F) \text{ if and only if } imp(\ulcorner g_0 \urcorner, \ulcorner s \urcorner) \in provs_2(G_2) .$$

I now argue indirectly. Observe that $\ulcorner g_0 \urcorner$ is a fixed number. If $provs_2(G_2)$ were recursive then so would be the set of all x such that

$$x \in sentence_2 \text{ and } imp(\ulcorner g_0 \urcorner, x) \in provs_2(G_2)$$

which is $provs_2(F)$.

Obviously, before being able to make any applications of this Corollary, I will need at least one example of a finite, radically undecidable set G_0.

4. Uniformization

This section is devoted to the application, of arithmetized deducibility relations, to some more advanced properties recursive functions. They will not be used further in this book.

Throughout this section, I consider an arithmetical situation with L and a primitive recursive arithmetization L_A, and such that the injection δ is primitive recursive. I further assume a set G of L-formulas whose image G in L_A is primitive recursive. For every R^{k+2} and every a let R_a^{k+1} be the (k+1)-ary relation $R^{k+2}(a,-)$.

I shall begin by stating a weak *Uniformization Theorem* :

THEOREM 5 For every $k>0$ there is a primitive recursive relation H^{k+2} which parametrizes the functions f^k with G-formally definable graphs :

If a function f^k has a G-formally definable graph then there exists a number e such that

(i) the relation H_e^{k+1} is full
(ii) for every λ in ω^k : $f^k(\lambda) = cro_0 \circ \mu H_e^{k+1}(\lambda)$.

In particular, every function f^k arising this way is recursive.

I first translate the situation into the arithmetization L_A of L. Let x_0, x_1, ... be a fixed enumeration of the variables of L and let y_0, y_1, ... be the corresponding enumeration of the variables of L_A; set $\psi_k = \langle y_0, \dots, y_{k-1} \rangle$. Observe that for codes m of L-deductions β there holds $m = g_S(\beta) = g_A(\alpha)$ for $\alpha = g_F \cdot \beta$; keeping the notation $[\![m]\!]$ for β, I write $[\![m]\!]_A$ for the L_A-deduction α. Since G is primitive recursive, also the 2-ary relation $ded(G)$ is primitive recursive. Since δ is primitive recursive, the Uniformization Lemma 2, stated for the case k+1 instead of k, says that

For any sequence $\langle \psi_k, y_k \rangle$, I can find a (k+2)-ary primitive recursive function ϑ_{k+1} such that for all L_A-formulas $u = u(\psi_k, y_k)$:

for all a and all λ in ω^k : $\vdash_{GA} u(\delta \cdot \lambda, \delta_a)$ if, and only if, $\vartheta_{k+1}(u, \lambda, a) \in prov(G)$

i.e.

for all a and all λ in ω^k : $\vdash_{GA} u(\delta \cdot \lambda, \delta_a)$ if, and only if,
there is an m such that $\langle \vartheta_{k+1}(u, \lambda, a), m \rangle \in ded(G)$.

By an arithmetical trick I define the relations

$$Q^{k+3} = ded(G)\,[\,\vartheta_{k+1} \circ \langle p_0^{k+3}, \dots, p_{k+1}^{k+3} \rangle, p_{k+2}^{k+3}\,]$$
$$H^{k+2} = Q^{k+3}[\,p_0^{k+2}, \dots, p_k^{k+2}, cro_0 \circ p_{k+1}^{k+2}, cro_1 \circ p_{k+1}^{k+2}\,]$$

which together with $ded(G)$ and ϑ_{k+1} are primitive recursive. Their raison d'être is that now, for any u, the three statements

$$\langle \vartheta_{k+1}(u, \lambda, cro_0(p)), cro_1(p) \rangle \in ded(G)$$
(1) $$\langle u, \lambda, cro_0(p), cro_1(p) \rangle \in Q^{k+3}$$
$$\langle u, \lambda, p \rangle \in H^{k+2}$$

are equivalent, and for any formula $u = u(\psi_k, y_k)$ they are equivalent to

(2) $[\![cro_1(p)]\!]_A$ proves $\vdash_{GA} u(\delta \cdot \lambda, \delta_a)$ with $a = cro_0(p)$.

Having set the stage, consider a function f^k whose graph is G-formally defined. By Lemma 4.1, I may assume that this is done with a formula $v(\xi_k, x_k)$ such that $f^k(\lambda) = a$ is equivalent to $\vdash_G v(\kappa \cdot \lambda, \kappa_a)$. If $e = \ulcorner v(x_0, \xi_k) \urcorner = u(y_0, \psi_k)$ in L_A, then $f^k(\lambda) = a$ also implies $\vdash_{GA} u(\delta \cdot \lambda, \delta_a)$ with some proof $[\![m]\!]_A$. Hence

$$\langle e, \lambda, cau(a, m) \rangle \in H^{k+2} .$$

by (2). Consequently, H_e^{k+1} is full. Also by (1), and since $\langle u, \lambda, p \rangle$ is in H^{k+2} for every p, $[\![cro_1(p)]\!]$ proves $\vdash_G v(\kappa \cdot \lambda, \kappa_a)$ with $a = cro_0(p)$, and this implies $f^k(\lambda) = a$. In particular, this holds for $p = \mu H_e^{k+1}(\lambda)$ whence also $f^k(\lambda) = cro_0(\mu H_e^{k+1}(\lambda))$. Finally, H^{k+2} and H_e^{k+1} being (primitive) recursive, a function defined by (ii) will be μ-recursive under the hypothesis (i). This concludes the proof. – In particular, I obtain a weak form of KLEENE's Uniformization Theorem as

COROLLARY 4 In addition to the hypotheses of Theorem 5, assume G to be such that every recursive function has a G-formally definable graph.

For every $k>0$ there is a primitive recursive relation H^{k+2} such that a function f^k is recursive if, and only if, there exists a number e such that (i) and (ii) hold.

Once H^{k+2} has been found, the axiom system G does not appear in this parametrization of recursive functions, and various axioms systems satisfying the Corollary's hypothesis will be exhibited in the next Chapter.

A number e satisfying (i), (ii) is called an *h-index* of the function f^k, and the Corollary may be read as saying that every recursive function has an h-index. Observe that e is not at all unique: if $v(\xi,x)$ formally defines the graph of f^k then so do the iterated conjunctions of $v(\xi,x)$, determining an infinite sequence of h-indices of f^k.

On the other hand, not every number e occurs as an h-index of some f^k, because the condition (i) will be violated for at least one e. In order to see this, I argue indirectly: if (i) would hold for every e then the function

(U) $h^{k+1}(e,\lambda) = cro_0 \circ \mu H^{k+2}(e,\lambda)$

would be defined on all of ω^{k+1} and, formed by minimization, would be recursive. Thus the class **FRF** of recursive functions would contain h^{k+1} as a *universal function* for its k-ary functions: for every f^k in **FRF** there is an e such that $h^{k+1}(e,-) = f^k$. That this is impossible follows from the

LEMMA 6 A class **F** of functions cannot contain a function h^{k+1} which is universal for the **k**-ary functions in **F** if

(u) $g^k \in \mathbf{F}$ implies $csg \circ g^k \in \mathbf{F}$

(uu) $h^{k+1} \in \mathbf{F}$ implies $g^k \in \mathbf{F}$ for the function g^k which arises by contraction of identical 1st and 2nd arguments of h^{k+1}:

$$g^k(\lambda) = h^{k+1}(\lambda(0), \lambda) .$$

Because assume that h^{k+1} is universal and in **F**. Define g^k by (uu) and $f^k = csg \circ g^k$. Then g^k and f^k are in **F** such that f^k has an index e with respect to h^{k+1}. Hence for every λ

$$h^{k+1}(e,\lambda) = f^k(\lambda) = csg \circ g^k(\lambda) = csg \circ h^{k+1}(\lambda(0), \lambda) .$$

Choosing λ such that $\lambda(0) = e$, this implies the contradiction $h^{k+1}(e,\lambda) = csg \circ h^{k+1}(e, \lambda)$.

COROLLARY 5 For every $k>0$ there is a primitive recursive relation V^{k+2} which parametrizes the G-formally definable relations R^k in the following sense :

A relation R^k is G-formally definable if, and only if, there exists a number e such that $R^k = \exists V_e^{k+1}$.

In particular, if G is such that the recursive relations are G-formally definable then V^{k+2} parametrizes the recursive relations.

The Uniformization Lemma 2 shows that, depending on k, I can find a (k+1)-ary primitive recursive function ϑ_k such that, for every relation R^k, G-formally definable by an L_A-formula $u(\psi_k)$, $\vdash_{GA} u(\delta\cdot\lambda)$ is equivalent to $\vartheta_k(u, \lambda) \epsilon \mathit{prov}(G)$, i.e.

$\lambda \epsilon R^k$ if, and only if, there exists m such that $<\vartheta_k(u, \lambda), m> \epsilon \mathit{ded}(G)$

and

$\lambda \epsilon R^k$ if, and only if, there exists m such that $<u, \lambda, m> \epsilon Q^{k+2}$

where $Q^{k+2} = \mathit{ded}(G)\, [\vartheta_k \circ <p_0^{k+2}, \dots, p_k^{k+2}>,\ p_{k+1}^{k+2}]$ is primitive recursive. Together with Q^{k+2} also the relation V^{k+2} defined by

$<u, \lambda, m> \ \epsilon\ V^{k+2}$ if and only if $<u, \lambda, m> \ \epsilon\ Q^{k+2}$ and $u \ \epsilon\ \mathit{form}$ and
for all $z<u$: if $<z,u> \epsilon\ \mathit{occfree}$ then $z = y_0$ or ... or $z = y_{k-1}$

is primitive recursive, and there holds

$<u, \lambda, m> \ \epsilon\ V^{k+2}$ if and only if
u is a formula $u(\psi)$ and $[m]_A$ proves $\vdash_{GA} u(\delta\cdot\lambda)$.

Thus if R^k is G-formally definable by $v(\xi)$ and $e = \ulcorner v(\xi) \urcorner = u(\psi)$ then $\lambda \epsilon R^k$ implies $\vdash_G v(\kappa\cdot\lambda)$ with some proof $[m]$ whence $<\lambda, m> \epsilon V_e^{k+1}$ and $R^k \subseteq \exists V_e^{k+1}$.

Conversely, $\lambda \epsilon \exists V_f^{k+1}$ implies $<d, \lambda, m> \epsilon V^{k+2}$ where, by definition of V, d is some L_A-formula $w(\psi)$ and where, by definition of Q^{k+2}, $[m]_A$ proves $\vdash_{GA} w(\delta\cdot\lambda)$. Hence $\lambda \epsilon P^k$ for the relation P^k which is G-formally defined by both $w(\psi)$ and $v(\xi)$ with $d = \ulcorner v(\xi) \urcorner = u(\psi)$. Consequently, $\exists V_e^{k+1} \subseteq P^k$.

Starting from R^k and setting $d = e$, I find $R^k = P^k$ and thus $R^k = \exists V_e^{k+1}$. Starting from d and setting $e = d$, I find $P^k = R^k$ and thus $P^k = \exists V_e^{k+1}$

COROLLARY 6 Assume that the recursive relations are G-formally definable. For every $k>0$ there is a primitive recursive relation W^{k+2} which parametrizes the recursively enumerable relations R^k in the following sense:

A relation R^k is recursively enumerable if, and only if, there exists a number e such that $R^k = \exists W_e{}^{k+1}$.

Let Q^{k+3} be defined for recursive relations R^{k+1} as Q^{k+2} was defined in the proof of the last Corollary. Thus if R^{k+1} is G-formally defined by a formula $u = u(\psi_k, y_k)$ then

$$<\lambda, a> \epsilon R^{k+1} \text{ if, and only if, there exists m such that } <u, \lambda, m> \epsilon\, Q^{k+2}.$$

Define $W^{k+2} = Q^{k+3}[p_0^{k+2}, \ldots, p_k^{k+2}, cro_0 \circ p_{k+1}^{k+2}, cro_0 \circ p_{k+1}^{k+2}]$. Then W^{k+2} is primitive recursive, and every $\exists W_e{}^{k+1}$ is recursively enumerable. Conversely, if R^k is recursively enumerable then let R^{k+1} be recursive such that $R^k = \exists R^{k+1}$, and let R^{k+1} be G-formally defined by the L_A-formula e. Then R^k is $\exists W_e{}^{k+1}$ because $\lambda \epsilon R^k$ is equivalent to each of the statements

there exists a such that $<\lambda, a> \epsilon\, R^{k+1}$	since $R^k = \exists R^{k+1}$
there exist a and m such that $<e, \lambda, a, m> \epsilon\, V^{k+3}$	by (3)
there exists p such that $<e, \lambda, p> \epsilon\, W^{k+2}$	with $p = cau(a,m)$, $a = cro_0(p)$, $m = cro_1(p)$.

In Theorem 5 and its Corollaries, the uniformization relations H^{k+2}, V^{k+2} and W^{k+2} were primitive recursive, in consequence of the primitive recursiveness of the arithmetization L_A and of the set G, in particular of the functions *repf*, ϑ_1 and of the relation *dedu*(G). If an elementary arithmetization L_A is given and if G is assumed to be elementary (which will be the case for the examples to be studied in the following Chapter) then the uniformization relations also will be elementary. In its present form, Corollary 6 provides another proof for the statement **RER** = **ER(FPF)** contained in Theorem 6.5; if W^{k+2} can be chosen as elementary, then Corollary 6 also implies **RER** = **ER(FEF)**.

It follows from **FRF** = **F(RFR)** and **RFR** = **R(FRF)** in Theorem 6.4 that the recursive functions f are precisely those for which G(f) is recursive and that the recursive relations R are precisely those for which χ_R is recursive. But if $G(\chi_R)$ is G-formally defined by $v(\xi, x)$ then $\lambda \epsilon R^k$ is equivalent to $\vdash v(\lambda \cdot \kappa, \kappa_1)$ and thus R^k is G-formally defined by $v(\xi, \kappa_1)$. Consequently, G makes the graphs G(f) of recursive functions G-formally definable if, and only if, G makes the recursive relations G-formally definable: the hypotheses on G in Corollaries 4 and 6 are the same.

Returning to Theorem 5, the following observations can be made.

(a) If e is arbitrary, then the relation $H^{k+2}(e,-)$ may not be full. The equivalences (1) and (2) hold for any $u = u(\psi_k, y_k)$; thus $cro_0 \circ \mu H^{k+2}(u,-)$ will always be a *partial* function, assigning to λ the value a such that *cau*(a,m) is minimal for the property that $[m]_A$ proves $\vdash u(\kappa \cdot \lambda, \kappa_a)$. In particular, if the formula $u(\psi_k, y_k)$ G-formally defines the graph of a partial function g^k, then the number a is uniquely determined by λ, and thus $g^k = cro_0 \circ \mu H^{k+2}(u,-)$.

(b) Even if $cro_0 \circ \mu H^{k+2}(e,-)$ *is* a total, hence recursive function f^{k+1}, and if recursive functions happen to be G-formally definable, then the number e may be totally unrelated to a any L_A-formula formally defining f^{k+1} (and in Corollary 5, I avoided the analogous problem by narrowing the relation Q^{k+2} to V^{k+2}).

A partial function is called *partial recursive* if its graph is recursive.

Assume for the moment that G makes recursive relations G-formally definable. Then Theorem 5 and Corollary 4 remain valid for partial recursive functions. Further, it can be shown that, for any recursive R^{k+2}, also the partial function μR^{k+2} is partial recursive in the above sense (cf. e.g. FELSCHER 93, Theorem 23.2); consequently, for every $k>0$,

(U) $h^{k+1} = cro_0 \circ \mu H^{k+2}$

will now be a partial recursive function, universal for all partial recursive functions g^k. I thus obtain a *family* $\langle h^{k+1} \mid 0<k<\omega \rangle$ of universal functions.

In connection with the uses made of recursive functions, the reader may well be tempted consider the study of partial recursive function as a generalization for generalization's sake only. That this is not so may be seen from the surprising usefulness of the Corollary 7 below, in which I shall use partial recursive functions in order to show that the (total) ACKERMANN-PETER function is recursive. But it is correct that the true motivation for the investigation of partial functions, as well as its usefulness, becomes apparent only in the study of *Computability* which is beyond the scope of this book.

There are, in the context of computability, various other ways by which to construct families $\langle h^{k+1} \mid 0<k<\omega \rangle$ of functions universal for partial recursive functions. By a theorem of ROGERS any two such families are in a certain sense *isomorphic*, provided they satisfy the following *SMN-property* :

for all positive m, n there exists an (m+1)-ary recursive function S^m_n such that for every e and every α in ω^m there holds

(SMN) $h^{m+n}(e, \alpha, -) = h^n(S^m_n(e, \alpha), -)$.

For this reason, it obviously would be desirable to secure the SMN-property for a universal family of functions, defined in the present setting of arithmetizations. Unfortunately, the information contained in the h-indices e used for the relation H^{k+2} does not suffice to permit this task. However, a slight revision of Theorem 5, adding stronger hypotheses, will provide me with a universal family for which I can prove the so-called SMN-property. To this end, observe first that the set $form_k$ of all u with

$$u \in form \text{ and for all } z<u : \text{if } <z,u> \in occfree \text{ then } z = y_0 \text{ or } \ldots \text{ or } z = y_k$$

is [elementary] recursive (as was used already in the proof of Corollary 5).

THEOREM 6

Assume that L contains a predicate symbol $\leq$. Assume that G is consistent and that it proves

(INEQ) if $\vdash \kappa_n \equiv \kappa_m$ then $n = m$

(MINn) $\vdash x \leq \kappa_n \rightarrow (x \equiv \kappa_0 \vee x \equiv \kappa_1 \vee \ldots \vee x \equiv \kappa_n)$

(CMPn) $\vdash x \leq \kappa_n \vee \kappa_n \leq x$.

Assume that G is such that recursive relations are representable.

For every $k>0$ there is a primitive recursive relation K^{k+2} which parametrizes the partial recursive functions f^k :

For every partial recursive function f^k there exists a number e such that

$$\text{for every } \lambda \text{ in } \omega^k : f^k(\lambda) = cro_0 \circ \mu K_e^{k+1}(\lambda)$$

and every f^k arising in this way is (partial) recursive.

For every $k>0$, define a partial recursive function U^{k+1} by cases as

$$U^{k+1}(e,\lambda) = cro_0 \circ \mu K^{k+2}(e,\lambda) \quad \text{if } e \in form_k$$
$$U^{k+1}(e,\lambda) = \text{undefined} \qquad \text{otherwise} .$$

Then there is a primitive recursive function S^m_n with which the family $<U^{k+1} \mid 0<k<\omega>$ of universal functions has the SMN-property:

$$\text{for every e and every } \alpha : \quad U^{m+n}(e, \alpha, -) = U^n(S^m_n(e, \alpha), -) .$$

The first part of the proof repeats that of Theorem 5. However, instead of proofs of $\vartheta_{k+1}(u, \lambda, a) = u(\delta \cdot \lambda, \delta_a)$, I now shall consider proofs of the formula $u(\delta \cdot \lambda, y_{k+1}) \leftrightarrow y_{k+1} \equiv \delta_a$. This I obtain from $u = u(\psi_k, y_k)$ by sending it first into

$$\Psi_k(u) = u(\psi, y_{k+1}) \leftrightarrow y_{k+1} \equiv y_k ,$$

where k refers to the last variable in the sequence $<\psi_k, y_k>$ specified for u, and then applying the map

$$\phi_k(-, \lambda, a) = repf(y_k, \delta_a \mid repf(y_{k-1}, \delta_{\lambda(k-1)} \mid \ldots$$
$$\mid repf(y_1, \delta_{\lambda(1)} \mid repf(y_0, \delta_{\lambda(0)} \mid -))) \ldots)) .$$

Here Ψ_k is primitive recursive in view of the construction of my arithmetization, and together with *repf* and δ also $\phi_k(-,-,-)$ is primitive recursive. Thus $u(\delta\cdot\lambda, y_{k+1}) \leftrightarrow y_{k+1} \equiv \delta_a$ becomes a formula $\Theta_k(u, \lambda, a)$ with the primitive recursive function $\Theta_k = \phi_k(\Psi_k(-), -, -)$.

Instead of Q^{k+3}, therefore, I form the primitive recursive relations

$$^*Q^{k+3} = ded(G)\,[\Theta_{k+1}\circ\langle p_0^{k+3},\dots, p_{k+1}^{k+3}\rangle,\ p_{k+2}^{k+3}]$$

and

$$K^{k+2} = {}^*H^{k+2} = {}^*Q^{k+3}[p_0^{k+2},\dots, p_k^{k+2}, cro_0\circ p_{k+1}^{k+2}, cro_1\circ p_{k+1}^{k+2}]\ .$$

Again, for any u, the three statements

$$\begin{array}{ll} & \langle\Theta_{k+1}(u, \lambda, cro_0(p)), cro_1(p)\rangle \in ded(G) \\ (3) & \langle u, \lambda, cro_0(p), cro_1(p)\rangle \in {}^*Q^{k+3} \\ & \langle u, \lambda, p\rangle \in K^{k+2} \end{array}$$

are equivalent, and for any formula $u = u(\psi_k, y_k)$ they are equivalent to

(4p) $[cro_1(p)]_A$ proves $\vdash_{GA} u(\delta\cdot\lambda, y_{k+1}) \leftrightarrow y_{k+1} \equiv \delta_a$ with $a = cro_0(p)$.

For $u = u(\psi_k, y_k)$ then $\langle u, \lambda, q\rangle \in K^{k+2}$ is equivalent to

(4q) $[cro_1(q)]_A$ proves $\vdash_{GA} u(\delta\cdot\lambda, y_k) \leftrightarrow y_{k+1} \equiv \delta_b$ with $b = cro_0(q)$.

Since $\vdash_{GA} \delta_a \equiv \delta_a$, (4p) implies $\vdash_{GA} u(\delta\cdot\lambda, \delta_a)$. Hence (4p) together with (4q) implies $\vdash_{GA} \delta_a \equiv \delta_b$ whence by (INEQ)

(5) if $u = u(\psi_k, y_k)$ and $\langle u, \lambda, p\rangle \in K^{k+2}$ and $\langle u, \lambda, q\rangle \in K^{k+2}$ then $cro_0(p) = cro_0(q)$.

Consider now a partial function f^k whose graph is recursive. From (MINn), (CMPn) it follows by Lemma 4.5 that there is a formula $u = u(\psi_k, y_k)$ which represents $G(f^k)$ and functionally represents f^k. Thus $f^k(\lambda) = a$ implies $\vdash_{GA} u(\delta\cdot\lambda, y_k) \leftrightarrow y_k \equiv \delta_a$ with some proof $[m]_A$ such that (4p) holds with $p = cau(a,m)$. If $q = \mu K_u{}^{k+1}(\lambda)$ is the smallest q such that $\langle u,\lambda,q\rangle \in K^{k+2}$ then (5) implies $a = cro_0(p) = cro_0(q)$. But $cro_0(q) = cro_0(\mu K_u{}^{k+1}(\lambda))$ whence $f^k(\lambda) = a$ implies $cro_0(\mu K_u{}^{k+1}(\lambda)) = a$.

Conversely, $cro_0(\mu K_u{}^{k+1}(\lambda)) = a$ implies $\langle u,\lambda,p\rangle \in K^{k+2}$, $a = cro_0(p)$ for $p = \mu K_u{}^{k+1}(\lambda)$. Thus (4p) holds whence $\vdash_{GA} u(\delta\cdot\lambda, \delta_a)$. Now $u = u(\psi_k, y_k)$ represents $G(f^k)$, and as G was assumed to be consistent, u also G-formally defines $G(f^k)$. Hence $cro_0(\mu K_u{}^{k+1}(\lambda)) = a$, via $\vdash_{GA} u(\delta\cdot\lambda, \delta_a)$, implies $f^k(\lambda) = a$. Thus the functions f^k and $cro_0(\mu K_u{}^{k+1}(\lambda))$ are the same.

In the second part of the proof, I construct the function S_n^m. Let there be given $u = u(\psi_{m+n}, y_{m+n})$ and $\langle\alpha,\lambda,a\rangle$ in ω^{m+n+1}. The primitive recursive function

$$\phi_m(-, \alpha) = repf(y_{m-1}, \delta_{\alpha(m-1)}|\ \cdots\ |\ repf(y_0, \delta_{\alpha(0)}|-) \cdots)$$

sends

$$\begin{array}{lll} u & \text{into} & u(\delta\cdot\alpha, y_m, \dots, y_{m+n}) = v(y_m, \dots, y_{m+n})\text{ , and} \\ \Psi_{m+n}(u(\psi_{m+n}, y_{m+n})) & \text{into} & w(y_m, \dots, y_{m+n+1}) = \\ & & \quad v(y_m, \dots, y_{m+n-1}, y_{m+n+1}) \leftrightarrow y_{m+n+1} \equiv y_{m+n} \end{array}$$

whence

$$\begin{array}{ccc} u & \longrightarrow & \Psi_{m+n}(u) \\ \Big\downarrow \phi_m(-,\alpha) & & \Big\downarrow \phi_m(-,\alpha) \\ v & \longrightarrow & \Psi_{m+n}(v)=w \end{array}$$

commutes. The primitive recursive function $tot_{m+2n+3}(x,x\,|\,-)$ sends

$$\begin{array}{ll} v(y_m,\dots,y_{m+n}) & \text{into a formula } v'(y_m,\dots,y_{m+n})\text{ , and} \\ w(y_m,\dots,y_{m+n+1}) & \text{into } w'(y_m,\dots,y_{m+n+1}) = \\ & \quad v'(y_m,\dots, y_{m+n-1}, y_{m+n+1}) \leftrightarrow y_{m+n+1} \equiv y_{m+n} \end{array}$$

where the new formulas are provably equivalent to the old ones and do not contain bound variables from $y_0, \dots, y_{m+2n+3}$. The primitive recursive function $q_n^m(-)$ defined as

$$repf(y_{m+2n+3}, y_{n+1}\,|\; \dots\; repf(y_{m+n+2}, y_0\,|\; repf(y_{m+n+1}, y_{m+2n+3}\,|\; \dots \\ repf(y_m, y_{m+n+2}\,|\,-) \dots))\dots)$$

does not affect any bound variables in v' or w' and replaces $y_m, \dots, y_{m+n+1}$ by $y_0, \dots, y_{n+1}$; it thus sends

$$\begin{array}{ll} v'(y_m,\dots,y_{m+n}) & \text{into a formula } g(\psi_n, y_n) \\ w'(y_m,\dots,y_{m+n+1}) & \text{into a formula } h(\psi_n, y_n, y_{n+1}) = \\ & \quad g(\psi_n, y_{n+1}) \leftrightarrow y_{n+1} \equiv y_n . \end{array}$$

The function

$$\phi_n(-,\lambda,a,) = repf(y_n, a\,|\, repf(y_{n-1}, \delta_{\lambda(n-1)}\,|\; \dots\; repf(y_0, \delta_{\lambda(0)}\,|\,-) \dots))$$

sends

$$\begin{array}{ll} g(\psi_n, y_n) & \text{into } g(\delta\cdot\lambda, \delta_a)\text{ , and} \\ h(\psi_n, y_n, y_{n+1}) & \text{into } h(\delta\cdot\lambda, \delta_a, y_{n+1}) = g(\delta\cdot\lambda, y_{n+1}) \leftrightarrow y_{n+1} \equiv \delta_a \end{array}$$

which is $\Theta_n(g, \lambda, a)$. Hence again the diagram

$$\begin{array}{ccc} v' & \xrightarrow{\Psi_{m+n}} & w' \\ \Big\downarrow q_n^m & & \Big\downarrow q_n^m \\ g & \xrightarrow{\Psi_{m+n}} & h \\ \Big\downarrow \phi_n(-,\lambda,a) & & \Big\downarrow \phi_n(-,\lambda,a) \\ g(\delta\cdot\lambda,\ \nu_a) & & h(\delta\cdot\lambda,\ \delta_a,\ y_{n+1}) \end{array}$$

commutes. Here the composed map $\phi_n(-, \lambda, a)\circ q_n^m$ is

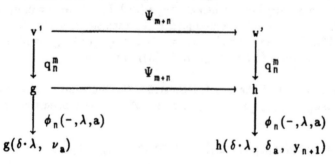

$$\begin{aligned}&\mathit{repf}(y_n, a \mid \mathit{repf}(y_{n-1}, \delta_{\lambda(n-1)} \mid \;\cdots\; \mathit{repf}(y_o, \delta_{\lambda(o)} \mid \\ &\qquad \mathit{repf}(y_{m+2n+3}, y_{n+1} \mid \;\cdots\; \mathit{repf}(y_{m+n+2}, y_o \mid \\ &\qquad \mathit{repf}(y_{m+n+1}, y_{m+2n+3} \mid \;\cdots\; \mathit{repf}(y_m, y_{m+n+2} \mid -) \cdots) \cdots)) \cdots)).\end{aligned}$$

None of the variables here is quantified in the argument $v'(y_m, \dots, y_{m+n})$; hence the functions *repf* act homomorphically also on quantified subformulas and act as term homomorphisms on the arguments of atomic subformulas. Thus I can verify directly that, for $m \leq i \leq m+n$, the replacements within the last map send

first y_{m+i} into $y_{m+i+n+2}$ on the third line (no value being an argument),
next $y_{m+i+n+2}$ into y_i on the second line (no value being an argument),
next y_i into $\delta_{\lambda(i)}$ for $i<n$, y_n into a, y_{n+1} into y_{n+1} on the first line.

Hence $g(\delta \cdot \lambda, \delta_a)$ and $h(\delta \cdot \lambda, \delta_a, y_{n+1})$ are also the images of the formulas $v'(y_m, \dots, y_{m+n})$ and $w'(y_m, \dots, y_{m+n+1})$ under the map

$$\phi_{m+n}(-, \lambda, a) = \mathit{repf}(y_{m+n}, a \mid \mathit{repf}(y_{m+n-1}, \delta_{\lambda(n-1)} \mid \;\cdots\; \mathit{repf}(y_m, \delta_{\lambda(o)} \mid -) \cdots)) ,$$

i.e.

$$\begin{array}{ccc} v' & \xrightarrow{\quad \Psi_{m+n} \quad} & w' \\ \Big\downarrow {\scriptstyle \phi_{m+n}(-,\lambda,\alpha)} & & \Big\downarrow {\scriptstyle \phi_{m+n}(-,\lambda,\alpha)} \\ g(\delta \cdot \lambda,\ \delta_a) & & h(\delta \cdot \alpha,\ \delta_a, y_{n+1}) \end{array}$$

Since v and v', as well as w and w', were provably equivalent, applications of $\phi_{m+n}(-, \lambda, a)$ lead to provably equivalent formulas again. But the application to $w(y_m, \dots, y_{m+n+1}) = \phi_m(\Psi_{m+n}(u), \alpha)$ produces

$$\begin{aligned}&\mathit{repf}(y_{m+n}, a \mid \mathit{repf}(y_{m+n-1}, \delta_{\lambda(n-1)} \mid \;\cdots\; \mathit{repf}(y_m, \delta_{\lambda(o)} \mid \phi_m(\Psi_{m+n}(u), \alpha)) \cdots)) \\ &= \mathit{repf}(y_{m+n}, a \mid \mathit{repf}(y_{m+n-1}, \delta_{\lambda(n-1)} \mid \;\cdots\; \mathit{repf}(y_m, \delta_{\lambda(o)} \mid \\ &\qquad \mathit{repf}(y_{m-1}, \delta_{\alpha(m-1)} \mid \;\cdots\; \mid \mathit{repf}(y_o, \delta_{\alpha(o)} \mid \Psi_{m+n}(u)) \cdots)) \cdots)) \\ &= u(\delta \cdot \alpha, \delta \cdot \lambda, y_{m+n+1}) \leftrightarrow y_{m+n+1} \equiv \delta_a .\end{aligned}$$

Consequently, $h(\delta \cdot \lambda, \delta_a, y_{n+1})$ and $\phi_{m+n}(\phi_m(\Psi_{m+n}(u), \alpha), \lambda, a)$ are provably equivalent, i.e.

(6a) $g(\delta \cdot \lambda, y_{n+1}) \leftrightarrow y_{n+1} \equiv \delta_a$ is provably equivalent to
$u(\delta \cdot \alpha, \delta \cdot \lambda, y_{m+n+1}) \leftrightarrow y_{m+n+1} \equiv \delta_a$.

Up to this point, I have done nothing but to repeat the proof of Lemma 4.1, though now with repeated singular replacements instead of simultaneous ones. However, I now have several primitive recursive functions, giving rise to the primitive recursive function $S^m_n(e, \alpha)$ defined as

$q^m_n \circ \mathit{tot}_{m+2n+3}(x, x \mid -) \circ \phi_m(e, \alpha)$	if $e \in \mathit{form}_{m+n}$
e'	with e' not in form_n otherwise

such that, for any $u = u(\psi_{m+n}, y_{m+n})$ and any $<\alpha,\lambda,a>$ in ω^{m+n+1}, there is a formula $S^m_n(u,\alpha) = g(\psi_n, y_n)$ with the property

(6b) $u(\delta\cdot\alpha,\ \delta\cdot\lambda,\ y_{m+n+1}) \longleftrightarrow y_{m+n+1} \equiv \delta_a$ and
$S^m_n(u,\alpha)(\delta\cdot\lambda,\ y_{n+1}) \longleftrightarrow y_{n+1} \equiv \delta_a$ are provably equivalent.

It remains to verify the SMN-property, and if e is not in $form_{m+n}$ then both of its sides give the nowhere defined function. So let now e be a u $form_{m+n}$, $u = (\psi_{m+n}, y_{m+n})$.

For the function U^{m+n}, the value $U^{m+n}(u,\ \alpha,\ \lambda) = a$ is defined if $a = cro_0(q)$ for the smallest q such that $<u,\ \alpha,\ \lambda,\ q> \ \epsilon\ K^{m+n+2}$. By (4p) this is equivalent to

(7q) $[cro_1(q)]_A$ proves $\vdash_{GA} u(\delta\cdot\alpha,\ \delta\cdot\lambda,\ y_{m+n+1}) \longleftrightarrow y_{m+n+1} \equiv \delta_a$
with $a = cro_0(q)$,

and (5) implies that then also $a = cro_0(p)$ for any p satisfying (7p).

For the function U^n, the value $U^{m+n}(S^m_n(u,\alpha),\ \lambda) = b$ is defined if $b = cro_0(r)$ for the smallest r such that $<S^m_n(u,\alpha),\ \lambda,\ r> \ \epsilon\ K^{m+n+2}$. By (4p) this is equivalent to

(8r) $[cro_1(r)]_A$ proves $\vdash_{GA} S^m_n(u,\alpha)\ (\delta\cdot\lambda,\ y_{n+1}) \longleftrightarrow y_{n+1} \equiv \delta_b$
with $b = cro_0(r)$,

and (5) implies that then also $b = cro_0(p)$ for any p satisfying (8p).

If (7q) holds then (6b) implies that $S^m_n(u,\alpha)(\delta\cdot\lambda,\ y_{n+1}) \longleftrightarrow y_{n+1} \equiv \delta_a$ has a proof $[m]_A$. Hence $p = cau(m,a)$ satisfies (8p) with $b = a$ whence also $a = cro_0(r)$ for the minimal r in (8r). Consequently, $U^{m+n}(u,\ \alpha,\ \lambda) = a$ implies $U^{m+n}(S^m_n(u,\alpha),\ \lambda) = a$.

If (8r) holds then (6b) implies that $u(\delta\cdot\alpha,\ \delta\cdot\lambda,\ y_{m+n+1}) \longleftrightarrow y_{m+n+1} \equiv \delta_a$ has a proof $[m]_A$. Hence $p = cau(m,b)$ satisfies (7p) with $a = b$ whence also $b = cro_0(q)$ for the minimal q in (7q). Consequently, $U^{m+n}(S^m_n(u,\alpha),\ \lambda) = b$ implies $U^{m+n}(u,\ \alpha,\ \lambda) = b$. This concludes the proof of Theorem 6.

The indices e with respect to the universal functions U^k I shall call *u-indices*. The following is KLEENE 's (second) Recursion Theorem :

COROLLARY 7 Assume the hypotheses of Theorem 6. If f^{k+1} is a partial recursive function then there exists an e such that $g^k = f^{k+1}(e,-)$ has itself the u-index e.

Making use of the function S^k_1, I define the recursive function h^{k+1} by

$$h(x_0,\ \xi) = f^{k+1}(S^k_1(x_0,\ x_0),\ \xi)\ .$$

If $U^{k+1}(q,-) = h^{k+1}$ then $h^{k+1}(q,-)$ is $U^k(S^k_1(q,q),-)$ by the SMN-property.

But $h^{k+1}(q,-) = f^{k+1}(S_1^k(q,q),-)$ by definition of h^{k+1}, and so I choose $e = S_1^k(q,q)$.

As an application of this Corollary, I shall show that (under the hypotheses of Theorem 6) the ACKERMANN-PETER function P mentioned at the end of Chapter 6 is recursive. I reformulate the equations defining P as

$$P(0,n) = n+1 \quad , \quad P(m,0) = P(m\dot{-}1,1) \text{ for } m>0 \quad ,$$
$$P(m,n) = P(m\dot{-}1,\ P(m,n\dot{-}1)) \text{ for } m>0,\ n>0 .$$

Mathematical induction shows immediately that P is a total function, uniquely determined by these equations. I define the primitive recursive function *ife*(x,y,z) satisfying

$$ife(x,y,z) = y \text{ for } x=0 \quad , \quad ife(x,y,z) = z \text{ for } x\neq 0$$

by $ife(0,y,z) = p_0^2(y,z)$, $ife(n+1,y,z) = p_1^4(y,z,n,ife(n,y,z))$. Then the equations defining P are equivalent to the one recursion equation

$$P(y,z) = ife(y,\ z+1,\ ife(z,\ P(y\dot{-}1,1),\ P(y\dot{-}1,\ P(y,\ z\dot{-}1))))\ .$$

Assume, for the moment, that P is indeed recursive. Then P would have a u-index x such that $P(y,z) = U^2(x,y,z)$, and thus the right side of my recursion equation would have the form

$$h(x,y,z) = ife(y,\ z+1,\ ife(z,\ U^2(x,\ y\dot{-}1,1),\ U^2(x,\ y\dot{-}1,\ U^2(x,y,\ z\dot{-}1))))\ .$$

Forgetting about my assumption, h^3 *is* a partial recursive function, and now Corollary 7 shows that there is a u-index e such that $U^2(e,y,z)$ is $h(e,y,z)$. Consequently

$$U^2(e,y,z) = ife(y,\ z+1,\ ife(z,\ U^2(e,\ y\dot{-}1,1),\ U^2(e,\ y\dot{-}1,\ U^2(e,y,\ z\dot{-}1))))$$

whenever $U^2(e,-,-)$ is defined. Hence $U^2(e,-,-)$ is a partial recursive solution of the recursion equation for P. Since P is the unique solution of that equation, it follows that $U^2(e,-,-)$ is P , and thus P is recursive.

Chapter 9. Axioms for Arithmetic

In this Chapter I shall consider three axioms systems, PC, PB and PQ. The first two employ, for every natural number n, a constant κ_n, and the axioms of PB, say, express essentially that $\kappa_n + \kappa_m$ is κ_{n+m} and that $\kappa_n \cdot \kappa_m$ is κ_{nm}. So PB employs 2-ary terms formed with + and · , while PC expresses the same facts with help of 3-ary predicates. Clearly, both PC and PB have infinitely many axioms. The system PQ, introduced by R.M.ROBINSON 50, uses only seven axioms which, instead of mirroring the tables of addition and multiplication, express their recursive definitions with help of a constant κ_0 and a function symbol s for successor. All three axiom systems turn out to be radically undecidable. Being finite, PQ then permits to apply an idea of CHURCH 36 in order to show that also the empty axiom system, i.e. elementary quantifier logic, is undecidable.

On the way to the undecidability results, more insights are be obtained on the arithmetical relations describable under the axioms systems. Under PC, the relations of the class **RAB**, basic for the developments in Chapter 6, become those represented by *bounded* formulas, i.e. in which quantifiers $\forall x$ and $\exists x$ are applied only to formulas of the form $x \leq y \rightarrow v$ and $x \leq y \wedge v$ respectively, where v is either atomic or a propositional combination of formulas bounded already. Under PB the situation is analogous, only that now the variable x may be bounded by a term t instead of a variable y. In Section 3 the representability of recursive relations, both for PC and PB, is proved by a rapid recourse upon the recursion-free characterisation of recursive relations in Corollary 6.6. In Section 4 it is shown that, under a weak consistency hypothesis, recursive relations are represented by Σ_1-formulas and that the recursively enumerable relations are those formally defined by such formulas. Section 5 gives a short glance at formal definability by Diophantine formulas.

References

A. Church: A note on the Entscheidungsproblem. J.Symb.Logic 1 (1936) 40–41 and 101–102

R.M. Robinson: An essentially undecidable axiom system. Proc.Int.Congr.of Mathematicians 1950. Vol. 1 (1952) 729–730

A. Tarski, A. Mostowski, R.M.Robinson: Undecidable Theories. Amsterdam 1953

1. The Arithmetic PC

Consider an equality language L_{1o} which contains

for every n in ω a constant κ_n,
two 3-ary predicate symbols *plu* and *tim* (in addition to $\equiv$).

Let PC_0 be the axiom system

(PC0) $plu(\kappa_n, \kappa_m, z) \longleftrightarrow z \equiv \kappa_{n+m}$

(PC1) $tim(\kappa_n, \kappa_m, z) \longleftrightarrow z \equiv \kappa_{nm}$

(PC2) $\neg\, \kappa_n \equiv \kappa_m$ for $n \neq m$.

Provability shall refer to provability from PC_0 .

Let A_{1o} be the L_{1o}-structure over ω in which the graphs $G(+)$ and $G(\cdot)$ interpret *plu* and *tim*, and in which the numbers n interpret the constants κ_n. Then (PC0), (PC1) express that the functions $+$ and $\cdot$ are functionally represented by *plu*(x,y,z) and *tim*(x,y,z), and in view of (PC2) their graphs are represented by these formulas. By equality logic and (PC2), the identity E^2 on ω is represented by the formula $x \equiv x$.

I begin with two observations the usefulness of which will become clear in a moment. Recall that, for a sequence ξ of pairwise different variables, $v(\xi)$ denotes a formula having its free variables among those in ξ . In what follows, I set $\xi = \langle x_0, \ldots, x_{k-1} \rangle$ and $\psi = \langle y_0, \ldots, y_{m-1} \rangle$ (without assumptions about the relationships between $im(\xi)$ and $im(\psi)$).

(1) Let $v(\xi)$ represent R^k and let $u(\xi)$ represent S^k. Then $v(\xi) \wedge u(\xi)$, $v(\xi) \vee u(\xi)$, $\neg\, v(\xi)$ represent $R^k \cap S^k$, $R^k \cup S^k$, $-R^k$ respectively .

Consider, for instance, $v(\xi) \wedge u(\xi)$. Then $\lambda \epsilon R^k \cap S^k$ implies both $\vdash v(\kappa \cdot \lambda)$ and $\vdash u(\kappa \cdot \lambda)$ whence $\vdash v(\kappa \cdot \lambda) \wedge u(\kappa \cdot \lambda)$. And *not* $\lambda \epsilon R^k \cap S^k$ implies $\vdash \neg v(\kappa \cdot \lambda)$ or $\vdash \neg u(\kappa \cdot \lambda)$ whence $\vdash \neg v(\kappa \cdot \lambda) \vee \neg u(\kappa \cdot \lambda)$ and $\vdash \neg(v(\kappa \cdot \lambda) \wedge u(\kappa \cdot \lambda))$. Observe that classical logic is used here.

(2) Let k be decomposed into two disjoint subsets I and J and let π be a map sending I into m and J into ω. Let η be defined for the variables x_i, $i<k$, by $\eta(x_i) = y_{\pi(i)}$ for $i \epsilon I$ and $\eta(x_i) = \kappa_{\pi(i)}$ for $i \epsilon J$; let $g^m_{\pi(i)}$ be $p^m_{\pi(i)}$ for $i \epsilon I$ and $c^m_{\pi(i)}$ for $i \epsilon J$. Let $v(\xi)$ represent R^k and let w be the formula $sub(\eta \mid v)$.

Then $w = w(\psi)$ represents $S^m = R^k[g^m_{\pi(o)}, \ldots, g^m_{\pi(k-1)}]$.

Observe that always $\vdash v \longleftrightarrow tot(\chi \mid v)$. Thus I may assume that v does not contain bound variables from $im(\psi)$ whence $w = rep(\eta \mid v)$. As $v(\xi)$ represents R^k, I know that $\beta \epsilon R^k$ implies $\vdash v(\kappa \cdot \beta)$ and that $\beta \epsilon - R^k$ implies $\vdash \neg v(\kappa \cdot \beta)$. For $\beta \epsilon \omega^k$ and $\alpha \epsilon \omega^m$ there holds by definition

$v(\kappa\cdot\beta) = rep(\vartheta_{\kappa\beta} \mid v)$ for the map $\vartheta_{\kappa\beta}$ such that $\vartheta_{\kappa\beta}(x_i) = \kappa_{\beta(i)}$

$w(\kappa\cdot\alpha) = rep(\vartheta_{\kappa\alpha} \mid w)$ for the map $\vartheta_{\kappa\alpha}$ such that $\vartheta_{\kappa\alpha}(y_j) = \kappa_{\alpha(j)}$.

For $\alpha\epsilon\omega^m$ I define $\gamma\epsilon\omega^k$ by $\gamma(i) = \alpha_{\pi(i)}$ for $i\epsilon I$ and $\gamma(i) = \pi(i)$ for $i\epsilon J$. Again by definition

$v(\kappa\cdot\gamma) = rep(\vartheta_{\kappa\gamma} \mid v)$ for the map $\vartheta_{\kappa\gamma}$ such that $\vartheta_{\kappa\gamma}(x_i) = \kappa_{\gamma(i)}$.

By definition of w

$$w(\kappa\cdot\alpha) = rep(\vartheta_{\kappa\alpha} \mid rep(\eta \mid v)) = rep(\chi \mid v)$$

for the map χ such that $\chi(x_i) = \vartheta_{\kappa\alpha}(y_{\pi(i)}) = \kappa_{\alpha\pi(i)} = \kappa_{\gamma(i)}$ for $i\epsilon I$ and $\chi(x_i) = \vartheta_{\kappa\alpha}(\kappa_{\pi(i)}) = \kappa_{\pi(i)} = \kappa_{\gamma(i)}$ for $i\epsilon J$. Hence χ and $\vartheta_{\kappa\gamma}$ coincide on $im(\xi)$, whence $w(\kappa\cdot\alpha) = v(\kappa\cdot\gamma)$.

By definition of S^m, there holds $\alpha\epsilon S^m$ if, and only if, $\gamma\epsilon R^k$. Thus $\alpha\epsilon S^m$ implies $\gamma\epsilon R^k$, whence $\vdash v(\kappa\cdot\gamma)$ and $\vdash w(\kappa\cdot\alpha)$. And not $\alpha\epsilon S^m$ implies not $\gamma\epsilon R^k$, whence $\vdash \neg v(\kappa\cdot\gamma)$ and $\vdash \neg w(\kappa\cdot\alpha)$.

While a relation R^k, formally defined by a formula $v(\xi)$, is uniquely determined by that formula, the situation is slightly different for a relation represented by a formula. For example, *plu*(x,y,z) represents G(+), but if a relation R^3 is given as represented by *plu*(x,y,z), then in order to conclude from $\lambda\epsilon R^3$, hence $\vdash$ *plu*$(\kappa\cdot\lambda)$, upon $\lambda\epsilon G(+)$, I would have to know that *not* $\vdash \neg$ *plu*$(\kappa\cdot\lambda)$, because then *not* $\lambda\epsilon -G(+)$. Uniqueness of the relations, represented by formulas $v(\xi)$ of some type, could therefore be secured by assuming PC_0 to be consistent with respect to the sentences $v(\kappa\cdot\lambda)$ arising from these $v(\xi)$.

Rather than adding such assumptions (which, for open formulas $v(\xi)$, actually could be proved, though by means somewhat stronger than those offered by PC_0), I shall define the relations *canonically represented* by open L_{10}-formulas by saying

(RC0) The relations G(+), G(·), E^2 are canonically represented by the atomic formulas *plu*(x,y,z), *tim*(x,y,z) and $x \equiv y$.

(RC1) If in (1) the relations R^k and S^k are canonically represented by the formulas $v(\xi)$, $u(\xi)$ then so are $R^k \cap S^k$, $R^k \cup S^k$, $-R^k$ by $v(\xi) \wedge u(\xi)$, $v(\xi) \vee u(\xi)$, $\neg v(\xi)$.

(RC2) If in (2) the relation R^k is canonically represented by $v(\xi)$ then so is S^m by $w(\psi)$.

It follows from (1) and (2) that these canonically represented relations are actually represented by the respective formulas.

An *open sentence* shall be a sentence arising from an open formula by replacing its free variables with constants.

Let $\mathbf{RAB}_0$ be the smallest class of relations which contains the identity E^2

and the graphs of both + and $\cdot$, and which is Boolean closed and closed under superpositions with projections and constant functions.

LEMMA 1 Every open L_{10}-formula canonically represents a relation in $\mathbf{RAB}_0$.

For every relation R in $\mathbf{RAB}_0$, I can find an open L_{10}-formula which canonically represents R .

PC_0 is *complete* for open sentences s: there holds $\vdash s$ or $\vdash \neg s$.

By (RC0), the atomic formulas without constants canonically represent relations from $\mathbf{RAB}_0$, and it then follows from (1) and (2) by induction on formulas that any open L_{10}-formula canonically represents a relation which, by definition of $\mathbf{RAB}_0$, still belongs to that class. Conversely, the initial relations of $\mathbf{RAB}_0$ are canonically represented by atomic formulas, and induction on the construction of $\mathbf{RAB}_0$ shows how to find canonically representing formulas for newly formed relations, by making use of (1) and (2).

Consider an open sentence s of the form $\kappa_a \equiv \kappa_b$. Since $x \equiv x$ represents E^2, I have $\vdash s$ if a = b and $\vdash \neg s$ otherwise. Consider $s = \mathit{plu}(\kappa_a,\kappa_b,\kappa_c)$. Since $\mathit{plu}(x,y,z)$ represents + , I have $\vdash s$ if c = a+b and $\vdash \neg s$ otherwise. The case of $\mathit{tim}(x,y,z)$ is analogous. If ($\vdash s$ or $\vdash \neg s$) then also ($\vdash \neg s$ or $\vdash \neg\neg s$) by classical logic. If ($\vdash s_0$ and $\vdash s_1$) then $\vdash s_0 \wedge s_1$; if not ($\vdash s_0$ and $\vdash s_1$) then ($\vdash \neg s_0$ or $\vdash \neg s_1$) by inductive hypothesis, hence $\vdash \neg s_0 \vee \neg s_1$ and $\vdash \neg(s_0 \wedge s_1)$. The case of $s_0 \vee s_1$ is analogous. This concludes the proof of Lemma 1 .

The first two statements of Lemma 1 may be expressed by saying that the relations in $\mathbf{RAB}_0$ are those which are canonically representable by open formulas of L_{10}. If PC_0 is assumed to be consistent for open sentences then canonical representability and representability coincide for open formulas.

I now embed my language L_{10} into a language L_1 containing a further predicate symbol $\lesssim$.

Let PC be the axiom system consisting of PC_0 together with

(PC3) $x \lesssim \kappa_n \rightarrow (x \equiv \kappa_0 \vee x \equiv \kappa_1 \vee \ldots \vee x \equiv \kappa_n)$

(PC4) $x \lesssim \kappa_n \vee \kappa_n \lesssim x$.

Provability shall refer to provability from PC. There then holds

(PC5) If $n \lesssim m$ then $\vdash \kappa_n \lesssim \kappa_m$, if not $n \lesssim m$ then $\vdash \neg \kappa_n \lesssim \kappa_m$.

For n = m the first statement follows from (PC4). If n<m then

for every $i \leq n : \vdash \neg \kappa_m \equiv \kappa_i$ — by (PC2)
$\vdash (\neg \kappa_m \equiv \kappa_0 \wedge \neg \kappa_m \equiv \kappa_1 \wedge \cdots \neg \kappa_m \equiv \kappa_n)$
$\vdash \neg (\kappa_m \equiv \kappa_0 \vee \kappa_m \equiv \kappa_1 \vee \cdots \vee \kappa_m \equiv \kappa_n)$
$\vdash \neg \kappa_m \leq \kappa_n$ — by (PC3)
$\vdash \kappa_n \leq \kappa_m$ — by (PC4) .

If not $n \leq m$ then $m < n$, and thus by the above $\vdash \neg \kappa_n \leq \kappa_m$.

Let A_1 be the L_1–structure arising from A_{10} by interpreting the predicate $\leq$ through the usual relation $\leq^A$ on ω. It follows from (PC5) that $x \leq y$ represents the relation $\leq^A$. – I now extend the my previous observations by

(3) Let $v(\xi,x)$ represent R^{k+1}, and let y be distinct from x.

If y does not occur in ξ then $u(\xi,y) = \forall x\, (x \leq y \rightarrow v(\xi,x))$ represents $\forall_{\leq} R^{k+1}$.

If y occurs in ξ as x_i then $u(\xi,y)$ represents $\forall^i_{\leq} R^{k+1}$.

Assume first that y does not occur in ξ . Then $<\alpha,a> \epsilon \forall_{\leq} R^{k+1}$ implies $<\alpha,b> \epsilon R^{k+1}$ for every $b \leq a$. Hence

(3a) for every $b \leq a : \vdash v(\kappa \cdot \alpha, \kappa_b)$ — assumption
$\vdash v(\kappa \cdot \alpha, z) \wedge x \equiv z \rightarrow v(\kappa \cdot \alpha, x)$ — equality logic
for every $b \leq a : \vdash v(\kappa \cdot \alpha, \kappa_b) \wedge x \equiv \kappa_b \rightarrow v(\kappa \cdot \alpha, x)$
for every $b \leq a : \vdash x \equiv \kappa_b \rightarrow v(\kappa \cdot \alpha, x)$ — using the assumption
$\vdash (x \equiv \kappa_0 \rightarrow v(\kappa \cdot \alpha,x)) \wedge (x \equiv \kappa_1 \rightarrow v(\kappa \cdot \alpha,x)) \wedge \ldots$
$\wedge (x \equiv \kappa_a \rightarrow v(\kappa \cdot \alpha,x))$
$\vdash (x \equiv \kappa_0 \vee x \equiv \kappa_1 \vee \ldots x \equiv \kappa_a) \rightarrow v(\kappa \cdot \alpha, x)$
$\vdash x \leq \kappa_a \rightarrow v(\kappa \cdot \alpha, x)$ — making use of (PC3)
$\vdash \forall x\, (x \leq \kappa_a \rightarrow v(\kappa \cdot \alpha, x))$
$\vdash u(\kappa \cdot \alpha, \kappa_a)$ — since y is not in $v(\xi,x)$

And not $<\alpha,a> \epsilon \forall_{\leq} R^{k+1}$ implies

(3b) for some b such that $b \leq a : \vdash \neg v(\kappa \cdot \alpha, \kappa_b)$
$\vdash \kappa_b \leq \kappa_a \wedge \neg v(\kappa \cdot \alpha, \kappa_b)$ — making use of (PC5)
$\vdash \exists x\, (x \leq \kappa_a \wedge \neg v(\kappa \cdot \alpha,x))$
$\vdash \neg \forall x\, (x \leq \kappa_a \rightarrow v(\kappa \cdot \alpha,x))$.

The case that y occurs in ξ is analogous.

Observe that if $v(\xi,x)$ represents R^{k+1} and if x does not occur in v then (d) $\forall_{\leq} R^{k+1} = R^{k+1}$, (dd) $v(\xi)$ represents $S^k = \exists R^{k+1}$ *and* $R^{k+1} = S^k[p_0^{k+1}, \ldots, p_{k-1}^{k+1}]$, (ddd) $\forall^i_{\leq} R^{k+1} = S^k$.

Recall that **RAB** is the smallest AB–closed class of relations, i.e. the smallest class which contains the graphs of + and · and is trivially closed, Boolean closed and closed under bounded quantifications $\forall_{\leq}$.

I define the set of *bounded formulas* of L_1 as follows:

(FC0) The atomic formulas of L_1 are bounded formulas .

(FC1) Propositional compositions of bounded formulas are bounded .

(FC2) If v is bounded then for all x, y with $x \neq y$ also $\forall x\,(x \leq y \rightarrow v)$ and $\exists x\,(x \leq y \wedge v)$ are bounded .

I define relations *canonically represented* by bounded L_1-formulas by adding to (RC0) that $\leq^A$ be canonically represented by the formula $x \leq y$, keeping (RC1), (RC2), and adding

(RC3) If the relation R^{k+1} is canonically represented by $v(\xi,x)$ and if $y \neq x$, then, depending on whether y does not occur in ξ or does,

$\forall_{\leq} R^{k+1}$, respectively $\forall^i_{\leq} R^{k+1}$, is canonically represented by $\forall x(x \leq y \rightarrow v(\xi,x))$,

and $-\forall_{\leq} - R^{k+1}$, respectively $-\forall^i_{\leq} - R^{k+1}$, is canonically represented by $\exists x(x \leq y \wedge v(\xi,x))$.

Again, canonically represented relations are actually represented because $\exists x(x \leq y \wedge v(\xi,x))$ is provably equivalent to $\neg\forall x\,(x \leq y \wedge \neg v(\xi,x))$, which by (3) and (2) represents $-\forall_{\leq} - R^{k+1}$ or $-\forall^i_{\leq} - R^{k+1}$.

THEOREM 1 Every bounded L_1-formula canonically represents a relation in **RAB** .

For every relation R in **RAB**, I can find a bounded L_1-formula which canonically represents R .

PC is *complete* for bounded sentences s : there holds $\vdash s$ or $\vdash \neg s$.

The proof of the first two statements is a straightforward extension of that of Lemma 1, making use of the fact that E^2 and $\leq^A$ are in **RAB** by 5.(AB3) and that together with R^{k+1} also $\forall_{\leq} R^{k+1}$ and $\forall^i_{\leq} R^{k+1}$ are in **RAB**. - As for bounded sentences, the open ones can be handled as in Lemma 1. Consider now a sentence s arising from $\forall x\,(x \leq y \rightarrow v(\xi,x))$ as $\forall x\,(x \leq \kappa_a \rightarrow v(\kappa \cdot \alpha,\, x)$ (and here it does not matter whether y is in ξ , i.e. κ_a is some $\kappa_{\alpha(i)}$, or not). Making the assumption that for every $b \leq a$ there holds $\vdash v(\kappa \cdot \alpha,\, \kappa_b)$, I arrive, as in (3a), at the provability of s . If the assumption does not hold, then there is some b such that $b \leq a$ and not $\vdash v(\kappa \cdot \alpha,\, \kappa_b)$. Hence $\vdash \neg v(\kappa \cdot \alpha,\, \kappa_b)$ by inductive hypothesis, and I arrive as in (3b) at the provability of $\neg s$.

The first two statements of Theorem 1 may be expressed by saying that the relations in **RAB** are those which are canonically representable by bounded formulas of L_1. If PC is assumed to be consistent for bounded sentences then canonical representability and representability coincide for bounded formulas .

2. The Arithmetic PB

I consider an equality language L_{2o} which contains

for every n in ω a constant κ_n ,
the operation symbols + and $\cdot$,
the predicate symbol $\equiv$.

Let PB_0 be the axiom system

(PB0) $\kappa_n+\kappa_m \equiv \kappa_{n+m}$

(PB1) $\kappa_n\cdot\kappa_m \equiv \kappa_{nm}$

(PB2) $\neg\,\kappa_n \equiv \kappa_m$ for $n\neq m$

such that (PB2) coincides with (PC2). Provability shall refer to provability from PB_0.

Let A_{2o} be the obvious L_{2o}-structure over ω. Then the operations + and $\cdot$ are represented by the terms x+y and x$\cdot$y; hence their graphs are functionally represented by the formulas $x+y \equiv z$ and $x\cdot y \equiv z$. In view of (PB2), the graphs then are represented by these formulas.

I recall that the class POF of *polynomial functions* was defined as the smallest class which contains + , $\cdot$, the projections and the constant functions c^k_n, and which is closed under superposition.

I consider again term objects $t(\xi)$ where t is a term and $\xi = <x_0,\dots, x_{k-1}>$ is such that $occ(t) \subseteq im(\xi)$. For every homomorphism h sending the term algebra T of L into A_{2o}, the value h(t) is uniquely determined by the values of h on occ(t); hence for a term object $t(\xi)$ it is uniquely determined by the sequence $h\cdot\xi$. A function f^k is said to be *induced by* $t(\xi)$ if

for every α in ω^k : $f^k(\alpha) = h_\alpha(t(\xi))$ for the homomorphism h_α such that $h_\alpha\cdot\xi = \alpha$.

I now collect some observations on the use of polynomial functions:

(4) The functions induced by term objects are precisely the polynomial functions.

Observe first that every term object induces a polynomial function. Because $x_i(\xi)$ with $\xi = <x_0,\dots, x_{k-1}>$ induces p^k_i. Also, $\kappa_n(\xi)$ induces c^k_n. If $t(\xi)$, $s(\xi)$ induce f^k and g^k then $(t+s)(\xi)$ and $(t\cdot s)(\xi)$ induce $+\circ<f^k, g^k>$ and $\cdot\circ<f^k, g^k>$. Conversely, + , $\cdot$, the p^k_i and the c^k_n are induced by term objects. Assume now that f^k is induced by $t(\xi)$ and that functions g^m_i, $i<k$, are induced by terms $s_i(\psi)$ with $\psi = <y_0,\dots, y_{m-1}>$. Let d be the endomorphism (substitution) of the term algebra sending x_i into $s_i(\psi)$, and set r =

$d(t(\xi))$. Then $occ(r)$ is contained in $im(\psi)$, and $r(\psi)$ induces the superposition $f^k \circ <g_0^m, \ldots, g_{k-1}^m>$: if $\beta \epsilon \omega^m$ and $\gamma = <g_0^m(\beta), \ldots, g_{k-1}^m(\beta)>$ then

$$\begin{aligned} g_i^m(\beta) &= h_\beta(s_i(\psi)) && \text{since } s_i(\psi) \text{ induces } g_i^m \\ &= h_\beta(d(x_i)) && \text{by definition of d} \end{aligned}$$

such that the homomorphisms $h_\beta \cdot d$ and h_γ coincide on $im(\xi)$. Hence

$$\begin{aligned} f^k(\gamma) &= h_\gamma(t(\xi)) && \text{since } t(\xi) \text{ induces } f^k \\ &= h_\beta(d(t(\xi))) && \text{since } t(\xi) \text{ contains at most the variables from } \xi \\ &= h_\beta(r(\psi)) && \text{by definition of r .} \end{aligned}$$

(5) Every term object represents the function which it induces.

The proof is by induction on the structure of terms. $x_i(\xi)$ represents p_i^k since $x_i(\kappa \cdot \alpha) = \kappa_{\alpha(i)}$ and $\vdash \kappa_{\alpha(i)} \equiv \kappa_{\alpha(i)}$. $\kappa_n(\xi)$ represents c_n^k since $\kappa_n(\kappa \cdot \alpha) = \kappa_n$. Also, $x+y$ and $x \cdot y$ represent $+$ and $\cdot$ by (PB0), (PB1). Consider now $t(\xi) = t_0(\xi) + t_1(\xi)$ and let f^k, f_0^k, f_1^k be the functions induced by these terms; I know already that $f^k = + \circ <f_0^k, f_1^k>$. If $t_0(\xi)$, $t_1(\xi)$ represent f_0^k, f_1^k then $f_0^k(\alpha) = a$, $f_1^k(\alpha) = b$ imply

$$\begin{aligned} &\vdash t_0(\kappa \cdot \alpha) \equiv \kappa_a \quad , \quad \vdash t_1(\kappa \cdot \alpha) \equiv \kappa_b \\ &\vdash t_0(\kappa \cdot \alpha) + t_1(\kappa \cdot \alpha) \equiv \kappa_a + \kappa_b && \text{by equality logic} \\ &\vdash t(\kappa \cdot \alpha) \equiv \kappa_a + \kappa_b && \text{by definition of t} \\ &\vdash t(\kappa \cdot \alpha) \equiv \kappa_{a+b} && \text{by (PB0)} \end{aligned}$$

and $f^k(\alpha) = a+b$ since $f^k = + \circ <f_0^k, f_1^k>$. The case of $t(\xi) = t_0(\xi) \cdot t_1(\xi)$ is analogous.

(6) Let h_0, h_1 be endomorphisms of the term algebra. Then for every term t: if for every $x \epsilon occ(t)$ there holds $\vdash h_0(x) \equiv h_1(x)$ then also $\vdash h_0(t) \equiv h_1(t)$.

This follows immediately from the axioms of equality logic. – In the following, let $\mathcal{P}$ be a 2-ary predicate symbol (observe that, at present, I only have $\equiv$).

(7) Let $t_0(\xi) \mathcal{P} t_1(\xi)$ represent a relation R^k, $\xi = <x_0, \ldots, x_{k-1}>$, and let η be a map from variables to terms. Let m be sufficiently large such that with $\psi = <y_0, \ldots, y_{m-1}>$ there holds $\eta(x_i) = s_i(\psi)$. Let g_i^m, $i<k$, be the polynomial function induced, hence represented, by $s_i(\psi)$.

Then $h_\eta(t_0)(\psi) \mathcal{P} h_\eta(t_1)(\psi)$ PB_0-represents $S^m = R^k[g_0^m, \ldots, g_{k-1}^m]$.

Let β be in ω^m and set $\gamma = <g_0^m(\beta), \ldots, g_{k-1}^m(\beta)>$; hence $\beta \epsilon S^m$ if, and only if, $\gamma \epsilon R^k$. Let $h_{\kappa\gamma}$ be the homomorphism mapping x_i to $\kappa_{\gamma(i)}$, and let $h_{\kappa\beta}$ be the homomorphism, mapping y_j to $\kappa_{\beta(j)}$. Let p be 0 or 1. I claim that

(x) $\vdash h_{\kappa\beta}(h_\eta(t_p)) \equiv h_{\kappa\gamma}(t_p)$.

Because the variables of t_p are in ξ, and for x_i in ξ I have $h_{\kappa\beta}(h_\eta(x_i)) = h_{\kappa\beta}(s_i(\psi)) = s_i(\kappa \cdot \beta)$ and $h_{\kappa\gamma}(x_i) = \kappa_{\gamma(i)}$. But $\vdash s_i(\kappa \cdot \beta) \equiv \kappa_{\gamma(i)}$ since s_i represents g_i^m. Thus (x) follows from (6).

Now $\beta\epsilon S^m$ gives $\gamma\epsilon R^k$, hence $\vdash t_o(\kappa\cdot\gamma)\mathscr{P}t_1(\kappa\cdot\gamma)$, i.e. $\vdash h_{\kappa\gamma}(t_o)\mathscr{P}h_{\kappa\gamma}(t_1)$. Thus by (x) and equality logic also $\vdash h_{\kappa\beta}(h_\eta(t_o)\mathscr{P}h_{\kappa\beta}(h_\eta(t_1)$, meaning $\vdash (h_\eta(t_o)(\kappa\cdot\beta)\mathscr{P}(h_\eta(t_1)(\kappa\cdot\beta)$. And *not* $\beta\epsilon S^m$ implies *not* $\gamma\epsilon R^k$ from which there follows the provability of the negated formulas.

Observe that (7) can be generalized to arbitrary formulas, making use of suitable substitutions.

I define the relations *canonically represented* by open L_{2_0}–formulas by (RB1) = (RC1), (RB2) = (RC2) and

(RB0) Let $t_o(\xi)$, $t_1(\xi)$ terms inducing functions f_0^k, f_1^k . Then the relation $E^2[f_0^k, f_1^k]$ is canonically represented by $t_o(\xi) \equiv t_1(\xi)$.

Again, canonically represented relations are represented. There holds

(8) Let the open formula $v(\xi)$ canonically represent R^k, and let $g_0^m, \ldots, g_{k-1}^m$ be polynomial functions. Then I can find an open L_{2_0}–formula of the same complexity as $v(\xi)$ which canonically represents $R^k[g_0^m, \ldots, g_{k-1}^m]$.

It follows from (4), (5) that there are terms $s_i(\psi)$ representing the g_i^m, $i<k$; let h_η be the endomorphism of the term algebra mapping x_i to $s_i(\psi)$. I now proceed by induction on the complexity of $v(\xi)$, and if $v(\xi)$ is atomic then my claim follows from (7). If $v(\xi)$ is, say, $u(\xi)\wedge w(\xi)$ then $u(\xi)$ and $w(\xi)$ canonically represent relations U^k and W^k such that $R^k = U^k \cap W^k$. Now

$$R^k[g_0^m, \ldots, g_{k-1}^m] = U^k[g_0^m, \ldots, g_{k-1}^m] \cap W^k[g_0^m, \ldots, g_{k-1}^m]$$

since there hold the equivalences

$$\begin{aligned}\alpha\epsilon R^k[g_0^m, \ldots, g_{k-1}^m] \quad &\text{if and only if} \quad <g_0^m(\alpha), \ldots, g_{k-1}^m(\alpha)>\epsilon R^k\\ &\text{if and only if} \quad <g_0^m(\alpha), \ldots, g_{k-1}^m(\alpha)>\epsilon U^k\cap W^k\\ &\text{if and only if} \quad \alpha\epsilon U^k[g_0^m, \ldots, g_{k-1}^m] \text{ and}\\ &\qquad\qquad\qquad \alpha\epsilon W^k[g_0^m, \ldots, g_{k-1}^m] .\end{aligned}$$

By inductive hypothesis, there are open formulas $u'(\xi)$, $w'(\xi)$, of the complexities of $u(\xi)$, $w(\xi)$, which canonically represent $U^k[g_0^m, \ldots, g_{k-1}^m]$ and $W^k[g_0^m, \ldots, g_{k-1}^m]$. Hence $u'(\xi)\wedge w'(\xi)$ canonically represents the superposition $R^k[g_0^m, \ldots, g_{k-1}^m]$.

I should mention that it is this last proof in which the notion of relations *canonically* represented by a formula plays a decisive role. As it has been set up, the proof proceeds by induction on the representing formula $v(\xi)$. If $v(\xi) = u(\xi)\wedge w(\xi)$ represents a given R^k, and if $u(\xi)$, $w(\xi)$ represent U^k and W^k, then *without* uniqueness hypotheses (e.g. consistency) I *cannot* conclude that $R^k = U^k \cap W^k$ as the proof requires.

I define the class $\mathbf{RAB}_0^*$ to be the smallest class which contains E^2 , is Boolean closed and is closed under superposition with polynomial functions.

LEMMA 2 Every open L_{2_0}-formula canonically represents a relation in $\mathbf{RAB}_0^*$.

For every relation R in $\mathbf{RAB}_0^*$, I can find an open L_{2_0}-formula which canonically represents R.

PB_0 is *complete* for open sentences s: there holds $\vdash s$ or $\vdash \neg s$.

The first statement is proved by induction on formulas. It holds for atomic formulas since they have the form $t_0(\xi) \equiv t_1(\xi)$ and, by the definition (RB0), canonically represent relations arising from E^2 by superposition with polynomial functions f_0^k, f_1^k. The induction then follows making use of (RB1). The second statement is proved by induction on the construction of $\mathbf{RAB}_0^*$. It holds for the initial relation E^2, is preserved under Boolean constructions by (1), and is preserved under superpositions with polynomial functions by (8). The third statement is handled as in Lemma 1.

The first two statements of Lemma 2 may be expressed by saying that the relations in $\mathbf{RAB}_0^*$ are those which are canonically representable by open formulas of L_{2_0}. If PB_0 is assumed to be consistent for open sentences then canonical representability and representability coincide for open formulas.

I now embed my language L_{2_0} into a language L_2 containing a further predicate symbol $\leq$.

Let PB be the axiom system consisting of PB_0 together with

(PB3) $x \leq \kappa_n \rightarrow (x \equiv \kappa_0 \vee x \equiv \kappa_1 \vee \ldots \vee x \equiv \kappa_n)$

(PB4) $x \leq \kappa_n \vee \kappa_n \leq x$

which coincide with (PC3), (PC4). Provability shall refer to provability from PB. Then (PC5) remains in effect with the same proof.

Let A_2 be the obvious L_2-structure over ω. Again, it follows from (PC5) that $\leq$ represents the relation $\leq^A$ of A_2. - I extend my previous observations by

(9) Let $v(\xi,x)$ represent R^{k+1}, and let $t(\xi)$ not contain x. Let f^k be induced by $t(\xi)$. Then $u(\xi) = \forall x\,(x \leq t(\xi) \rightarrow v(\xi,x))$ represents $V^k = (\forall_{\leq} R^{k+1})[p_0^k, \ldots, p_{k-1}^k, f^k]$.

It is $\alpha \in V^k$ equivalent to $<\alpha, f^k(\alpha)> \in \forall_{\leq} R^{k+1}$. Thus $\alpha \in V^k$ implies $<\alpha, b> \in R^{k+1}$ for every $b \leq f^k(\alpha)$. With $a = f^k(\alpha)$ I find as in (3a) $\vdash \forall x(x \leq \kappa_a \rightarrow v(\kappa \cdot \alpha, x))$. But $\vdash t(\kappa \cdot \alpha) \equiv \kappa_a$ by (5), and so $\vdash \forall x(x \leq t(\kappa \cdot \alpha) \rightarrow v(\kappa \cdot \alpha, x))$, i.e. $\vdash u(\kappa \cdot \alpha)$. Conversely, *not* $\alpha \in V^k$ implies $\vdash \neg \forall x(x \leq \kappa_a \rightarrow v(\kappa \cdot \alpha, x))$ by (3b), i.e. $\vdash \neg u(\kappa \cdot \alpha)$.

I define the set of *p-bounded formulas* of L_2 by the clauses (FB0) = (FC0), (FB1) = (FC1) (with p-bounded in place of bounded) and, instead of (FC2),

(FB2) If v is p-bounded then for all x and all terms t with *not* $x \epsilon occ(t)$ also $\forall x\ (x \leq t \rightarrow v)$ and $\exists x\ (x \leq t \wedge v)$ are p-bounded .

Up to this point, it did suffice to discuss k-ary relations (canonically) represented by formulas $v(\xi)$ which had their free variables in ξ. Already (RC3), however, required a slight distinction, and for p-bounded formulas a more careful distinction has to be made. Of course, I always can choose ξ so large that in a formula $\forall x\ (x \leq t \rightarrow v)$ all variables of t, or free in v, occur in $<\xi,x>$, writing it as

$$\forall x(x \leq t(\xi) \rightarrow v(\xi,x)) .$$

However, studying the relations represented in such situations will only give me internally bounded quantifications, but not the standard bounded quantifications. This explains the distinction made below in (RB3).

I define the relations *canonically represented* by p-bounded L_2-formulas as follows. In (RB0) I add that the atomic formulas $t_0(\xi) \leq t_1(\xi)$ canonically represent $\leq^A[f_0^k, f_1^k]$; by (7) they also represent these relations. I keep (RB1), (RB2), and I add, strengthening (RB3),

(RB3) If the relation R^{k+1} is canonically represented by $v(\xi,x)$ and if y does not occur in $<\xi,x>$

then $\forall_{\leq} R^{k+1}$ is canonically represented by $\forall x(x \leq y \rightarrow v(\xi,x))$,

and $-\forall_{\leq} - R^{k+1}$ is canonically represented by $\exists x(x \leq y \wedge v(\xi,x))$.

If R^{k+1} is canonically represented by $v(\xi,x)$, if $t(\xi)$ does not contain x, and if f^k is induced by $t(\xi)$,

then $(\forall_{\leq} R^{k+1})[p_0^k, \ldots, p_{k-1}^k, f^k]$ is canonically represented by $\forall x(x \leq t(\xi) \rightarrow v(\xi,x))$,

and $-((\forall_{\leq} - R^{k+1})[p_0^k, \ldots, p_{k-1}^k, f^k])$ is canonically represented by $\exists x(x \leq t(\xi) \wedge v(\xi,x))$.

It follows from (9), together with provable equivalence of $\exists x\ (x \leq t \wedge v)$ and $\neg \forall x\ (x \leq t \rightarrow \neg v)$, that also in the new case canonical representability implies representability.

I now extend (8) to p-bounded formulas:

(10) Let the p-bounded formula $v(\xi)$ canonically represent R^k, and let $g_0^m, \ldots, g_{k-1}^m$ be polynomial functions. Then I can find a p-bounded L_2-formula of the same complexity as $v(\xi)$ which canonically represents $R^k[g_0^m, \ldots, g_{k-1}^m]$.

As (8) was proved by induction on the complexity of formulas, it remains to consider a p-bounded formula

$u(\xi) = \forall x\,(x \leq t(\xi) \rightarrow v(\xi,x))$ or
$u(\xi) = \forall x\,(x \leq y \rightarrow v(\xi,x))$, y not in $<\xi,x>$

under the assumption that (10) already holds for formulas at most of the complexity of $v(\xi,x)$. I shall consider the first case, the second being even simpler. Let W^k be canonically represented by $u(\xi)$; I have to find a formula $u_0(\psi)$ of the complexity of $u(\xi)$ which represents the relation $S^m = W^k[g_0^m,\dots, g_{k-1}^m]$.

It follows from (RB3) that W^k is $(\forall^{k+1}{}_{\leq}R^{k+1})[p_0^k,\dots, p_{k-1}^k, f^k]$ where R^{k+1} is canonically represented by $v(\xi,x)$ and f^k is induced by $t(\xi)$. I shall show that

$$\text{(y)} \qquad S^m = (\forall^{m+1}{}_{\leq}(W^{k+1}[e_0^{m+1},\dots, e_{k-1}^{m+1}, p_m^{m+1}]))\,[p_0^m,\dots, p_{m-1}^m, h^m]$$

for $h^m = f^k \circ <g_0^m,\dots, g_{k-1}^m>$ and the polynomial functions e_i^{m+1} such that $e_i^{m+1}(\beta,c) = g_i^m(\beta)$. Now $\beta \epsilon S^m$ is equivalent to $\gamma \epsilon W^k$ with $\gamma = <g_0^m(\beta),\dots, g_{k-1}^m(\beta)>$. Hence each of the following statements is equivalent to $\beta \epsilon S^m$:

for every $b \leq f^k(\gamma)$: $<g_0^m(\beta),\dots, g_{k-1}^m(\beta), b> \epsilon R^{k+1}$

for every $b \leq f^k(\gamma)$: $<e_0^{m+1}(\beta, b),\dots, e_{k-1}^{m+1}(\beta, b), p_m^{m+1}(\beta, b)> \epsilon R^{k+1}$

for every $b \leq h^m(\beta)$: $<\beta, b> \epsilon R^{k+1}[e_0^{m+1},\dots, e_{k-1}^{m+1}, p_m^{m+1}]$

$\beta \epsilon (\forall^{m+1}{}_{\leq}(R^{k+1}[e_0^{m+1},\dots, e_{k-1}^{m+1}, p_m^{m+1}]))\,[p_0^m,\dots, p_{m-1}^m, h^m]$.

By inductive hypothesis, there is a formula $v_0(\psi,x)$ of the same complexity as $v(\xi,x)$ which canonically represents $R^{k+1}[e_0^{m+1},\dots, e_{k-1}^{m+1}, p_m^{m+1}]$. So if $q(\psi)$ is a term representing h^m, then it follows from (y) and (RB3) that S^m is represented by

$$u_0(\psi) = \forall x\,(x \leq q(\psi) \rightarrow v_0(\psi,x)) \quad .$$

I define the class **RAB*** to be the smallest class of relations which is AB-closed and is closed under superpositions with polynomial functions. Making use of Theorem 6.0, I could conclude that **RAB*** coincides with **RAB**, but as observed earlier, this Theorem will not be proven in this book. A p-bounded sentence shall arise from a p-bounded formula by replacing its free variables with constants.

THEOREM 2 Every p-bounded L_2-formula canonically represents a relation in **RAB*** .

For every relation R in **RAB***, I can find a p-bounded L_1-formula which canonically represents R .

PB is *complete* for p-bounded sentences s: there holds $\vdash s$ or $\vdash \neg s$.

The first statement is proved by induction on formulas. It holds for atomic formulas since they are of the forms $t_0(\xi) \equiv t_1(\xi)$ and $t_0(\xi) \leq t_1(\xi)$ which, by the amended (RB0), canonically represent relations arising from E^2 and $\leq^A$ by superposition with polynomial functions f_0^k, f_1^k. The induction then follows making use of (RB1) and (RB3). The second statement is proved by induction on the construction of $\mathbf{RAB}_0^*$. It holds for the initial relations $G(+)$ and $G(\cdot)$ since, say, $G(+) = E^2[+\circ<p_0^3, p_1^3>, p_2^3]$ is canonically represented by $x+y \equiv z$. It is preserved under Boolean constructions by (RB1), and is preserved under superpositions with polynomial functions by (10). The third statement is handled as in Theorem 1. Because let s arise from $\forall x\, (x \leq t(\xi) \rightarrow v(\xi, x))$, replacing ξ by $\kappa \cdot \alpha$:

$$s = \forall x\, (x \leq t(\kappa \cdot \alpha) \rightarrow v(\kappa \cdot \alpha, x)) .$$

If $t(\xi)$ induces f^m and $a = f^m(\beta)$ then $\vdash t(\kappa \cdot \alpha) \equiv \kappa_a$, and thus s is provably equivalent to the sentence $s' = \forall x\, (x \leq \kappa_a \rightarrow v(\kappa \cdot \alpha, x))$ which can be handled as in the earlier proof.

The first two statements of Theorem 2 may be expressed by saying that the relations in $\mathbf{RAB}^*$ are those which are canonically representable by p-bounded formulas of L_2. If PB is assumed to be consistent for p-bounded sentences then canonical representability and representability coincide for p-bounded formulas.

3. Representability of Recursive Relations I

Let L be L_1 or L_2 and let PD be PC or PB. Provability shall refer to PD.

There are various reasons which make it desirable to know that recursive relations and functions are PD-representable, one of them being the radical undecidability of PD to be discussed in a moment. For most such applications, it would suffice to know this only for primitive recursive relations (and for many of them, even elementary relations would suffice). The statements expressible in L, however, as well as the means provided by the axioms PD, make it convenient to attack representability of recursive relations in its full generality, whence primitive recursive relations become representable as a special case. (Actually, recursive relations may be considered to have been introduced precisely for this reason, isolating in the machinery of Chapters 5 and 6 the difficulties caused by recursively defined objects.)

There now are two roads on which to approach representability. The first one leads directly to the

THEOREM 3 Every recursive relation is PD-representable by an L-formula, and every recursive function is functionally PD-representable.

The proof uses the results developed in the last section of Chapter 6 which employed total minimization: it is a special case of Corollary 6.6 that the class **RFR** of recursive relations is the smallest class S which contains E^2, $G(+)$, $G(\cdot)$, which is trivially closed, is Boolean closed, is closed under superpositions with functions from **F**(**S**), and is closed under total minimization. Thus I have to show that the class of PD-representable relations is a class S having these properties.

Now E^2, $G(+)$, $G(\cdot)$ are already representable in PC_0. Also, the class of relations representable by arbitrary L-formulas is both trivially and Boolean closed. As for closure under superpositions, let R^k be represented by $v(\xi)$, and let $g_0^m, \ldots, g_{k-1}^m$ be functions with representable graphs. Now the hypotheses (MINn), (CMPn) of Lemma 4.4 are available, because they coincide with (PD3) and (by classical logic) with (PD4). Consequently, the functions g_i^m are functionally represented by formulas $w_i(\psi, x_i)$, $i<k$. Then $R^k[g_0^m, \ldots, g_{k-1}^m]$ is represented by

$$u(\xi) = \exists x_0 \ldots \exists x_{k-1} (w_0(\xi, x_0) \wedge \ldots \wedge w_{k-1}(\xi, x_{k-1}) \wedge v(\xi)) .$$

Because if $\mu \epsilon \omega^m$ and $\lambda(i) = g_i^m(\mu)$ and $\lambda = <\lambda(i) \mid i<k>$ then $\vdash w_i(\kappa \cdot \mu, x_i) \longleftrightarrow x_i \equiv \kappa_{\lambda(i)}$ implies by equality logic that

$$\vdash u(\kappa \cdot \mu) \longleftrightarrow \exists x_0 \ldots \exists x_{k-1} (x_0 \equiv \kappa_{\lambda(0)} \wedge \ldots \wedge x_{k-1} \equiv \kappa_{\lambda(k-1)} \wedge v(\xi)) .$$

Applying k times that $\vdash \exists x_i (x_i \equiv \kappa_i \wedge v) \longleftrightarrow rep(x_i, \kappa_i \mid v)$, there then results

$$\vdash u(\kappa \cdot \mu) \longleftrightarrow v(\kappa \cdot \lambda) .$$

From $\mu \in R^k[g_0^m, \ldots, g_{k-1}^m]$ there follows $\vdash v(\kappa \cdot \lambda)$ by definition of λ, hence $\vdash u(\kappa \cdot \mu)$. If μ is *not* in $R^k[g_0^m, \ldots, g_{k-1}^m]$ then λ is *not* in R^k, and therefore $\vdash \neg v(\kappa \cdot \lambda)$ implies $\vdash \neg u(\kappa \cdot \mu)$.

Finally, the class of PD-representable relations is closed under total minimization, meaning that, for every full relation R^{k+1}, also the graph νR^{k+1} of the function μR^{k+1} belongs to this class. This follows immediately from Lemma 4.4 which, from a formula $v(\xi, x)$ representing R^{k+1} (even if R^{k+1} is not full), constructs a formula $u(\xi, x)$ functionally representing the (partial) function with the graph νR^{k+1}. Again, the hypotheses (MINn), (CMPn) of this Lemma are available here.

As for recursive functions f, Lemma 4.4 implies that they are functionally PD-representable since their graphs are PD-representable. This concludes the proof of Theorem 3.

Both my languages L_1 and L_2 have signatures which can be elementarily presented in the sense of Chapter 7; hence they have recursively arithmetized copies. Under these arithmetizations, both axioms systems PC and PB have elementary images *PC* and *PB*. Taking A to be the structure with $\omega = u(A)$ and the relations and operations interpreting the language L, there exists in both cases an *arithmetical situation* in the sense of Chapter 8.

For instance, a recursively arithmetized copy L_A of L_2, built as in the proof of Theorem 7.1, may use as variables and constants

$$g_T(x_n) = 2^{2^{n+2}} \quad , \quad \gamma(n) = g_T(\kappa_n) = 2^{3^{3n}} \; ,$$

use 2^{3^4} and $2^{3^{12}}$ as characteristic factors for + and · and use 2^{5^4}, $2^{5^{12}}$ as the characteristic factors for ≡ and ≤ ; thus $arity_a(m) = \varepsilon(1, \varepsilon(0,m))$ and $arity_b(m) = \varepsilon(2, \varepsilon(0,m))$. Then the axioms (PB0) correspond to the numbers h such that

$|h| = 3$ and $\varepsilon(0,h) = 5^4$ and $|\varepsilon(1,h)| = 3$ and $\varepsilon(0, \varepsilon(1,h)) = 3^4$ and there exist $n \leq h$ and $m \leq h$ such that $\varepsilon(1, \varepsilon(1,h)) = 2^{3^{3n}}$ and $\varepsilon(2, \varepsilon(1,h)) = 2^{3^{3m}}$ and $\varepsilon(2,h)) = 2^{3^{3n+m}}$

and taking x in (PB4) to be the first variable, these axioms correspond to the numbers h such that

$|h| = 3$ and $\varepsilon(0,h) = 7^2$ and $|\varepsilon(1,h)| = |\varepsilon(2,h)| = 3$ and $\varepsilon(0, \varepsilon(1,h)) = \varepsilon(0, \varepsilon(1,h)) = 5^{12}$ and there exists $n \leq h$ and $\varepsilon(1, \varepsilon(1,h)) = \varepsilon(2, \varepsilon(2,h)) = 2^{2^2}$ and $\varepsilon(2, \varepsilon(1,h)) = \varepsilon(1, \varepsilon(2,h)) = 2^{3^{3n}}$.

The other three types of axioms being translated analogously, the set *PB*, characterized by the five conditions, will be elementary. Also, the injection γ is elementary.

It follows from Theorem 3 that my arithmetical situation has the property 7.(B_1). Thus I may apply the results of Chapter 7 and obtain the

COROLLARY 1 Precisely the recursive relations are PD–representable.

PD is radically undecidable.

If PD is consistent then it is incomplete, its theory is undecidable, and and both *prov(PD)* and *provs(PD)* are examples of recursively enumerable sets which are not recursive.

If the structure A on ω is model of PD then none of the sets *true*(A) and *trues*(A) is semantically definable in A, and there is no recursively enumerable set G of axioms which would axiomatize *trues*(A) .

Incompleteness of PD can also be concluded from Theorem 4.6 whose hypotheses (MINn), (CMPn) coincide with (PD3) and (PD4).

It is, of course, not the incompleteness but the radical undecidability of PD which makes these insights important. That PD is incomplete cannot surprise because it is a very weak system of axioms, and making use of models it is quite easy to find sentences s such that neither s nor $\neg$s can be provable from, say, PB - for instance the sentence $\forall x(x+\kappa_0 \equiv x)$. To this end, consider a structure A with u(A) obtained from ω by putting a new element a on top, i.e. $n \leq a$ for all $n \in \omega$. Extending addition and multiplication arbitrarily to the case that one, or both, of the summands (factors) is a, I still will obtain a model of PB. In particular, defining $a+0=0$, my sentence s will not hold and thus cannot be provable. Defining $a+0=a$, the sentence $\neg$s will not hold.

4. Representability of Recursive Relations II

Let L be L_1 or L_2 and let PD be PC or PB. Provability shall refer to PD. In both cases, bounded formulas have been defined. Let A be the obvious L-structure over ω.

There is a second approach to representability which uses only the description **RFR** = **QE**(**RAB**) of recursive relations (and thus avoids the use of minimization). Referring to projections in **E**(**RAB**), however, a certain hypothesis of consistency with respect to formulas with existential quantifiers will be required, rewarding me, in turn, with stronger forms of representability.

A formula which is provably equivalent to a bounded formula is called a Δ_0-*formula*.

A formula is called existentially bounded, or EB, if it is of the form $\exists x w$ where w is bounded. A formula which is Δ_0 or is provably equivalent to an EB-formula is called a Σ_1-*formula*.

In Chapter 4 I defined the property that an axiom system G be ω-*consistent*. Specializing this to PD and ω, the axiom system PD is ω-*consistent* if $\vdash \neg\forall y\, u(y)$ implies that *not* $\vdash u(\kappa_m)$ for at least one m in ω. Making use of classical logic, PD is ω-consistent if, and and only if, $\vdash \exists y\, u(y)$ implies that *not* $\vdash \neg u(\kappa_m)$ for at least one m in ω. Specializing further to the case that u(y) is bounded, I define

> PD is ω-consistent for EB-sentences if, for any bounded u(y), $\vdash \exists y\, u(y)$ implies that *not* $\vdash \neg u(\kappa_m)$ for at least one m in ω

and conclude from the completeness statements in Theorems 1 and 2

If PD is ω-consistent for EB-sentences then, for any bounded u(y), $\vdash \exists y\, u(y)$ implies that $\vdash u(\kappa_m)$ for at least one m in ω .

Observe that ω-consistency for EB-sentences implies consistency, hence, in particular, consistency for bounded sentences.

Recall that the relation formally defined by a formula $v(\psi)$ consists of all β such that $\vdash v(\kappa\cdot\beta)$; recall that the relation semantically defined in A by $v(\psi)$ consists of all β such that $v(\kappa\cdot\beta)$ is true. Recall that consistency, hence also ω-consistency, has the effect that representability entails formal definability.

Let G be an axiom system in one of my languages. I shall say that a formula $u(\xi)$ is A-*sound* if the relation formally defined by $u(\xi)$ coincides with the relation semantically defined by $u(\xi)$ in A.

LEMMA 3 Let A be model of G. If $v(\xi,x)$ is A-sound then so is $w(\xi) = \exists x\, v(\xi,x)$.

By hypothesis, for the relation R^{k+1} formally defined by $v(\xi,x)$, the three statements

(1v) $\vdash v(\kappa\cdot\lambda, \kappa_a)$ (2v) $<\lambda,a>\epsilon R^{k+1}$ (3v) $v(\kappa\cdot\lambda, \kappa_a)$ is true

are equivalent. Observe that the relation S^k semantically defined by $w(\xi) = \exists x\, v(\xi,x)$ is $\exists R^{k+1}$, i.e.

$\lambda\epsilon S^k$ if and only if there exists a such that $<\lambda,a>\epsilon R^{k+1}$
if and only if $\exists x v(\kappa\cdot\lambda,x)$ is true .

Thus of

(1w) $\vdash \exists x\, v(\kappa\cdot\lambda,x)$ (2w) $\lambda \epsilon S^k$ (3w) $\exists x\, v(\kappa\cdot\lambda,x)$ is true

(2w) and (3w) are equivalent. Since A is a model of G, (1w) implies (3w). But

if $\exists x\, v(\kappa\cdot\lambda,x)$ is true
then there exists a such that $<\lambda,a>\epsilon R^{k+1}$ i.e. (2v)
then there exists a such that $\vdash v(\kappa\cdot\lambda, \kappa_a)$ i.e. (1v)
then $\vdash \exists x\, v(\kappa\cdot\lambda,x)$.

Hence (3w) implies (1w).

LEMMA 4 If A is a model of PD then every EB-formula is A-sound, and PD then is ω-consistent for EB-sentences.

I did prove earlier that every bounded formula v canonically represents a relation R. Inspection of the definition of canonically represented formulas, beginning with the atomic formulas, shows that R then also is the relation semantically defined by v. If A is a model of PD then R is formally defined by v. Hence every bounded formula is A-sound, and by Lemma 3 then also EB-formulas are A-sound. If an EB-sentence $\exists y u(y)$ is provable then it is true. Hence $u(\kappa_m)$ is true for some m, and by soundness (only of the bounded formula u(y)) then $u(\kappa_m)$ is provable.

THEOREM 4 Assume that PD is ω-consistent for EB-sentences.

For every relation R in **RER**, I can find an EB-formula which formally defines R .

Every relation, formally defined by a Σ_1-formula, is in **RER** .

If R^k is in **RAB** then it follows from Theorems 1 or 2 that R^k is canonically representable by a bounded formula. If R^k is in **RER** but not in **RAB** then it is still in **E(RAB)**, hence $R^k = \exists S^{k+1}$ with S^{k+1} in **RAB** and thus representable by a bounded formula $v(\xi,x)$. Then $\exists x\, v(\xi,x)$ formally defines R^k, because it is equivalent that

$\alpha \in R^k$
there is an m such that $<\alpha,m>\in S^{k+1}$ since $R^k = \exists S^{k+1}$
there is an m such that $\vdash v(\kappa\cdot\alpha, \kappa_m)$ since S^{k+1} is also formally defined by $v(\xi,x)$
$\vdash \exists x\, v(\kappa\cdot\alpha, x)$ making use of ω-consistency when going up.

Conversely, let R^k be formally defined by an EB-formula $\exists x\, v(\xi,x)$. Then $v(\xi,x)$ is bounded, and since PD is consistent for bounded sentences, $v(\xi,x)$ represents, hence formally defines, a relation S^{k+1} in **RAB**. So there holds $R^k = \exists S^{k+1}$, because it is equivalent that

$\alpha \in R^k$
$\vdash \exists x\, v(\kappa\cdot\alpha, x)$ since R^k is formally defined by $\exists x\, v(\xi,x)$
there is an m such that $\vdash v(\kappa\cdot\alpha, \kappa_m)$ making use of ω-consistency when going down
there is an m such that $<\alpha,m>\in S^{k+1}$ since S^{k+1} is formally defined by $v(\xi,x)$.

Thus R^k is in **E(RAB)** = **RER**. This concludes the proof. If PD is assumed to have A as a model then I can argue more rapidly. Because then the relations in **RAB** are semantically defined by bounded formulas, whence the relations in **E(RAB)** are those semantically defined by EB-formulas. EB-formulas being A-sound, they also formally define these relations.

THEOREM 5 Assume that PD is ω-consistent for EB-sentences.

For every recursive relation in **RFR** = **QE(RAB)**, I can find an EB-formula which PD-represents R .

In particular, for every recursive function f^k, I can find an EB-formula representing its graph $G(f^k)$.

If R^k is recursive then both R^k and $-R^k$ are recursively enumerable. Hence (ENB) implies that $R^k = \exists S_0^{k+1}$ and $-R^k = \exists S_1^{k+1}$ where S_0^{k+1}, S_1^{k+1} are representable by bounded formulas $v_0(\xi, x)$, $v_1(\xi, x)$. Then also

$$w(\xi, x) = v_0(\xi, x) \wedge \forall y (y \leq x \rightarrow \neg v_1(\xi, y))$$

is bounded, and I shall show that R^k is represented by the EB-formula $u(\xi) = \exists x\, w(\xi, x)$.

If $\alpha \epsilon R^k$ then there is an a such that $\langle \alpha, a \rangle \epsilon S_0^{k+1}$, hence

(a) $\vdash v_0(\kappa \cdot \alpha, \kappa_a)$.

But $\alpha \epsilon R^k$ also implies *not* $\alpha \epsilon -R^k$, hence for every b *not* $\langle \alpha, b \rangle \epsilon S_1^{k+1}$ and thus $\vdash \neg v_1(\kappa \cdot \alpha, \kappa_b)$. This, in particular, holds for every $b \leq a$. Making use of (PD3), the argument employed in (3a) now gives

(b) $\vdash \forall y (y \leq \kappa_a \rightarrow \neg v_1(\kappa \cdot \alpha, y)))$.

From (a) and (b) I obtain $\vdash w(\kappa \cdot \alpha, \kappa_a)$, and this implies $\vdash u(\kappa \cdot \alpha)$.

If not $\alpha \epsilon R^k$ then $\alpha \epsilon -R^k$, and then there is an a such that $\langle \alpha, a \rangle \epsilon S_1^{k+1}$. Thus

$\vdash v_1(\kappa \cdot \alpha, \kappa_a)$
$\vdash \kappa_a \leq x \rightarrow (\kappa_a \leq x \wedge v_1(\kappa \cdot \alpha, \kappa_a))$
$\vdash \kappa_a \leq x \rightarrow \exists y (y \leq x \wedge v_1(\kappa \cdot \alpha, y))$
$\vdash \kappa_a \leq x \rightarrow (v_0(\kappa \cdot \alpha, x) \rightarrow \exists y (y \leq x \wedge v_1(\kappa \cdot \alpha, y)))$
(c) $\vdash \forall x (\kappa_a \leq x \rightarrow (v_0(\kappa \cdot \alpha, x) \rightarrow \exists y (y \leq x \wedge v_1(\kappa \cdot \alpha, y))))$.

But *not* $\alpha \epsilon R^k$ also implies that *not* $\langle \alpha, b \rangle \epsilon S_0^{k+1}$ for every b. Therefore $\vdash \neg v_0(\kappa \cdot \alpha, \kappa_b)$. This, in particular, holds for every $b \leq a$, and making use of (PD3), the argument employed in (3b) now gives

$\vdash \forall x (x \leq \kappa_a \rightarrow \neg v_0(\kappa \cdot \alpha, x))$
$\vdash \forall x (x \leq \kappa_a \rightarrow (\neg v_0(\kappa \cdot \alpha, x) \vee \exists y (y \leq x \wedge v_1(\kappa \cdot \alpha, y))))$
(d) $\vdash \forall x (x \leq \kappa_a \rightarrow (v_0(\kappa \cdot \alpha, x) \rightarrow \exists y (y \leq x \wedge v_1(\kappa \cdot \alpha, y))))$.

From (c) and (d) I obtain

$$\vdash \forall x (\kappa_a \leq x \vee x \leq \kappa_a \rightarrow (v_0(\kappa \cdot \alpha, x) \rightarrow \exists y (y \leq x \wedge v_1(\kappa \cdot \alpha, y))))$$

and so by (PD4)

$\vdash \forall x (v_0(\kappa \cdot \alpha, x) \rightarrow \exists y (y \leq x \wedge v_1(\kappa \cdot \alpha, y)))$
$\vdash \neg \exists x (v_0(\kappa \cdot \alpha, x) \wedge \forall y (y \leq x \rightarrow \neg v_1(\kappa \cdot \alpha, y)))$

i.e. $\vdash \neg u(\kappa \cdot \alpha)$. – The attentive reader will notice the close relationship between this proof and that of Theorem 4.6 .

COROLLARY 2 Assume that PD is ω-consistent for EB-sentences.

Every PD-formally definable relation is PD-formally definable by an EB-formula.

The PD-representable relations are precisely the recursive relations, and every PD-representable relation is PD-representable by an EB-formula.

If R^k is PD-formally defined then it is recursively enumerable by Theorem 8.2 ; hence Theorem 3 implies that it is PD-formally definable by an EB-formula. Also, PD-representable relations are recursive by Theorem 8.2, and Theorem 4 implies that these are PD-representable by EB-formulas.

Given a formula $v(\xi)$, formally defining a relation R^k, the Corollary tells me that *there is* an EB-formula $u(\xi)$ formally defining the same R^k. The problem how to actually find $u(\xi)$ can be attacked as follows. Consider the recursively arithmetized copy L_A, and let *PD* be the image of the axioms of PD. By Theorem 8.2, $\lambda \epsilon R^k$ is equivalent to $\vartheta_0(\lambda) \epsilon$ *prov(PD)* where $\vartheta_0 = \vartheta_1(-, u)$ is primitive recursive, u is the arithmetization of $v(\xi)$, and *prov(PD)* is $\exists$*ded(PD)* with *ded(PD)* being primitive recursive. Thus $\lambda \epsilon R^k$ is equivalent to

(x) there exists z such that $\vartheta_0(\lambda) = cro_0(z)$ and $< cro_0(z), cro_1(z) > \epsilon$ *ded(PD)*.

Hence R^k is a projection of a primitive recursive relation, i.e. in **E(FPF)**. Making use of **E(FPF)** = **E(RAB)**, I know that there *exists* a relation S^{k+1} in **RAB** having R^k as projection, and hence there *exists* a bounded formula representing S^{k+1} which gives rise to an EB-formula formally defining R^k. From these inconstructive existence statements, I can extract a construction of S^k as follows. Consider the recursive definitions of

cro_0, cro_1,

ϑ_0, reduced to *repf* and *rept* and the recursively defined operations of $Fm(L_A)$, reduced to the recursively defined operations of $T(L_A)$, reduced to the recursively defined exponentiation,

dedu(PD), reduced to the recursively defined set *axiom*, reduced to the recursively defined functions *exp* and *prino* and so again to exponentiation .

This leads to a schema of nested primitive recursions, starting from relations in **RAB**. Employing the technique of the proof of Theorem 6.1, each of these recursions can be replaced by the construction of a graph H of the desired function. By 6.(AB14), these constructions (Boolean, bounded quantification and superposition) remain in **E(RAB)**; if the graphs of the input functions of a recursion were formally defined by EB-formulas, then the construction steps translate into the formation of new EB-formulas, the last of which will formally define the output function. In the same way then, my schema of nested recursions produces EB-formulas which formally define cro_0, cro_1, ϑ_0, *dedu(PD)* and *ded(PD)*, and from them I arrive at an EB-formula formally defining the set characterized in (x) above.

5. Formal Definability of Recursive Relations

There is a third approach, this time to formal definability, which uses MATIYASEVIC's theorem **RER** = **E*(DIO)** mentioned in Chapter 6.

Consider the language L_{20} and the axiom system PB_0. Let A be obvious structure on ω.

Recall that by (7) every relation R^k in DIO, being of the form $E^2[f_0^k, f_1^k]$ with polynomial functions f_0^k, f_1^k, is represented by the atomic formula $t_0(\xi) \equiv t_1(\xi)$ where $t_0(\xi)$, $t_1(\xi)$ are terms inducing f_0^k, f_1^k. Observe further that R^k then also is the relation semantically defined by this atomic formula. Assuming consistency of PB_0 for atomic sentences, all relations representable by atomic formulas are of this form, and then they are also formally defined by the atomic formulas. In particular then, all atomic formulas are A-sound.

Recall that **E*(DIO)** was defined as the union of the classes $\mathbf{E}_n(\mathbf{DIO})$ with $\mathbf{E}_0(\mathbf{DIO}) = \mathbf{DIO}$ and $\mathbf{E}_{n+1}(\mathbf{DIO}) = \mathbf{E}(\mathbf{E}_n(\mathbf{DIO}))$.

I define the set of ED_0-formulas to consist of all atomic formulas, and I define the set of ED_{n+1}-formulas to consist of all formulas $\exists x v$ where v is an ED_n-formula. The set of ED*-formulas shall be the union of all sets ED_n. - I define

> PB_0 is ω-α-consistent for ED*-sentences if it is consistent for atomic sentences and if, for any ED*-formula u(y), $\vdash \exists y\, u(y)$ implies that $\vdash u(\kappa_m)$ for at least one m in ω.

LEMMA 5 If A is a model of PB_0 then every ED*-formula is A-sound, and PB_0 is ω-α-consistent for ED*-sentences.

Because PB_0 then is consistent for atomic sentences. Thus all atomic formulas are A-sound. But then Lemma 3 implies that every ED*-formula is A-sound. If $\exists y u(y)$ is provable then it is true. Hence $u(\kappa_m)$ is true for some m, and by soundness of the ED*-formula u(y) then $u(\kappa_m)$ is provable.

THEOREM 6 Assume that PB_0 is ω-α-consistent for ED^*-sentences.

For every relation R in $\mathbf{E}_n(\mathbf{DIO})$, I can find an ED_n-formula which formally defines R .

Every relation, formally defined by an ED_n-formula, is in $\mathbf{E}_n(\mathbf{DIO})$.

I remarked above that the relations in $\mathbf{DIO} = \mathbf{E}_0(\mathbf{DIO})$ are formally defined by the atomic ED_0-formulas, and since PB_0 is consistent for atomic sentences, the soundness of atomic formulas implies that the relations, formally defined by them, are in **DIO**. Assume the statements to be proven for a fixed n. If R^k is in $\mathbf{E}_{n+1}(\mathbf{DIO})$ then $R^k = \exists S^{k+1}$ with S^{k+1} in $\mathbf{E}_n(\mathbf{DIO})$ and thus formally defined by an ED_n-formula $v(\xi,x)$. Then $\exists x\ v(\xi,x)$ formally defines R^k, because it is equivalent that

$\alpha \in R^k$
there is an m such that $<\alpha,m> \in S^{k+1}$ since $R^k = \exists S^{k+1}$
there is an m such that $\vdash v(\kappa\cdot\alpha, \kappa_m)$ since S^{k+1} is formally defined by $v(\xi,x)$
$\vdash \exists x\ v(\kappa\cdot\alpha,x)$ making use of ω-α-consistency when going up.

Conversely, let R^k be formally defined by a ED_{n+1}-formula $v(\xi,x)$. Then $v(\xi,x)$ is an ED_n-formula, formally defining a relation S^{k+1} in $\mathbf{E}_{n+1}(\mathbf{DIO})$. Then there holds $R^k = \exists S^{k+1}$, because it is equivalent that

$\alpha \in R^k$
$\vdash \exists x v(\kappa\cdot\alpha,x)$ since R^k is formally defined by $\exists x\ v(\xi,x)$
there is an m such that $\vdash v(\kappa\cdot\alpha, \kappa_m)$ making use of ω-α-consistency when going down
there is an m such that $<\alpha,m> \in S^{k+1}$ since S^{k+1} is formally defined by $v(\xi,x)$.

This concludes the proof. If PB_0 is assumed to have A as a model then I can argue more rapidly, as I did already in the proof of Theorem 4.

Under the hypothesis that PB_0 be ω-α-consistent for ED^*-sentences, it follows from Theorem 6 that the relations in $\mathbf{E}^*(\mathbf{DIO})$ are precisely those which are formally definable by ED^*-formulas. Thus MATIYASEVIC's theorem $\mathbf{RER} = \mathbf{E}^*(\mathbf{DIO})$ implies these are the recursively enumerable relations.

Again I now have an arithmetical situation, and this time, recursive relations being recursively enumerable, it satisfies the condition **5**.(B_0). In analogy to Corollary 1, I obtain the

COROLLARY 3 Assume that PB_0 is ω-α-consistent for ED^*-sentences.

Precisely the relations in **RER** are formally definable by ED^*-formulas.

PB_0 is radically undecidable.

PD_0 is incomplete, its theory is undecidable, and and both *prov(PD)* and *provs(PD)* are examples of recursively enumerable sets which are not recursive.

In particular, the Corollary holds if PB_0 is assumed to have A as a model. Clearly, the axiom system PB_0 is much weaker than are PC or PB, but in contrast to Corollary 1, I here need the additional assumption of ω-α-consistency.

6. The Arithmetic PQ

I consider an equality language L which contains

a constant κ_0 ,
a 1-ary operation symbol s and the 2-ary operation symbols $+$ and $\cdot$,
no predicate symbols besides $\equiv$.

Let PQ be the axiom system

(PQ0) $\neg x \equiv \kappa_0 \rightarrow \exists y(x \equiv s(y))$

(PQ1) $s(x) \equiv s(y) \rightarrow x \equiv y$

(PQ2) $\neg s(x) \equiv \kappa_0$

(PQ3) $x+\kappa_0 \equiv x$

(PQ4) $x+s(y) \equiv s(x+y)$

(PQ5) $x\cdot\kappa_0 \equiv \kappa_0$

(PQ6) $x\cdot s(y) \equiv x\cdot y+x$.

PQ was studied by R.M.ROBINSON 50 and is often called *Robinson Arithmetic.* Observe that PQ, in contrast to PC and PB, contains only finitely many (namely seven) axioms. Apart from being esthetically pleasant, this fact will have a surprising consequence below.

The missing constants κ_n with $n>0$ can be introduced (a) by expanding L to a language L' containing such constants, and enlarging PQ to PQ' by adding, for every $n>0$, the sentence

(PQ7n) $\kappa_n \equiv s(\kappa_{n-1})$,

or (b) be defining κ_n as an abbreviation for the constant term $s(\kappa_{n-1})$ in which case (PQ7N) is provable by equality logic. Likewise, L' can be embedded into a language L" containing a predicate symbol $\leq$, and PQ' then can be embedded into an axiom system PQ" by adding

(PQ8) $x \leq y \leftrightarrow \exists z(z+x \equiv y)$.

It follows from the developments about equality logic that PQ' is conservative over PQ and that PQ" is conservative over PQ'. Also, every L"-formula is provably equivalent to an L-formula whence, in particular, PQ"-representability of relations and functions by L"-formulas is equivalent to their PQ-representability by L-formulas. In view of this situation, I shall, in the following, continue to speak of provability with respect to PQ while freely using L"-formulas and referring to the defining axioms (PQ7n) and (PQ8) for the κ_n and for $\leq$. Also, I shall speak for short about bounded formulas with respect to PQ when what is meant are bounded L"-formulas with respect to PQ".

LEMMA 6 The axioms of PB are provable from PQ .

The formulas (PB0) are proved by induction on m. For $m = 0$, I use (PQ3). Then

$\vdash \kappa_n+\kappa_m \equiv \kappa_{n+m} \rightarrow s(\kappa_n+\kappa_m) \equiv s\kappa_{n+m}$
$\vdash s(\kappa_n+\kappa_m) \equiv \kappa_n+s\kappa_m$ by (PQ4)
$\vdash s\kappa_m \equiv \kappa_{m+1}$ by (PQ7)
$\vdash s\kappa_{n+m} \equiv \kappa_{(n+m)+1}$ by (PQ7)
$\vdash \kappa_n+\kappa_m \equiv \kappa_{n+m} \rightarrow \kappa_n+\kappa_{m+1} \equiv \kappa_{(n+m)+1}$

and $\kappa_{(n+m)+1} = \kappa_{n+(m+1)}$. The formulas (PB1) are proved analogously from (PQ5) and (PQ6).

As for (PB2), I may assume $n<m$. I then prove the arising statement *for all* m by induction on n. If $0<m$ then $\vdash \kappa_m \equiv s\kappa_{m-1}$ by (PQ7), $\vdash \neg s\kappa_{m-1} \equiv \kappa_0$ by (PQ2) , hence $\vdash \neg \kappa_m \equiv \kappa_0$. If $\vdash \neg \kappa_n \equiv \kappa_m$ for all m with $n<m$ then $n+1<m$ implies $n<m-1$ whence $\vdash \neg \kappa_n \equiv \kappa_{m-1}$. This, together with (PQ1), gives $\vdash \neg s\kappa_n \equiv \kappa_{m-1}$, and with (PQ7) I obtain $\vdash \neg \kappa_{n+1} \equiv \kappa_m$.

Also (PB3) is proved by induction on n. For $n = 0$ I start from

$\vdash \neg s(z+y) \equiv \kappa_0$ by (PQ2)
$\vdash \neg z+sy \equiv \kappa_0$ by (PQ4)
$\vdash \neg (x \equiv sy \wedge z+x \equiv \kappa_0)$
$\vdash x \equiv sy \rightarrow \neg z+x \equiv \kappa_0$
$\vdash \exists y(x \equiv sy) \rightarrow \neg z+x \equiv \kappa_0$
$\vdash z+x \equiv \kappa_0 \rightarrow \neg \exists y(x \equiv sy)$
$\vdash \exists z(z+x \equiv \kappa_0) \rightarrow \neg \exists y(x \equiv sy)$

but

$\vdash x \leq \kappa_0 \rightarrow \exists z(z+x \equiv \kappa_0)$ by (PQ8)
$\vdash \neg x \leq \kappa_0 \rightarrow \neg\neg \exists y(x \equiv sy)$ by (PQ0)
$\vdash \neg \exists y(x \equiv sy) \rightarrow \neg\neg x \equiv \kappa_0$

whence

$\vdash x \leq \kappa_0 \rightarrow x \equiv \kappa_0$.

The induction step now is

$\vdash y \leq \kappa_n \rightarrow (y \equiv \kappa_0 \vee \ldots \vee y \equiv \kappa_n)$ by inductive [hypothesis

$\vdash y \leq \kappa_n \rightarrow (sy \equiv s\kappa_0 \vee \ldots \vee sy \equiv s\kappa_n)$

$\vdash y \leq \kappa_n \rightarrow (sy \equiv \kappa_1 \vee \ldots \vee sy \equiv \kappa_{n+1})$ by (PQ7)

$\vdash sy \equiv x \wedge y \leq \kappa_n \rightarrow (x \equiv \kappa_1 \vee \ldots \vee x \equiv \kappa_{n+1})$.

Also

$\vdash sy \equiv x \wedge z+x \equiv \kappa_{n+1} \rightarrow sy \equiv x \wedge z+x \equiv s\kappa_n$ by (PQ7)

$\vdash sy \equiv x \wedge z+x \equiv s\kappa_n \rightarrow sy \equiv x \wedge s(z+y) \equiv s\kappa_n$ by (PQ4)

$\vdash sy \equiv x \wedge z+x \equiv s\kappa_n \rightarrow sy \equiv x \wedge z+y \equiv \kappa_n$ by (PQ1)

$\vdash sy \equiv x \wedge z+x \equiv s\kappa_n \rightarrow sy \equiv x \wedge \exists(z+y \equiv \kappa_n)$

$\vdash sy \equiv x \wedge \exists(z+x \equiv s\kappa_n) \rightarrow sy \equiv x \wedge \exists(z+y \equiv \kappa_n)$

$\vdash sy \equiv x \wedge x \leq s\kappa_n \rightarrow sy \equiv x \wedge y \leq \kappa_n$

whence

$\vdash sy \equiv x \wedge x \leq \kappa_{n+1} \rightarrow (x \equiv \kappa_1 \vee \ldots \vee x \equiv \kappa_{n+1})$

$\vdash \exists y(sy \equiv x) \wedge x \leq \kappa_{n+1} \rightarrow (x \equiv \kappa_1 \vee \ldots \vee x \equiv \kappa_{n+1})$

$\vdash \neg x \equiv \kappa_0 \wedge x \leq \kappa_{n+1} \rightarrow (x \equiv \kappa_1 \vee \ldots \vee x \equiv \kappa_{n+1})$ by (PQ0)

$\vdash x \leq \kappa_{n+1} \rightarrow (x \equiv \kappa_0 \vee x \equiv \kappa_1 \vee \ldots \vee x \equiv \kappa_{n+1})$.

The propositional argument used to arrive at the last line is to conclude from $\neg a \wedge b \rightarrow c$ upon $\neg a \wedge b \rightarrow a \vee c$, hence $(a \wedge b) \vee (\neg a \wedge b) \rightarrow a \vee c$ since trivially $a \wedge b \rightarrow a \vee c$, and thus $(a \vee \neg a) \wedge b \rightarrow a \vee c$.

As for (PB4), I begin with some auxiliary provabilities :

(a) $\vdash \kappa_0+\kappa_m \equiv \kappa_m$ from the already proven (PB0)

(b) if $i \leq m$ then $\vdash \kappa_i \leq \kappa_m$ because $\vdash \kappa_{m-i}+\kappa_i \equiv \kappa_m$ by (PB0)

(c) $x \leq \kappa_n \rightarrow x \leq \kappa_{n+1}$

$\vdash x \equiv \kappa_i \rightarrow (\kappa_i \leq \kappa_n \rightarrow x \leq \kappa_n)$ by equality logic

for $i \leq n$ $\vdash x \equiv \kappa_i \rightarrow x \leq \kappa_{n+1}$ by (b)

$\vdash (x \equiv \kappa_0 \vee \ldots \vee x \equiv \kappa_n) \rightarrow x \leq \kappa_{n+1}$

Now I apply the already proven (PB3) .

(d) $\vdash sx+\kappa_n \equiv x+s\kappa_n$

This holds for $n = 0$ since $\vdash sx \equiv s(x+\kappa_0)$ by (PQ3) whence $\vdash sx \equiv x+s\kappa_0$ by (PQ4) , but also $\vdash sx+\kappa_0 \equiv sx$ by (PQ3). If (d) holds for n, then I continue by

$\vdash sx+s\kappa_n \equiv s(sx+\kappa_n)$ by (PQ4)

$\vdash s(sx+\kappa_n) \equiv s(x+s\kappa_n)$ by inductive hypothesis

$\vdash s(x+s\kappa_n) \equiv x+ss\kappa_n$ by (PQ4)

$\vdash sx+s\kappa_n \equiv x+ss\kappa_n$

(e) $\vdash \kappa_n \leq x \rightarrow (x \equiv \kappa_n \vee \kappa_{n+1} \leq x)$

$\vdash z+\kappa_n \equiv x \rightarrow z+\kappa_n \equiv x \wedge (z \equiv \kappa_0 \vee \neg z \equiv \kappa_0)$

$\vdash z+\kappa_n \equiv x \rightarrow (z+\kappa_n \equiv x \wedge z \equiv \kappa_0) \vee (z+\kappa_n \equiv x \wedge \neg z \equiv \kappa_0)$

$$\begin{array}{ll}
\vdash z+\kappa_n \equiv x \wedge z \equiv \kappa_0 \;\rightarrow\; \kappa_0+\kappa_n \equiv x & \\
\vdash z+\kappa_n \equiv x \wedge z \equiv \kappa_0 \;\rightarrow\; \kappa_n \equiv x & \text{by (a)} \\
\vdash z+\kappa_n \equiv x \;\rightarrow\; \kappa_n \equiv x \vee (z+\kappa_n \equiv x \wedge \neg z \equiv \kappa_0) . &
\end{array}$$

Further

$$\begin{array}{ll}
\vdash z+\kappa_n \equiv x \wedge sy \equiv z \;\rightarrow\; sy+\kappa_n \equiv x & \\
\vdash z+\kappa_n \equiv x \wedge sy \equiv z \;\rightarrow\; y+\kappa_{n+1} \equiv x & \text{by (d)} \\
\vdash z+\kappa_n \equiv x \wedge sy \equiv z \;\rightarrow\; \exists y(y+\kappa_{n+1} \equiv x) & \\
\vdash z+\kappa_n \equiv x \wedge \exists y(sy \equiv z) \;\rightarrow\; \exists y(y+\kappa_{n+1} \equiv x) & \\
\vdash z+\kappa_n \equiv x \wedge \neg z \equiv \kappa_0 \;\rightarrow\; \kappa_{n+1} \leq x & \text{by (PQ0).}
\end{array}$$

Hence

$$\begin{array}{l}
\vdash z+\kappa_n \equiv x \;\rightarrow\; \kappa_n \equiv x \vee \kappa_{n+1} \leq x \\
\vdash \exists z(z+\kappa_n \equiv x) \;\rightarrow\; \kappa_n \equiv x \vee \kappa_{n+1} \leq x .
\end{array}$$

Now I can prove (PB4) by induction on n. For $n = 0$ there holds $\vdash \kappa_0 \leq y$ by (PQ3), (PQ8). But also

$$\begin{array}{ll}
\vdash x \equiv \kappa_n \;\rightarrow\; (\kappa_n \leq \kappa_{n+1} \rightarrow x \leq \kappa_{n+1}) & \text{by equality} \\
\vdash x \equiv \kappa_n \;\rightarrow\; x \leq \kappa_{n+1} & \text{by (b) [logic} \\
\vdash (x \equiv \kappa_n \vee \kappa_{n+1} \leq x) \;\rightarrow\; (x \leq \kappa_{n+1} \vee \kappa_{n+1} \leq x) & \\
\vdash \kappa_n \leq x \;\rightarrow\; (x \leq \kappa_{n+1} \vee \kappa_{n+1} \leq x) & \text{by (e)} \\
\vdash x \leq \kappa_n \;\rightarrow\; x \leq \kappa_{n+1} & \text{(c)} \\
\vdash (x \leq \kappa_n \vee \kappa_n \leq x) \;\rightarrow\; (x \leq \kappa_{n+1} \vee \kappa_{n+1} \leq x) . &
\end{array}$$

This concludes the Proof of Lemma 6.

It was observed in connection with Corollary 1 that $\forall x(x+\kappa_0 \equiv x)$ is not provable in PB. As it is provable in PQ by (PQ3), the axiom system PQ is strictly stronger than PB.

Since PB was already recognized as being radically undecidable, there follows from Theorem 8.4 the

COROLLARY 4 PQ is radically undecidable .

In particular, if PQ is consistent then it is incomplete. This, again, is not surprising since, for instance, none of the formulas

$$\begin{array}{ll}
x \leq x & \neg\, x \equiv s(x) \\
x \leq y \wedge y \leq x \rightarrow x \equiv y & \kappa_0+x \equiv x \\
\kappa_0 \cdot x \equiv \kappa_0 & x+y \equiv y+x \\
x \cdot y \equiv y \cdot x & x+(y+z) \equiv (x+y)+z \\
x \cdot y \equiv (x \cdot y) \cdot z & x \cdot (y+z) \equiv x \cdot y + x \cdot z
\end{array}$$

is provable from PQ. To this end, consider a structure A obtained from ω by adding two new elements a_0 and a_1 on top. The operations of A extended those on ω by setting, for x in ω and z in ω or $z = a_i$, $i = 0,1$,

$$s(a_i) = a_i \quad a_i + x = a_i \quad z + a_i = a_{1-i} \quad x \cdot a_i = a_i \quad a_i \cdot 0 = 0 \quad a_i \cdot z = a_{1-i}$$

for $z \neq 0$. Then (PQ0–3) and (PQ5) obviously hold. Also, $z + s(a_i) = a_{1+i} = s(a_{1+i}) = s(z + a_i)$ and $x \cdot s(a_i) = x \cdot a_i = a_i = a_i + x = x \cdot a_i + x$, $a_i \cdot s(a_i) = a_i \cdot a_i = a_{1-i} = a_{1-i} + a_i = a_i \cdot a_i + a_i$, $a_{1-i} \cdot s(a_i) = a_{1-i} \cdot a_i = a_i = z + a_{1-i} = a_{1-i}\, a_i + a_i$. Thus A is a model of PQ. On the other hand

not $a_0 \leq a_0$ since $z + a_0 \neq a_0$ $\qquad$ $a_0 = s(a_0)$
$a_0 \leq a_1$ and $a_1 \leq a_0$ $\qquad$ $\kappa_0 + a_0 = a_1$
$\kappa_0 \cdot a_0 = a_0$ $\qquad$ $a_0 + a_1 = a_0$, $a_1 + a_0 = a_1$
$a_0 \cdot a_1 = a_1$, $a_1 \cdot a_0 = a_0$
$z + (a_0 + a_1) = z + a_0 = a_1$, $(z + a_0) + a_1 = a_0$
$a_0 \cdot (a_0 \cdot a_1) = a_0 \cdot a_1 = a_1$, $(a_0 \cdot a_0) \cdot a_1 = a_1 \cdot a_1 = a_0$
$a_0 \cdot (a_0 + a_1) = a_0 \cdot a_0 = a_1$, $a_0 \cdot a_0 + a_0 \cdot a_1 = a_1 + a_1 = a_0$.

It was observed by ROBINSON 50 (cf. also TARSKI–MOSTOWSKI–ROBINSON 53) that none of the seven subsystems, arising from PQ by removing one of its axioms, is radically undecidable: each of them has a decidable extension.

Summarizing one of the aspects treated in this Chapter, I have, in the form of PB, PC and PQ, found axiom systems G of fragments of arithmetic for which the general incompleteness theorems of Chapter 4 hold, and once, at the beginning of the next Chapter, also full arithmetic PA will have been defined, it will be evident that also PA is such a system G. The question whether there are complete axiomatizations of arithmetic was historically the starting point of all these investigations, and already from the Theorems 8.1 and 8.3 we now know that the answer must be No if such axiomatizations are to have a minimum of practicability. The point then of studying PB, PC and PQ is *not* that they provide examples of theories which are incomplete – which in view of their weakness can easily be seen by semantical arguments – it is *rather* that these fairly weak theories are radically undecidable.

7. Application: The Undecidability of Elementary Logic

The sentences, provable by ccqt [by $\text{ccqt}_=$] without any further axioms, may be called provable *by* [*equality*] *logic*. For those sentences which contain only 1–ary predicate symbols a decision method for validity, hence also for provability by logic, was found already by LÖWENHEIM 15 and simplified by SKOLEM 19 , BEHMANN 22 , HILBERT-BERNAYS 34 and QUINE 45 ; a detailed exposition is given as Chapter 4 in ACKERMANN 54 . For the general case, however, there can be no decision method for provability by logic.

Making use of the radical undecidability of PQ, together with the fact that PQ has only a finite number of axioms, this can be seen quite easily. The language L of PQ has only finitely many symbols; hence every expansion L_2 of L with a recursively presented signature contains the signature of L in the recursive way required for Corollary 8.3. Since PQ is finite, I now may apply this Corollary with $G_0 = PQ$ to any recursively enumerable axiom system G_2 such that $PQ \cup G_2$ is consistent. In particular, G_2 may be chosen as empty (or as the set of axioms of elementary logic), and so I obtain (a variant of) CHURCH's 36 Undecidability Theorem:

THEOREM 7 Under the hypothesis that PQ is consistent, provability by equality logic is undecidable in any equality language at least as expressive as the language of PQ.

CHURCH's proof also used a reduction to a finitely axiomatized, but more complicated fragment of arithmetic. A different, and in a way more direct proof was found by KALMAR 56 , cf. also Chapter 8 in SURANYI 59 .

As for logic without equality, it suffices to observe the

LEMMA 7 Let L_2 be a language with a recursively presented signature. If provability by equality logic is undecidable then so is provability by logic.

It follows from the hypothesis on L_2 that the set EQ of equality axioms is recursive. By definition of the calculus $ccqt_=$, provability by equality logic means provability from EQ by ccqt. A sentence v, therefore, is provable by $ccqt_=$ if there is a finite subset F_v of EQ such that it is provable from F_v by ccqt, and F_v may be chosen uniquely as to consist of those equality axioms referring to the function symbols and predicate symbols occurring in v. As F_v consists of sentences, v then is provable by $ccqt_=$ if, and only if, $\bigwedge F_v \rightarrow v$ is provable by ccqt. Consequently, an algorithm deciding provability by ccqt will also decide provability by $ccqt_=$, and if χ and $\chi_=$ are the characteristic functions of the codes of sentences provable by ccqt and $ccqt_=$ respectively, then $\chi_= = \chi \circ \psi$ where ψ is the recursive function with $\psi(\ulcorner v \urcorner) = \ulcorner \bigwedge F_v \rightarrow v \urcorner$.

The language L of PQ contains one 2-ary predicate symbol, but also the constant κ_0 and the operation symbols s, + and · ; the latter may be replaced by 2- and 3-ary predicate symbols. So there still is a gap between L and the languages with only 1-ary predicate symbols for which provability *is* decidable. Yet there holds the

COROLLARY 5 Under the hypothesis that PQ is consistent, provability by logic is undecidable in any language containing at least a 2-ary predicate symbol.

This follows immediately from KALMAR's 32, 36 theorem that decidability of validity is reducible to the particular case of a formula containing only one 2-ary predicate symbol; cf. also Theorem XIV in SURANYI 59. It also can be shown by applying an idea of SZMIELEW-TARSKI 50 who proposed an interpretation of PQ in a fragment of a theory of finite sets.

References

W. Ackermann: Solvable Cases of the Decision Problem. Amsterdam 1954

H. Behmann: Beiträge zur Algebra der Logik, insbesondere zum Entscheidungsproblem. Math. Ann. 86 (1922) 163–229

A. Church: A note on the Entscheidungsproblem. J.Symb.Logic 1 (1936) 40–41 and 101–102

A. Church: An unsolvable problem of elementary number theory. American J.of Math. 58 (1936) 345–363

L. Kalmar: Zum Entscheidungsproblem der mathematischen Logik. Verhandl. des Internat. Math.Kongresses Zürich 1932, vol. 2, 337–338

L. Kalmar: Zurückführung des Entscheidungsproblems auf den Fall von Formeln mit einer einzigen binären Funktionsvariablen. Compositio Math. 4 (1936) 137–144

L. Kalmar: Ein direkter Beweis für die allgemein-rekursive Unlösbarkeit des Entscheidungsproblems des Prädikatenkalküls der ersten Stufe mit Identität. Zeitschr.Math. Logik Grundlagen d.Math 2 (1956) 1–14

D. Hilbert, P. Bernays: Grundlagen der Mathematik I. Berlin 1934

L. Löwenheim: Über Möglichkeiten im Relativkalkül. Math.Ann. 76 (1915) 137–148

W. van Orman Quine: On the Logic of Quantification. J.Symb.Logic 10 (1945) 1–12

Th. Skolem: Untersuchungen über die Axiome des Klassenkalkuls und über Produktations- und Summationsprobleme, welche gewisse Klassen von Aussagen betreffen. Skrifter Videnskab. Kristiana 3 (1919) 1–37

J. Surányi: Reduktionstheorie des Entscheidungsproblems im Prädikatenkalkul der ersten Stufe. Budapest 1959

W. Szmiliew, A. Tarski: Mutual interpretability of some essentially undecidable theories. Proc. Internat.Congress of Math. Cambridge 1950, vol. 1, 734

Chapter 10. Peano Arithmetic PA and its Expansion PR

The structure of this Chapter is simple; there is only one language being considered and only one axiom system, that of Peano arithmetic. In Section 1 a few mathematical statements will be derived from these axioms, and some of these proofs will be carried out using only intuitionistic logic, in which case Peano arithmetic is also called Heyting arithmetic HA.

Let L be the equality language used for PQ, containing a constant κ_0, a 1-ary operation symbol s, and the 2-ary operation symbols + and $\cdot$. Let PA be the axiom system by adding to the axioms

(PQ0) $\neg x \equiv \kappa_0 \rightarrow \exists y(x \equiv s(y))$
(PQ1) $s(x) \equiv s(y) \rightarrow x \equiv y$
(PQ2) $\neg s(x) \equiv \kappa_0$
(PQ3) $x+\kappa_0 \equiv x$
(PQ4) $x+s(y) \equiv s(x+y)$
(PQ5) $x\cdot\kappa_0 \equiv \kappa_0$
(PQ6) $x\cdot s(y) \equiv x\cdot y+x$.

of PQ, for every L-formula v(x), the universal closure of the formula

(ISA) $v(\kappa_0) \wedge \forall x(v(x) \rightarrow v(s(x))) \rightarrow \forall x\, v(x)$.

Thus PA has infinitely many axioms, and (ISA) is an axiom *schema*, called the *induction* schema. The set of all L-formulas being elementary, so is the set of all formulas (ISA); hence also PA is elementary, and since it contains QA it also is, if consistent, radically undecidable and incomplete.

PA is called *Peano-Arithmetic*, and the axioms (PQ0), (PQ1), (PQ2) and (ISA) (together with the equality axiom for s) are often called *Peano's* axioms for natural numbers. Historically this name is incorrect because (1) PEANO 89 and 91 did *not* use the language L with an axiom schema such as (ISA), but a language with *set* variables X and a single induction axiom for subsets of natural numbers, and (2) PEANOs characterization was already stated (and used) by DEDEKIND 88 . In Dedekind's set theoretical approach then (PQ3)-(PQ6) had not to be requested since the set theoretical frame permitted him to prove the existence (or the conservativness) of (adding function symbols for) functions defined by primitive recursion.

What is called *Heyting arithmetic* HA shall have the same axioms as PA, but shall differ from PA in that only the calculus of *intuitionistic* logic is used for its derivations. If not stated explicitly otherwise, all the derivations in Sections 1 and 2 work within HA. In the following, I shall denote as $\vdash$ the provability in an MP-calculus ciqt in the sense of Chapter 2.9.

As a first application of (ISA), observe that (PQ0) can be omitted. Because if v(x) is

$$\neg\, x \equiv \kappa_0 \rightarrow \exists y(x \equiv s(y))$$

then $\vdash \neg\, \kappa_0 \equiv \kappa_0$ implies $\vdash v(\kappa_0)$. Also, $\vdash s(x) \equiv s(x)$ implies

$$\vdash \exists y(s(x) \equiv s(y)) \;,$$

hence $\vdash v(s(x))$, $\vdash v(x) \rightarrow v(s(x))$ and $\vdash \forall x(v(x) \rightarrow v(s(x)))$, such that (ISA) gives the universal closure of (PQ0).

It follows from quantifier logic that (ISA) may be replaced by

(ISA') $\quad v(\kappa_0) \wedge \forall x(v(x) \rightarrow v(s(x))) \rightarrow v(y)$ with y *not* free in $v(\kappa_0)$,

and both HA and PA admit the *induction rule*

(IRA) $\quad v(\kappa_0),\, v(x) \rightarrow v(s(x)) \vdash v(y)$

from which, conversely, every instance of (ISA) can be derived v; the proof was carried out already in Section 2.4.

In the course of this Chapter, I shall repeatedly extend the language L of HA and PA by defined predicates and terms; it is important to notice that (ISA) and (IRA) become available also for all newly arising formulas of such extensions. Observe first that every such formula clearly contains only a finite number of new predicates and terms; so it will suffice to show that, if (ISA) can be derived for the formulas of an extension L' of L, then it can be derived for the formulas of an extension L" of L' by a defined predicate or term. But HA and PA, formulated for L", also contain the defining axiom u^f of that extension, and it follows from Lemma 2.10.8 that, for every L"-formula b, I can find an L'-formula b^σ provably equivalent to b under the axiom u^f and such that $fr(b) = fr(b^\sigma)$. Thus if (ISA) is available for L'-formulas, then it can be applied to an L"-formula v by applying it to the L'-formula v^σ.

In place of (ISA), weaker schemata (ISA_0), (ISA_δ), (ISA_σ) may be considered, obtained by restricting the formula v to be open or to be a Δ_0– or a Σ_1–formula in the sense of 9.4 for a language containing $\lesssim$ as basic predicate symbol; observe that for these schemata the rule corresponding to (IRA) remains admissible. Much of what follows can already be developed in the *fragments* of PA employing these weaker schemata; the reader will find a comprehensive discussion in HAJEK–PUDLAK 93, Chapter 1.

In Book 2, Chapter 10, I have shown that, quite generally, languages can always be expanded by a new function symbol *f* and a defining axiom for the arising *f*–terms such that this expansion is *conservative*: no old formula is provable with help of the new axiom, unless it was provable already without it. In Section 2 the Theorem 1 will say that, for every primitive recursive function f, my language L can be expanded in this manner such that the defining *recursion equations* of f, now formulated as free variable formulas with the function symbol *f*, become provable in this expansion of PA (or HA). This, clearly, is anything but obvious and amounts to the construction of an explicitly defining formula for a function given only recursively. As a matter of fact, in Chapter 1.16 the much simpler case was treated that in my language I can find a formula *semantically* defining such a function (in that case exponentiation) in a model of arithmetic, and the construction of this formula proceeded by an arithmetization of Dedekind's set theoretical proof of the existence of recursively defined functions. Here now, the uncontrolled, semantical statements about this arithmetization have to be formalized into syntactical provabilities, and it will be a sign of the strength of PA (and HA) that in presence of this axiom system such formal arithmetization can be carried out.

The third Section of this Chapter offers a first view of primitive recursive arithmetic PR, the expansion of PA obtained by adding primitive recursive terms, formed with function symbols for a family of (possibly all) primitive recursive functions. While conservative over PA, PR acquires its power and elegance from the provability of the free variable forms of recursion equations (and not only of their representing instances with numerical constants), and the presence of primitive recursive terms then makes it also possible to establish as provable the formulas expressing the properties of functions defined by of course of value recursion. Being a conservative expansion, PR does not actually say more than PA, but what it does say it says in a technically simpler manner – for instance, if R^n is a relation with a primitive recursive characteristic function χ_R, and if $T(\xi)$ is the term introduced for χ_R, then R^n in PR is represented by the simple atomic formula $T(\xi) \equiv \kappa_1$.

References

R. Dedekind: Was sind und was sollen die Zahlen . Braunschweig 1888

P. Hajek and P. Pudlak: Metamathematics of First-Order Arithmetic . Berlin 1993

D. Hilbert, P. Bernays: Grundlagen der Mathematik I . Berlin 1934

G. Peano: Arithmetices Principia Novo Methodo Exposita, Turin 1889

G. Peano: Sul concetto di numero, Rivist. di Mat. 1 (1891) 87–102)

1. Development of Basic Arithmetic

For reasons of convenience, I shall denote as L also the expansion of the language of PA by constants κ_n for $n>0$, connected with s by

(PQ7n) $\kappa_n \equiv s(\kappa_{n-1})$

and introduced as for PQ either by actual expansion or by explicit internal definition as $s(\kappa_{n-1})$. For PA, as well as for HA, many elementary facts from naive arithmetic can be proved, as long as they can be expressed in a language which does not use variables for sets. I begin with certain observations which will lead to an important variant of the induction schema. During the proofs, employing (ISA_0), I shall present the arguments in an abbreviated manner.

(a_0) $\vdash x+(y+z) \equiv (x+y)+z$.

Induction on z. If $z = \kappa_0$ then I use (PQ3); in the induction step I use (PQ4): $(x+y)+s(z) \equiv s((x+y)+z) \equiv s(x+(y+z)) \equiv x+s(y+z) \equiv x+(y+s(z))$.

(a_1) $\vdash \kappa_0+x \equiv x$.

Induction on x, making use of (PQ3) and (PQ4).

(a_2) $\vdash s(x)+y \equiv x+s(y)$.

Induction on y. First $s(x)+\kappa_0 \equiv s(x) \equiv s(x+\kappa_0) \equiv x+s(\kappa_0)$; in the induction step then $s(x)+s(y) \equiv s(s(x)+y) \equiv s(x+s(y)) \equiv x+ss(y)$.

(a_3) $\vdash x+y \equiv y+x$.

Induction on y. First (a_1); then (a_2) for $x+s(y) \equiv s(x+y) \equiv s(y+x) \equiv y+s(x) \equiv s(y)+x$.

(a_4) $\vdash x+y \equiv y \rightarrow x \equiv \kappa_0$.

Induction on y, concluding from $x+s(y) \equiv s(x+y) \equiv s(y)$ by (PQ1) that $x+y \equiv y$.

(a_5) $\vdash \neg(x \equiv \kappa_0) \rightarrow \neg(x+y \equiv y)$.

(a_6) $\vdash x+y \equiv \kappa_0 \rightarrow x \equiv \kappa_0 \quad , \quad x+y \equiv \kappa_0 \rightarrow y \equiv \kappa_0$.

I prove the second statement from which the first follows by (a_3). For the induction step I conclude from (PQ2) and $x+s(y) \equiv s(x+y)$ by *ex absurdo quodlibet*.

It follows from (a_1) that the reverse implication in (a_4) is provable, hence also the reverse implication in (a_5). – Next I define the predicate symbol $\leq$ by

(PQ8) $\quad x \leq y \leftrightarrow \exists z(z+x \equiv y)$

and obtain

(b_0) $\quad \vdash \kappa_0 \leq x$
(b_1) $\quad \vdash x \leq x$
(b_2) $\quad \vdash x \leq y \wedge y \leq z \rightarrow x \leq z$
(b_3) $\quad \vdash x \leq y \wedge y \leq x \rightarrow x \equiv y$.

Here (PQ3) implies (b_0) and (a_1) implies (b_1). For (b_2), I conclude from $z_0+x \equiv y$ and $z_1+y \equiv z$ that $(z_1+z_0)+x \equiv z_1+(z_0+x) \equiv z$ by (a_1). For (b_3), I conclude from $z_0+x \equiv y$ and $z_1+y \equiv x$ that $(z_1+z_0)+x \equiv x$, hence $z_1+z_0 \equiv \kappa_0$ by (a_4) whence $z_1 \equiv \kappa_0$ by (a_6). – I now define the predicate symbol $<$ by

(PQ9) $\quad x<y \leftrightarrow x \leq y \wedge \neg\, x \equiv y$,

and as its right hand formula is equivalent to each of

$\exists z\,(z+x \equiv y \wedge \neg\, x \equiv y)$
$\exists z\,(z+x \equiv y \wedge \neg\, z+x \equiv x)$
$\exists z\,(z+x \equiv y \wedge \neg\, z \equiv \kappa_0)$ by (a_5)
$\exists z\,(z+x \equiv y \wedge \exists z'(z \equiv s(z') \wedge \neg\, s(z') \equiv \kappa_0))$ by (PQ0)
$\exists z\,(z+x \equiv y \wedge \exists z'(z \equiv s(z')))$ by (PQ2)
$\exists z\, \exists z'\,(z+x \equiv y \wedge z \equiv s(z'))$
$\exists z'\,(s(z')+x \equiv y)$,

I arrive at

(b_4) $\quad \vdash x<y \leftrightarrow \exists z\,(z+x \equiv y \wedge \neg\, z \equiv \kappa_0)$,
$\quad \vdash x<y \leftrightarrow \exists z\,(s(z)+x \equiv y)$.

Applying this, I find

(b_5) $\quad \vdash x<y \wedge y \leq z \rightarrow x<z$,
$\quad \vdash x<y \wedge y \leq z \rightarrow x \leq z$,
$\quad \vdash x \leq y \wedge y<z \rightarrow x<z$,

since e.g. $s(z_0)+x \equiv y$, $z_1+y \equiv z$ imply $s(z_1+z_0)+x \equiv (z_1+s(z_0))+x \equiv z_1+$ $+(s(z_0)+x) \equiv z$; in the same manner $z_0+x \equiv y$, $s(z_1)+y \equiv z$ imply $s(z_1+$ $z_0)+x \equiv (z_1+s(z_0))+x \equiv (s(z_1)+z_0)+x$ making use of (a_2).

(b_6) $\quad \vdash x \leq y \leftrightarrow x < s(y)$,
(b_7) $\quad \vdash x<y \leftrightarrow s(x) \leq y$,
(b_8) $\quad \vdash x < s(x)$.

Because the formulas

$\exists z\ (z+x \equiv y)$
$\exists z\ (s(z+x) \equiv s(y))$ by (PQ1)
$\exists z\ (z+s(x) \equiv s(y))$
$\exists z\ (s(z)+x \equiv s(y))$ by (a_2)

are equivalent and imply (b_6) and also (b_7); hence also (b_8).

(b_9) $\vdash \neg\, x<x$,
(b_{10}) $\vdash \neg\, x<\kappa_0$,
(b_{11}) $\vdash x<y \;\rightarrow\; \neg\, y\leq x$,
(b_{12}) $\vdash y\leq x \;\rightarrow\; \neg\, x<y$.

Because $x<x$ is equivalent to

$\exists z\ (s(z)+x \equiv x)$
$\exists z\ (s(z)+x \equiv x \wedge x \equiv \kappa_0)$ by (a_4)
$\exists z\ (s(z) \equiv \kappa_0)$

thus $x<x$ implies $\exists z\ (s(z) \equiv \kappa_0)$, hence $\neg\, \exists z\ (s(z) \equiv \kappa_0) \rightarrow \neg\, x<x$; but $\vdash \forall x\ (\neg\ s(x) \equiv \kappa_0)$ by (PQ2), $\vdash \neg\ \exists z(s(z) \equiv \kappa_0)$, and thus ($b_9$). Also, $\kappa_0 \leq x \rightarrow (x<\kappa_0 \rightarrow x<x)$ by (b_5), hence $x<\kappa_0 \rightarrow x<x$ by (b_0), and so (b_{10}) by minimal contraposition from (b_9). Further, $x<y \rightarrow (y\leq x \rightarrow x<x)$ by (b_5), $x<y \rightarrow (y\leq x \rightarrow (x<x \wedge \neg\, x<x))$ by (b_9), hence (b_{11}) by *ex absurdo quodlibet*. Finally, (b_{12}) from (b_{11}) by minimal contraposition.

(b_{13}) $\vdash x \equiv \kappa_0 \vee \neg\, x \equiv \kappa_0$.

Induction on x with (PQ2) in the induction step.

(b_{14}) $\vdash x\leq y \leftrightarrow x<y \vee x \equiv y$.

This follows from the equivalences

$\exists z\ (z+x \equiv y)$
$\exists z\ (z \equiv \kappa_0 \wedge z+x \equiv y \;\vee\; \neg\, z \equiv \kappa_0 \wedge z+x \equiv y)$ by (b_{13})
$x \equiv y \vee \exists z(\neg\, z \equiv \kappa_0 \wedge z+x \equiv y)$
$x \equiv y \vee x<y$ by (b_4).

(b_{15}) $\vdash y<s(x) \leftrightarrow y<x \vee y \equiv x$.
(b_{16}) $\vdash x<y \vee x \equiv y \vee y<x$.
(b_{17}) $\vdash x\leq y \vee y\leq x$.

(b_{15}) follows from (b_6), (b_{14}). (b_{17}) will follow from (b_{14}) and (b_{16}). This now holds for $y = \kappa_0$ by (b_0) and (b_{14}). Now induction on y :

$x<y \vee x \equiv y \vee y<x \vdash x<s(y) \vee x<s(y) \vee y<x$ by (b_8), (b_5)
$\vdash x<s(y) \vee y<x$
$\vdash x<s(y) \vee s(y)\leq x$ by (b_7)
$\vdash x<s(y) \vee s(y) \equiv x \vee s(y)<x$ by (b_{14}).

(b_{18}) $\vdash x \equiv y \vee \neg\, x \equiv y$.
(b_{19}) $\vdash x<y \vee \neg\, x<y$.
(b_{20}) $\vdash x\leq y \vee \neg\, x\leq y$.

First, (b_{16}) implies (b_{18}) since $x<y \to \neg\, x \equiv y$ by (PQ9). Also, (b_{16}) implies (b_{19}) since (1) $x \equiv y \to \neg\, x<y$ from $x<y \to \neg\, x \equiv y$ by minimal contraposition, and (2) $y<x \to \neg\, x<y$ since $y<x \to \neg\, x \lesssim y$ by (b_{11}), as well as $\neg\, x \lesssim y \to \neg\, x<y$ by minimal contraposition from $x<y \to x \lesssim y$ in (PQ9). Finally, (b_{16}) implies $x \lesssim y \vee y<x$, hence (b_{20}) by (b_{11}).

In PA, obviously, the statements (b_{13}), (b_{18}), (b_{19}), (b_{20}) do not require mathematical proofs. Recalling the notion of *decidable* elements in a Heyting algebra from Chapter 2.6, (b_{18}), (b_{19}), (b_{20}) show that the atomic formulas $x \equiv y$, $x<y$ and $x \lesssim y$ determine decidable elements in the Heyting algebra associated to HA, and since the union of decidable elements is decidable again, (b_{20}) also follows from (b_{18}), (b_{19}) via (b_{14}). Decidable elements being also regular, classical contraposition rules become admissable for implications between them. In particular, the proof of Lemma 3 for PQ remains valid in HA, because its classical arguments remain admissible as $x \equiv \kappa_0$ first is regular and then is decidable.

(b_{21}) $\quad \vdash \neg\, y<x \to x \lesssim y$,
(b_{22}) $\quad \vdash \neg\, x \lesssim y \to y<x$.

It follows from (b_{14}) that $\vdash \neg x \lesssim y \leftrightarrow \neg x<y \wedge \neg x \equiv y$. There holds propositionally $\neg a \to (a \vee b \to b)$ since $\neg a \to (a \to a)$ trivially and $\neg a \to (a \to b)$ by 2.7.(i_0). Hence also $\vdash \neg a \wedge \neg b \to (a \vee b \vee c \to c)$ propositionally, and so (b_{16}) implies (b_{22}). From that follows (b_{21}) by admissible classical contraposition.

I now shall introduce several more induction principles, beginning with the schema

(ISO) $\quad \forall x\, (\forall y\, (y<x \to v(y)) \to v(x)) \to v(x)$

and, for v in which x is the only variable free, the rule

(IRO) $\quad \forall y\, (y<x \to v(y)) \to v(x) \vdash v(x)$

of *order induction*. Clearly, (IRO) is admissible if (ISO) is derivable. In order to derive (ISO), I set

$$w_0 = \forall x (\forall y (y<x \to v(y)) \to v(x))$$

$$w_1(x) = \forall y\, (y \lesssim x \to v(y)) \quad .$$

Since $\vdash w_0 \to (\forall y\, (y<\kappa_0 \to v(y)) \to v(\kappa_0))$ as a quantifier axiom, and since (b_{10}) implies $\vdash \forall y\, (y<\kappa_0 \to v(y))$, there follows propositionally

$$\vdash w_0 \to v(\kappa_0) \ .$$

Since $\vdash v(\kappa_0) \to w_1(\kappa_0)$ again by (b_{10}), there follows

(z0) $\quad \vdash w_0 \to w_1(\kappa_0)$.

Further, $\vdash w_0 \to \forall y\, (y<s(x) \to v(y)) \to v(s(x))$ as a quantifier axiom. But $\vdash \forall y\, (y<s(x) \to v(y)) \to \forall y\, (y \lesssim x \to v(y)$ by (b_6) , hence

$\vdash w_o \to (\forall y\, (y \le x \to v(y)) \to v(s(x)))$,
$\vdash w_o \to (w_1(x) \to v(s(x)))$.

Also by (b_6) now $\vdash w_1(x) \leftrightarrow \forall y\, (y < s(x) \to v(y))$, whence

(z1) $\quad \vdash w_o \to (w_1(x) \to \forall y\, (y < s(x) \to v(y)) \wedge v(s(x)))$.

By (b_{15}) also $\vdash y \le s(x) \leftrightarrow y < s(x) \vee y \equiv s(x)$, hence

$\vdash \forall y\, ((y < s(x) \vee y \equiv s(x)) \to v(y)) \wedge v(s(x)) \to \forall y\, (y \le s(x) \to v(y))$
$\vdash \forall y\, (y < s(x) \to v(y)) \wedge v(s(x)) \to \forall y\, (y \le s(x) \to v(y))$
$\vdash \forall y\, (y < s(x) \to v(y)) \wedge v(s(x)) \to w_1(s(x))$.

Thus by (z1)

$\vdash w_o \to (w_1(x) \to w_1(s(x)))$
$\vdash w_o \to \forall x\, (w_1(x) \to w_1(s(x)))$

and by (z0)

$\vdash w_o \to w_1(\kappa_0) \wedge \forall x\, (w_1(x) \to w_1(s(x)))$.

Making use of the axiom (ISA) for $w_1(x)$ in place of $v(x)$, the propositional chain rule gives

$\vdash w_o \to \forall x\, w_1(x)$

whence $\vdash w_o \to w_1(x)$. But $\vdash w_1(x) \to v(x)$ by (b_1), and so I arrive at (ISO): $\vdash w_o \to v(x)$. – It will be noticed that, if v in (ISO) is a Δ_o-formula, then only (ISA_δ) is needed in this derivation.

The principle of *symmetric double induction* concerns a formula $v(x,y)$ with free variables x,y which do not occur bound in v :

(IRDS) $\quad v(\kappa_0, y),\ v(x, \kappa_0),\ v(x,y) \to v(s(x), s(y)) \ \vdash\ v(x,y)$.

In the proof, I abbreviate as M the three premisses and, with a new variable z, I set

(x) $\quad w_1(z,y) = v(z, y+z) \quad$ and $\quad w_2(z,x) = v(x+z, z)$.

Employing (ISA) I first prove

$M \vdash w_1(z,y) \quad$ and $\quad M \vdash w_2(z,x)$

by induction on z. Indeed, there holds

$M \vdash w_1(\kappa_0, y)$ since $\vdash v(\kappa_0, y) \to v(\kappa_0, y+\kappa_0)$,
$M \vdash w_1(z,y) \to w_1(s(z),y)$ since $w_1(s(z), y) = v(s(z), y+s(z))$ and $\vdash v(s(z), s(y+z)) \to v(s(z), y+s(z))$,
$M \vdash w_2(\kappa_0, x)$ since $\vdash v(x, \kappa_0) \to v(x+\kappa_0, \kappa_0)$,
$M \vdash w_2(z,x) \to w_2(s(z),x)$ since $w_2(s(z), x) = v(x+s(z), s(z))$ and $\vdash v(s(x+z), s(z)) \to v(x+s(z), s(z))$.

Replacing variables in (x), I have

$M \vdash w_1(x,z) \quad$ and $\quad M \vdash w_2(y,z)$,

and so $\vdash x \leq y \rightarrow \exists z\,(z+x \equiv y)$ implies

$$M \vdash x \leq y \rightarrow \exists z\,(z+x \equiv y \wedge w_1(x,z))$$
$$M \vdash y \leq x \rightarrow \exists z\,(z+y \equiv x \wedge w_2(y,z))\ .$$

But $w_1(x,z) = v(x,\, z+x)$, $w_2(y,z) = v(z+y\, ,\, y)$ gives

$$\vdash \exists z\,(z+x \equiv y \wedge w_1(x,z)) \rightarrow v(x,y)\ ,\quad \vdash \exists z(z+y \equiv x \wedge w_2(y,z)) \rightarrow v(x,y)$$

whence

$$M \vdash x \leq y \rightarrow v(x,y)\quad ,\quad M \vdash y \leq x \rightarrow v(x,y)\ ,$$

such that (b_{17}) leads to $M \vdash v(x,y)$.

The last induction priciple to be considered concerns the well ordering of natural numbers. Making use of the abbreviation $E(v(x)) = v(x) \vee \neg v(x)$ it is stated as the *Principle of the Smallest Number*

$$\text{(NISA)}\qquad \forall x\, E(v(x)) \wedge \exists x\, v(x) \rightarrow \exists y(v(y) \wedge \forall z(z<y \rightarrow \neg v(z)))\ ;$$

obviously, the decidability hypothesis $\forall x\, E(v(x))$ can be omitted for PA. For a proof of (NISA) I set

$$u(y) = \forall z\,(z<y \rightarrow \neg v(z))$$
$$q = \exists y\,(v(y) \wedge u(y))$$
$$w(x) = \exists y\,(y<x \wedge v(y) \wedge u(y))\ .$$

There follow from the definitons

(p_0)	$\vdash w(x) \rightarrow q$	
(p_1)	$\vdash u(\kappa_0)$	by (b_{10})
(p_2)	$\vdash w(x) \rightarrow w(s(x))$	
(p_3)	$\vdash u(x) \wedge v(x) \rightarrow w(s(x))$	
(p_4)	$\vdash u(x) \wedge \neg v(x) \rightarrow u(s(x))$	
(p_5)	$\vdash u(s(x)) \rightarrow \neg v(x)$.	

Hence

(p_6)	$\vdash v(x) \wedge w(s(x)) \rightarrow q$	by (p_0)
	$\vdash v(x) \wedge u(s(x)) \rightarrow \neg v(x)$	by (p_5)
	$\vdash v(x) \wedge u(s(x)) \rightarrow v(x)$	
(p_7)	$\vdash v(x) \wedge u(s(x)) \rightarrow q$	ex absurdo quodlibet
(p_8)	$\vdash v(x) \wedge (w(s(x)) \vee u(s(x))) \rightarrow q$	by (p_6), (p_7).

Further

$$\vdash w(x) \rightarrow w(s(x)) \vee u(s(x)) \qquad \text{by } (p_2)$$
$$\vdash E(v(x)) \wedge u(x) \rightarrow w(s(x)) \vee u(s(x)) \qquad \text{by } (p_3),\ (p_4)$$
$$\vdash E(v(x)) \rightarrow (w(x) \vee u(x) \rightarrow w(s(x)) \vee u(s(x)))$$
$$\vdash \forall x\ E(v(x)) \rightarrow (w(x) \vee u(x) \rightarrow w(s(x)) \vee u(s(x)))$$
$$\vdash \forall x\ E(v(x)) \rightarrow \forall x\ (w(x) \vee u(x)) \rightarrow w(s(x)) \vee u(s(x)))\ .$$

Since also $\vdash \forall x\ E(v(x)) \rightarrow w(\kappa_0) \vee u(\kappa_0)$ by (p_1), there follows

$\vdash \forall x E(v(x)) \rightarrow (w(\kappa_0) \vee u(\kappa_0)) \wedge \forall x\,(w(x) \vee u(x)) \rightarrow w(s(x)) \vee u(s(x)))$.

Applying the chain rule to this formula and to (ISA) for $w(x) \vee u(x)$, I find

$$\vdash \forall x E(v(x)) \rightarrow \forall x\,(w(x) \vee u(x))\ .$$

Hence in particular

$$\vdash \forall x\, E(v(x)) \wedge v(x) \rightarrow (w(s(x)) \vee u(s(x))) \wedge v(x)$$

and so by (p_8)

$$\vdash \forall x\, E(v(x)) \wedge v(x) \rightarrow q$$

which proves (NISA). – It will be noticed again that, if v in (NISA) is a Δ_0-formula, then only (ISA_δ) is needed in this derivation.

Decidability implying regularity, under the hypothesis $\forall x\, E(v(x))$ contraposition gives from (b_{21})

$$\vdash (z<y \rightarrow \neg v(z)) \rightarrow (v(z) \rightarrow y \leq z)$$

and from (b_{11})

$$\vdash (v(z) \rightarrow y \leq z) \rightarrow (z<y \rightarrow \neg v(z))\ .$$

Hence, (NISA) is equivalent to

(NISA') $\vdash \forall x\, E(v(x)) \wedge \exists x\, v(x) \rightarrow \exists y\,(v(y) \wedge \forall z\,(v(z) \rightarrow y \leq z))$.

A characteristic application of (NISA') is the introduction of a term for the remainder under division. Let $v(x) = v(b,a,x)$ be $\exists y\,(b \equiv a \cdot y + x)$; since $\vdash b \equiv a \cdot \kappa_0 + b$ there follows $\vdash \exists x\, v(b,a,x)$. Making use of the formula $u(y) = u(b,a,y)$ introduced above, classical logic gives by (NISA')

$\vdash \forall b\, \forall a\, \exists y\,(v(b,a,y) \wedge u(b,a,y))$, where also
$\vdash u(b,a,y) \wedge u(b,a,y') \rightarrow y \equiv y'$ by (b_3) .

Thus the term MOD(a,b) can be introduced by the defining axiom

(mod) $\forall b\, \forall a\, (v(b,a,\mathrm{MOD}(a,b)) \wedge u(b,a,\mathrm{MOD}(a,b)))$.

But the use of classical logic and of (NISA') can be avoided. Observe first that

(a_7) $\vdash x+z \equiv y+z \rightarrow x \equiv y$,
(a_8) $\vdash x \cdot (y+z) \equiv x \cdot y + x \cdot z$,
(a_9) $\vdash x \cdot y \equiv y \cdot x$,
(a_{10}) $\vdash \kappa_1 \cdot z \equiv z$
(b_{23}) $\vdash x \leq x+y$,
(b_{24}) $\vdash \neg z \equiv \kappa_0 \rightarrow x \leq x \cdot z$.

Here (b_{23}) follows from (a_3), and (a_{10}) from (PQ6) and (a_9). The four other statements are proved by induction on z where (a_7), (a_8), (b_{24}) use (PQ1), (a_0), (b_{23}) respectively, and the case $z = \kappa_0$ in (b_{24}) follows by ex absurdo quodlibet. Taking $v_0(b,a,x,y)$ to be the formula

$$b \equiv a \cdot y + x \;\wedge\; (\neg\, a \equiv \kappa_0 \to x < a)$$

I first prove

(p_9) $\quad \vdash \exists x\, \exists y\; v_0(b,a,x,y)$

by induction on b. For $b = \kappa_0$ this follows by (b_0) and (PQ9) from $\vdash \kappa_0 \equiv a \cdot \kappa_0 + \kappa_0$ and $\vdash \neg\, a \equiv \kappa_0 \to \kappa_0 < a$. There further hold

$\vdash b \equiv a \cdot y + x \;\to\; s(b) \equiv a \cdot y + s(x)$

(p_{10}) $\quad \vdash v_0(b,a,x,y) \;\to\; s(b) \equiv a \cdot y + s(x)$

$\vdash v_0(b,a,x,y) \wedge a \equiv \kappa_0 \;\to\; s(b) \equiv a \cdot y + s(x) \wedge a \equiv \kappa_0$

(p_{11}) $\quad \vdash v_0(b,a,x,y) \wedge a \equiv \kappa_0 \;\to\; s(b) \equiv a \cdot y + s(x) \wedge (\neg\, a \equiv \kappa_0 \;\to\; s(x) < a)$

$\vdash v_0(b,a,x,y) \wedge a \equiv \kappa_0 \;\to\; v_0(s(b), a, s(x), y)$

where (p_{11}) arises via $\vdash a \equiv \kappa_0 \to (\neg\, a \equiv \kappa_0 \to s(x) < a)$ by ex absurdo quodlibet. So by quantifier logic

(p_{12}) $\quad \vdash v_0(b,a,x,y) \wedge a \equiv \kappa_0 \;\to\; \exists x' \exists y'\, v_0(s(b), a, x', y')$.

Also, from (p_{10}) and $\vdash s(x) < a \;\to\; (\neg\, a \equiv \kappa_0 \to s(x) < a)$ there follows

(p_{13}) $\quad \vdash v_0(b,a,x,y) \wedge s(x) < a \;\to\; \exists x' \exists y'\, v_0(s(b), a, x', y')$.

Finally, (p_{10}) also gives

$\vdash v_0(b,a,x,y) \wedge s(x) \equiv a \;\to\; s(b) \equiv a \cdot s(y)$

$\vdash v_0(b,a,x,y) \wedge s(x) \equiv a \;\to\; s(b) \equiv a \cdot s(y) + \kappa_0 \wedge (\neg\, a \equiv \kappa_0 \;\to\; \kappa_0 < a)$

$\vdash v_0(b,a,x,y) \wedge s(x) \equiv a \;\to\; v_0(s(b), a, \kappa_0, s(y))$

(p_{14}) $\quad \vdash v_0(b,a,x,y) \wedge s(x) \equiv a \;\to\; \exists x' \exists y'\, v_0(s(b), a, x', y')$

here again $\vdash \neg\, a \equiv \kappa_0 \to \kappa_0 < a$ by (b_0), (PQ9). But (b_7), (b_{14}) also imply $\vdash x < a \;\to\; (s(x) < a \vee s(x) \equiv a)$; thus

$\vdash v_0(b,a,x,y) \wedge a \equiv \kappa_0 \;\to\; x < a$ $\qquad$ by definition of v_0

$\vdash v_0(b,a,x,y) \wedge a \equiv \kappa_0 \;\to\; s(x) < a \vee s(x) \equiv a$

(p_{15}) $\quad \vdash v_0(b,a,x,y) \wedge a \equiv \kappa_0 \;\to\; \exists x' \exists y'\, v_0(s(b), a, x', y')$

by (p_{13}), (p_{14}). Consequently, (b_{13}) applied to (p_{12}) and (p_{15}) gives

$\vdash v_0(b,a,x,y) \;\to\; \exists x' \exists y'\, v_0(s(b), a, x', y')$,

$\vdash \exists x\, \exists y\, v_0(b,a,x,y) \;\to\; \exists x' \exists y'\, v_0(s(b), a, x', y')$,

and so (p_9) follows by (ISA). Actually, I need only (ISA_0) since in place of (p_9) I can work with $\exists x\, (x \le b \wedge \exists y\, (y \le b \wedge v_0(b,a,x,y)))$ which is a Δ_0-formula. - Next I show

(p_{16}) $\quad \vdash v_0(b,a,x,y) \wedge v_0(b,a,x',y') \;\to\; x \equiv x'$.

Abbreviating as v_1 the formula $v_0(b,a,x,y) \wedge v_0(b,a,x',y')$, it follows again from (b_{13}) that it will suffice to prove both

$$\vdash v_1 \wedge a \equiv \kappa_0 \rightarrow x \equiv x'$$
$$\vdash v_1 \wedge \neg a \equiv \kappa_0 \rightarrow x \equiv x' .$$

The first of these statements follows from $\vdash v_0(b,a,x,y) \wedge a \equiv \kappa_0 \rightarrow x \equiv b$ which is a consequence of (PQ5), (a_9), (a_1). The second one will follow from

(p_{17}) $\qquad \vdash v_1 \wedge \neg a \equiv \kappa_0 \rightarrow y \equiv y'$

since $\vdash a \cdot y + x \equiv a \cdot y + x' \rightarrow x \equiv x'$ by (a_7). Since (b_{16}) gives $\vdash x \equiv y' \vee y<y' \vee y'<y$, (p_{17}) will follow from

$$\vdash v_1 \wedge \neg a \equiv \kappa_0 \wedge y<y' \rightarrow y \equiv y'$$

since y, y' then may be interchanged. This follows by ex absurdo quodlibet from

(p_{18}) $\qquad \vdash v_1 \wedge \neg a \equiv \kappa_0 \wedge y<y' \rightarrow x<a \wedge \neg x<a .$

But $\vdash v_1 \wedge \neg a \equiv \kappa_0 \rightarrow x<a$ by definition of v_0, v_1, and so (p_{18}) will follow from

$$\vdash v_1 \wedge y<y' \rightarrow \neg x<a .$$

This follows by (b_{12}) from

$$\vdash b \equiv a \cdot y + x \wedge b \equiv a \cdot y' + x' \wedge y<y' \rightarrow a \leq x ,$$

hence also from

(p_{19}) $\qquad \vdash a \cdot y + x \equiv a \cdot y' + x' \wedge s(z) + y \equiv y' \rightarrow a \leq x .$

And (p_{19}) holds since

$$a \cdot y + x \equiv a \cdot y' + x' \wedge s(z) + y \equiv y'$$

permits to deduce

$a \cdot y + x \equiv a \cdot (s(z) + y) + x'$		
$a \cdot y + x \equiv a \cdot (y + s(z)) + x'$	by (a_3)
$a \cdot y + x \equiv (a \cdot y + a \cdot s(z)) + x'$	by (a_8)
$a \cdot y + x \equiv a \cdot y + (a \cdot s(z) + x')$	by (a_0)
$x \equiv a \cdot s(z) + x'$	by (a_3), (a_7)
$a \cdot s(z) \leq x$	by (b_{23})
$a \leq x$	by (b_{24}) .

Setting $v_1(b,a,x)$ for $\exists y\, v_0(b,a,x,y)$, generalization gives from (p_9)

(mod_0') $\qquad \vdash \forall b\, \forall a\, \exists x\, v_1(b,a,x)$,

and from (p_{16}) there follows

$\qquad \vdash v_1(b,a,x) \wedge v_0(b,a,x',y') \rightarrow x \equiv x'$

(mod_1') $\qquad \vdash v_1(b,a,x) \wedge v_1(b,a,x') \rightarrow x \equiv x'$

such that MOD(a,b) can be introduced with the defining axiom

(mod') $\qquad \forall b\, \forall a\, \exists y\, v_0(b,a,\mathrm{MOD}(a,b),y)$.

In particular,

(a_{11}) $\vdash \exists y\,(b \equiv a \cdot y + \mathrm{MOD}(a,b))$
(a_{12}) $\vdash \neg\, a \equiv \kappa_0 \rightarrow \mathrm{MOD}(a,b) < a$
(a_{13}) $\vdash \mathrm{MOD}(\kappa_0, b) \equiv b$.

2. Extensions by Recursive Terms

The aim of this section is to prove the

THEOREM 1 Let $G(\xi)$ and $R(x,y,\xi)$ be terms of a definitorial extension L' of the language of PA or HA. Then L' can be definitorially expanded by a term $F(x,\xi)$ such that the recursion equations

$$F(\kappa_0, \xi) \equiv G(\xi) \quad , \quad F(s(x), \xi) \equiv R(x, F(x,\xi), \xi)$$

are provable in PA respectively HA (together with the previous defining equations of the extension L').

The content of this theorem may be illustrated by comparing it with what has been shown for QA. (1) In QA every recursive function f is functionally representable, and as observed in Lemma 4.2, the language of QA then can be definitorially expanded by a term $F(\xi)$ such that $f(\lambda) = n$ implies $\vdash F(\kappa\lambda) \equiv \kappa_n$ and *not* $f(\lambda) = n$ implies $\vdash \neg F(\kappa\lambda) \equiv \kappa_n$, where provability is that of QA. In this manner the behaviour of F(–) for constant arguments is fully determined, but nothing is known about provable formulas containing F(–) with *free* variables. For instance, exponentiation a^b can be represented by a term exp(x,y), but the functional equation $\exp(x, y+y') \equiv \exp(x,y) \cdot \exp(x,y')$ *cannot* be proved in QA. Whereas in PA a term is available whose behaviour also with free variables is under control (and the provability of the recursion equations then will imply that of further functional equations). – (2) Recursive functions, viewed as usual mathematical objects, are *extensionally* determined (i.e. by their *being as such*), and one and the same function will have a multitude of descriptions. Whereas terms, with PA-provable recursion equations, are *intensionally* determined (i.e. by their *manner of being*), and the axioms of PA provide the provabilities which express how they have been constructed; the provability calculus for PA permits, therefore, to *control* these descriptions. (Simple minded people may view this as the *poverty* of PA which necessitates the intensional descriptions.)

In order to prove the theorem, I will have to replace by PA-derivations the informal mathematical arguments reducing primitive recursive definitions to explicit ones, carried out in the proof of Theorem 5.3 when showing that certain classes were closed under primitive recursion. Making use of MOD(a,b), I introduce terms $\sigma(x,k)$ and $\beta(c,d,p)$ by the explicit definitions

$$\sigma(x,k) \equiv s(s(x)\cdot k)$$
$$\beta(c,d,p) \equiv MOD(c, \sigma(p,d)) .$$

Given $G(\xi)$ and $R(x,y,\xi)$, let $w(p,q,x,z,\xi)$ be the formula

$$\beta(p,q,\kappa_0) \equiv G(\xi)$$
$$\wedge\ \forall y\,(y<x \rightarrow \beta(p,q,s(y)) \equiv R(y, \beta(p,q,y), \xi)) \ \wedge\ \beta(p,q,x) \equiv z$$

and let $v(x,z,\xi)$ be $\exists p\ \exists q\ w(p,q,x,z,\xi)$. I shall need the

LEMMA 1 $\vdash \exists k\ \exists p\ \forall y\ (y \leq x \rightarrow \beta(p,k,y) \equiv \beta(m,n,y)\ \wedge\ \beta(p,k, s(x)) \equiv y\,)$

whose proof I defer to a later stage. Theorem 1 then is proved in the following four steps.

Step 1: $\vdash v(\kappa_0,z,\xi) \leftrightarrow z \equiv G(\xi)$

Provability of the implication from left to right follows from the definition of the formula w. Provability of the reverse implication follows in the same manner, making use of $\vdash \exists p\ \exists q\ \beta(p,q,\kappa_0) \equiv z$ which follows from

$\vdash s(\kappa_0)\cdot q \equiv q$ by (ISA_0)–induction on q applying (PQ6), (PQ4)
$\vdash \sigma(\kappa_0, q) \equiv s(q)$
$\vdash z \equiv s(z)\cdot \kappa_0 + z$
$\vdash MOD(z, s(z)) \equiv z$
$\vdash \exists p\ \exists q\ MOD(p,\ \sigma(\kappa_0, q)) \equiv z$.

Step 2: $\vdash v(s(x),z,\xi) \leftrightarrow \exists z'\,(v(x,z',\xi) \wedge R(x,z',\xi) \equiv z\,)$.

I begin by listing some auxiliary facts from pure logic:

(q_0)	$\vdash ((u_0 \vee u_1) \rightarrow u) \leftrightarrow ((u_0 \rightarrow u) \wedge (u_1 \rightarrow u))$	propositionally
(q_1)	$\vdash \forall x(u_0 \wedge u_1) \leftrightarrow (\forall x u_0 \wedge \forall x u_1)$	
(q_2)	$\vdash \exists x(u_0 \wedge u_1) \leftrightarrow (\exists x u_0) \wedge u_1$	if *not* $x \epsilon fr(u_1)$
(q_3)	$u_0 \rightarrow (u_1 \rightarrow u_2) \vdash \forall x u_0 \rightarrow (\forall x u_1 \rightarrow \forall x u_2)$	by **2.9**.(a14) twice
(q_4)	$u_0 \rightarrow (u_1 \rightarrow u_2) \vdash \exists x u_0 \rightarrow (u_1 \rightarrow \exists x u_2)$	if *not* $x \epsilon fr(u_1)$ by **2.9**.(a15), **2.9**.(B1) .

The following observation is an immediate consequence of (b_{15}), (q_0), (q_1); it will be a basic tool in many of the following argumentations:

(q_5) $\vdash \forall y(y<s(x) \rightarrow u(y)) \leftrightarrow (\forall y(y<x \rightarrow u(y)) \wedge u(x)$.

It follows from (q_5) that $w(p,q, s(x),z,\xi)$ is deductively equivalent to

$$\beta(p,q,\kappa_0) \equiv G(\xi) \wedge \forall y\,(y<x \rightarrow \beta(p,q,\,s(y)) \equiv R(y,\,\beta(p,q,y),\,\xi))$$
$$\wedge\, \beta(p,q,\,s(x)) \equiv R(x,\,\beta(p,q,x),\,\xi) \wedge \beta(p,q,\,s(x)) \equiv z\ .$$

But the first conjunction here is equivalent to $w(p,q,x,\,\beta(p,q,x),\,\xi)$, hence

(r_0) $\vdash w(p,q,\,s(x),z,\xi) \longleftrightarrow w(p,q,x,\,\beta(p,q,x),\,\xi)$
$$\wedge\, R(x,\,\beta(p,q,x),\,\xi) \equiv z \wedge \beta(p,q,\,s(x)) \equiv z$$

$\vdash w(p,q,\,s(x),z,\xi) \longleftrightarrow \exists z'\,(w(p,q,x,z',\xi) \wedge R(x,z',\xi) \equiv z)$
$$\wedge\, \beta(p,q,\,s(x)) \equiv z$$

(r_1) $\vdash w(p,q,\,s(x),z,\xi) \rightarrow \exists z'\,(w(p,q,x,z',\xi) \wedge R(x,z',\xi) \equiv z)$

(r_2) $\vdash w(p,q,\,s(x),z,\xi) \rightarrow \exists z'\,(\exists p\,\exists q\, w(p,q,x,z',\xi) \wedge R(x,z',\xi) \equiv z\,)$

where (q_2) leads from (r_1) to (r_2). Now (r_2) implies the implication from left to right in Step 2.

In order to approach the reverse implication, I use $\vdash y\leq x \rightarrow y<x$ from (PQ9) and find for all terms $t_0(y)$, $t_1(y)$

$\vdash \forall y\,(y\leq x \rightarrow t_0(y) \equiv t_1(y)) \rightarrow \forall y\,(y<x \rightarrow t_0(y) \equiv t_1(y))$
$\vdash \forall y\,(y\leq x \rightarrow t_0(y) \equiv t_1(y)) \rightarrow (s(y)\leq x \rightarrow t_0(s(y)) \equiv t_1(s(y)))$
$\vdash \forall y\,(y\leq x \rightarrow t_0(y) \equiv t_1(y)) \rightarrow ((y<x \wedge s(y)\leq x)$
$$\rightarrow t_0(y) \equiv t_1(y) \wedge t_0(s(y)) \equiv t_1(s(y)))\ .$$

But $\vdash y<x \rightarrow s(y)\leq x$ by (b_6), thus

$\vdash \forall y\,(y\leq x \rightarrow t_0(y) \equiv t_1(y)) \rightarrow$
$$(y<x \rightarrow t_0(y) \equiv t_1(y) \wedge t_0(s(y)) \equiv t_1(s(y)))$$

(q_6) $\vdash \forall y\,(y\leq x \rightarrow t_0(y) \equiv t_1(y)) \rightarrow$
$$\forall y\,(y<x \rightarrow t_0(y) \equiv t_1(y) \wedge t_0(s(y)) \equiv t_1(s(y)))\ .$$

If $u(x_0,x_1)$ is a formula whose variables x_0,x_1 shall be replaced by terms, then by equality logic

$\vdash t_0(y) \equiv t_1(y) \wedge t_0{}'(y) \equiv t_1{}'(y) \rightarrow (u(t_0,t_0{}') \rightarrow u(t_1,t_1{}'))$
$\vdash (y<x \rightarrow t_0(y) \equiv t_1(y) \wedge t_0{}'(y) \equiv t_1{}'(y))$
$$\rightarrow (y<x \rightarrow u(t_0,t_0{}')) \rightarrow (y<x \rightarrow u(t_1,t_1{}'))$$

(q_7) $\vdash \forall y\,(y<x \rightarrow t_0(y) \equiv t_1(y) \wedge t_0{}'(y) \equiv t_1{}'(y))$
$$\rightarrow (\forall y\,(y<x \rightarrow u(t_0,t_0{}'))) \rightarrow \forall y\,(y<x \rightarrow u(t_1,t_1{}'))$$

by (q_3). In the particular case that $t_0{}'(y) = t_0(s(y))$, $t_1{}'(y) = t_1(s(y))$, (q_7) together with (q_6) gives

(q_8) $\vdash \forall y\,(y\leq x \rightarrow t_0(y) \equiv t_1(y)) \rightarrow$
$$(\forall y\,(y<x \rightarrow u(t_0(y),\,t_0(s(y))))) \rightarrow \forall y\,(y<x \rightarrow u(t_1(y),\,t_1(s(y))))\ .$$

The premiss of that implication is formally equivalent to

$$\forall y\,(y\leq x \rightarrow t_0(y) \equiv t_1(y)) \wedge t_0(\kappa_0) \equiv t_1(\kappa_0) \wedge t_0(x) \equiv t_1(x)\ ;$$

hence the conclusion of (q_8) may be strengthened to

$$(u_0(t_0(\kappa_0)) \wedge \forall y\,(y<x \rightarrow u(t_0(y),\,t_0(s(y)))) \wedge u_1(t_0(x))$$
$$\rightarrow (u_0(t_1(\kappa_0)) \wedge \forall y\,(y<x \rightarrow u(t_1(y),\,t_1(s(y)))) \wedge u_1(t_1(x)))\ .$$

I now choose for $u(x_0, x_1)$, $u_0(x_0)$, $u_1(x_0)$ the formulas $x_1 \equiv R(y, x_0, \xi)$, $x_0 \equiv G(\xi)$, $x_0 \equiv z$, and for $t_0(y)$, $t_1(y)$ the terms $\beta(p,q,y)$, $\beta(p',q',y)$, and obtain

$$\vdash \forall y\,(y \lesssim x \to \beta(p,q,y) \equiv \beta(p',q',y)) \;\to\; (w(p',q',x,z',\xi) \to w(p,q,x,z',\xi))$$

from where propositionally

$$(r_3) \qquad \begin{aligned} &\vdash \forall y\,(y \lesssim x \to \beta(p,q,y) \equiv \beta(p',q',y)) \wedge \beta(p,q,s(x)) \equiv z) \\ &\quad \to ((w(p',q',x,z',\xi) \wedge R(x,z',\xi) \equiv z) \\ &\qquad \to (w(p,q,x,z',\xi) \wedge R(x,z',\xi) \equiv z \wedge \beta(p,q,s(x)) \equiv z)). \end{aligned}$$

Now $\vdash w(p,q,x,z',\xi) \to z' \equiv \beta(p,q,x)$ by definition of w; hence it follows from (r_0) that the conjunction on the last line of (r_3) is deductively equivalent to $w(p,q,s(x),z,\xi)$. Thus

$$\begin{aligned} &\vdash \forall y\,(y \lesssim x \to \beta(p,q,y) \equiv \beta(p',q',y)) \wedge \beta(p,q,s(x)) \equiv z) \\ &\quad \to ((w(p',q',x,z',\xi) \wedge R(x,z',\xi) \equiv z) \to w(p,q,s(x),z,\xi)). \end{aligned}$$

So (q_4) and the definition of $v(s(x), z, \xi)$ imply

$$(r_4) \qquad \begin{aligned} &\vdash \exists p\,\exists q\,\forall y\,(y \lesssim x \to \beta(p,q,y) \equiv \beta(p',q',y)) \wedge \beta(p,q,s(x)) \equiv z) \\ &\quad \to ((w(p',q',x,z',\xi) \wedge R(x,z',\xi) \equiv z) \to v(s(x),z,\xi) \end{aligned}$$

if only p', q' have been chosen sufficiently new.

I next apply Lemma 1 to p', q', x, z in place of m, n, x, y, and I also write q instead of k. It then states that the upper line in (r_4) is provable. Hence so is the lower line, and thus I arrive at

$$\begin{aligned} &\vdash w(p',q',x,z',\xi) \wedge R(x,z',\xi) \equiv z \;\to\; v(s(x),z,\xi) \\ &\vdash \exists z'\,\exists p'\,\exists q'\,(w(p',q',x,z',\xi) \wedge R(x,z',\xi) \equiv z) \;\to\; v(s(x),z,\xi) \\ &\vdash \exists z'\,(v(x,z',\xi) \wedge R(x,z',\xi) \equiv z) \;\to\; v(s(x),z,\xi) \qquad \text{by } (q_2). \end{aligned}$$

This proves the reverse implication in Step 2.

Step 3: $\vdash \exists z\, v(x,z,\xi) \wedge \forall z_0\, \forall z_1\, (v(x,z_0,\xi) \wedge v(x,z_1,\xi) \to z_0 \equiv z_1)$.

I shall call this formula $u(x)$ and shall prove it by (IRA). First, $\vdash u(\kappa_0)$ follows from Step 1 since $\vdash \exists z\,(z \equiv G(\xi))$. Further

$$\begin{aligned} &\vdash u(x) \;\to\; \exists z'\, v(x,z',\xi) \\ &\vdash u(x) \;\to\; \exists z'\,(v(x,z',\xi) \wedge R(x,z',\xi) \equiv R(x,z',\xi)) \\ &\vdash u(x) \;\to\; \exists z\,\exists z'\,(v(x,z',\xi) \wedge R(x,z',\xi) \equiv z) \\ &\vdash u(x) \;\to\; \exists z\, v(s(x),z,\xi) \end{aligned}$$

by Step 2. Also by Step 2

$$\begin{aligned} &\vdash v(s(x),z_0,\xi) \wedge v(s(x),z_1,\xi) \\ &\quad \to \exists z_0'\,\exists z_1'\,(v(x,z_0',\xi) \wedge v(x,z_1',\xi) \wedge R(x,z_0',\xi) \equiv z_0 \wedge R(x,z_1',\xi) \equiv z_1) \\ &\vdash u(x) \;\to\; (v(x,z_0',\xi) \wedge v(x,z_1',\xi) \to z_0' \equiv z_1') \end{aligned}$$

$\vdash (z_0' \equiv z_1' \wedge R(x,z_0',\xi) \equiv z_0 \wedge R(x,z_1',\xi) \equiv z_1) \rightarrow z_0 \equiv z_1$
$\vdash u(x) \rightarrow (v(s(x),z_0,\xi) \wedge v(s(x),z_1,\xi) \rightarrow z_0 \equiv z_1)$

and this together leads to $\vdash u(x) \rightarrow u(s(x))$.

Step 4 :

It follows from the result of Step 3 that I can definitorially expand my language by a term $F(x,\xi)$ with the defining axiom

$$v(x, F(x,\xi), \xi) .$$

This axiom implies $\vdash F(\kappa_0,\xi) \equiv G(\xi)$ by the equivalence from Step 1. Applying the equivalence from Step 2, I obtain also

$\vdash \exists z' \; (v(x,z',\xi) \wedge R(x,z',\xi) \equiv F(s(x),\xi))$
$\vdash \exists z' \; (F(x,\xi) \equiv z' \wedge R(x,z',\xi) \equiv F(s(x),\xi))$
$\vdash R(x, F(x,\xi), \xi) \equiv F(s(x),\xi)$.

This completes the 4th Step and the proof of the Theorem. It only remains to prove the Lemma.

The content of this Lemma is the fundamental property of Gödel's β-function: given a number y and given numbers m,n which code a sequence of length x, then there are numbers p,k coding the prolongation of that sequence, obtained by adding the (x+1)-th member y. A proof by naive arithmetic was given in Chapter 1.16; I now will have to give a PA-proof, and that will have to avoid the arguments making use of finite sequences (particularly in the discussion of the Chinese Remainder Theorem). Also, the principle of the smallest number (used in the proof of 1.16.(a2)) can be avoided such that I actually arrive at an HA-proof. I beginn by developing some some more arithmetical tools.

The predicate $|$ of divisibility I introduce as usual by the axiom

$$x \,|\, y \leftrightarrow \exists z \, (z \leq y \wedge x \cdot z \equiv y) ;$$

properties such as reflexivity and transitivity I leave to the reader, but observe that $\vdash \kappa_0 \,|\, y \rightarrow y \equiv \kappa_0$ and $\vdash x \,|\, \kappa_0$.

I introduce the 3-ary predicate of being congruent with respect to a module as

$$x \simeq y \; (\textit{modulo } n) \leftrightarrow \exists p \, (x \equiv n \cdot p + y \vee y \equiv n \cdot p + x) ;$$

here reflexivity and symmetry are trivial, and the provability of the congruence properties for $+$ and $\cdot$ follows from (a_0), (a_8). For transitivity I use

$\vdash (x \equiv n \cdot p + y \wedge y \equiv n \cdot q + z) \rightarrow x \equiv n \cdot (p+q) + z$
$\vdash (y \equiv n \cdot p + x \wedge z \equiv n \cdot q + y) \rightarrow z \equiv n \cdot (q+p) + x$
$\vdash (x \equiv n \cdot p + y \wedge z \equiv n \cdot q + y)$
$\quad \rightarrow x \equiv z \vee \exists r \, (r+p \equiv q \wedge z \equiv n \cdot r + x) \vee \exists r \, (r+q \equiv p \wedge x \equiv n \cdot r + z)$

$$\vdash (y \equiv n \cdot p + x \wedge y \equiv n \cdot q + z) \\ \to x \equiv z \vee \exists r\,(r + p \equiv q \wedge x \equiv n \cdot r + z) \vee \exists r\,(r + q \equiv p \wedge z \equiv n \cdot r + x)$$

where (b_{16}) is used in the third and fourth case, and in the latter also (a_7).

(s_0) $\quad \vdash x \simeq y \ (\textit{modulo}\ \kappa_0) \to x \equiv y$

$\quad \vdash x \simeq \mathrm{MOD}(x,y) \ (\textit{modulo}\ y)$ $\quad$ by (a_{11})

(s_1) $\quad \vdash x \simeq r \ (\textit{modulo}\ y) \wedge r < y \to r \equiv \mathrm{MOD}(x,y)$

(s_2) $\quad \vdash x \simeq y \ (\textit{modulo}\ n) \wedge m \mid n \to x \simeq y \ (\textit{modulo}\ m)$.

I obtain (s_1) from

$\vdash x \equiv y \cdot p + r \wedge r < y \to r \equiv \mathrm{MOD}(x,y)$ $\quad$ by (mod_1')

$\vdash r \equiv y \cdot p + x \wedge p \equiv \kappa_0 \to r \equiv x$ $\quad$ from $\vdash x < y \to x \equiv \mathrm{MOD}(x,y)$

$\vdash r \equiv y \cdot p + x \wedge \neg\, p \equiv \kappa_0 \to y \lesssim r$ $\quad$ by (b_{24}) ,

and in the third case I conclude $r \equiv \mathrm{MOD}(x,y)$ from (b_{12}) by ex absurdo quodlibet. Next I introduce the 2–ary predicate of being coprime as

$$\Pi(a,b) \leftrightarrow \exists x\,(a \cdot x \simeq \kappa_1 \ (\textit{modulo}\ b))$$

where $\vdash \kappa_1 \equiv s(\kappa_0)$. I do not need to secure that this coincides with the familiar definition; what I do need are the consequences below.

(s_3) $\quad \vdash \Pi(\kappa_0, \kappa_1)$

(s_4) $\quad \vdash \Pi(\kappa_1, b)$

(s_5) $\quad \vdash \Pi(a, \kappa_0) \to a \equiv \kappa_1$

(s_6) $\quad \vdash \Pi(a,b) \wedge c \mid a \to \Pi(c,b)$

(s_7) $\quad \vdash \Pi(a,b) \wedge \Pi(c,b) \to \Pi(a \cdot c, b)$

(s_8) $\quad \vdash \Pi(a,b) \wedge a \simeq c \ (\textit{modulo}\ b) \to \Pi(c,b)$

(s_9) $\quad \vdash \Pi(a,b) \to \Pi(b,a)$.

Here (s_3) follows from $\vdash \kappa_0 \cdot x \simeq \kappa_1$ (mod κ_1), and (s_5) follows directly from (s_0). As for (s_6), I conclude from $a \equiv c \cdot d$ and $a \cdot x \simeq \kappa_1$ (*modulo* b) that $c \cdot (d \cdot x) \simeq \kappa_1$ (*modulo* b); for (s_7) I conclude from $a \cdot x \simeq \kappa_1$ (*modulo* b) and $c \cdot y \simeq \kappa_1$ (*modulo* b) that $(a \cdot c) \cdot (x \cdot y) \simeq \kappa_1$ (*modulo* b); for (s_8) I conclude from $a \cdot x \simeq \kappa_1$ (*modulo* b) and $a \simeq c$ (*modulo* b) that $a \cdot x \simeq a \cdot x \simeq \kappa_1$ (*modulo* b). More effort is required for (s_9), based on the distinction

$$\vdash b \equiv \kappa_0 \vee b \equiv \kappa_1 \vee \kappa_1 < b$$

coming from (b_7). If $b \equiv \kappa_0$ then $\Pi(b,a)$ by (s_5) and (s_3); if $b \equiv \kappa_1$ then I use (s_4). Dissolving $\Pi(a,b)$ according to the definitions, there results $\exists x\, \exists y\,(a \cdot x \equiv b \cdot y + \kappa_1 \vee \kappa_1 \equiv b \cdot y + a \cdot x)$. If $\kappa_1 \equiv b \cdot y + a \cdot x$ and $\kappa_1 < b$ then $b \cdot y < b$, thus $\neg\, b \lesssim b \cdot y$ by (b_{11}), hence $y \equiv \kappa_0$ by (b_{24}) and so $a \equiv \kappa_1$, $x \equiv \kappa_1$. Thus I obtain

$$\vdash \Pi(a,b) \wedge \kappa_1 < b \to \exists x\, \exists y\,(a \cdot x \equiv b \cdot y + \kappa_1)$$

which I shall improve to

$$\vdash \Pi(a,b) \wedge \kappa_1 < b \to \exists x\, \exists y\,(a \cdot x \equiv b \cdot y + \kappa_1 \wedge x < b \wedge y < a) \ .$$

Because assume $a \cdot x \equiv b \cdot y + \kappa_1$ and $x \equiv n \cdot b + p$, $y \equiv m \cdot a + q$, $p < b$, $q < a$

by (a_{11}), (a_{12}), thus $\neg\, a \equiv \kappa_0$ since $\kappa_1 \leq a\cdot x$, $\kappa_0 < a\cdot x$, $\neg\, a\cdot x \equiv \kappa_0$. There results $\vdash a\cdot n\cdot b + a\cdot p \equiv b\cdot m\cdot a + b\cdot q + \kappa_1$, and now I apply the distinction $\vdash n \leq m \vee m \leq n$ flowing from (b_{17}). If $n \leq m$, $n+z \equiv m$, then (a_7) leads to $a\cdot p \equiv b\cdot z\cdot a + b\cdot q + \kappa_1$, hence $b\cdot z\cdot a \leq a\cdot p$; but $p<b$ gives $a\cdot p < a\cdot b$, hence $(a\cdot b)\cdot z < a\cdot b$, $\neg\, a\cdot b \leq (a\cdot b)\cdot z$, $z \equiv \kappa_0$ by (b_{24}); thus $n \equiv m$ and $a\cdot p \equiv b\cdot q + \kappa_1$. If $m \leq n$, $m+z \equiv n$, then (a_7) leads to $a\cdot z\cdot b + a\cdot p \equiv b\cdot q + \kappa_1$, hence $a\cdot z\cdot b \leq b\cdot q + \kappa_1$; but $\kappa_1 < b$ gives $a\cdot z\cdot b < b\cdot q + b$, and since $b\cdot q + b \equiv b\cdot(q+\kappa_1)$ and $q+\kappa_1 \leq a$ this implies $(a\cdot b)\cdot z < a\cdot b$ from where I can conclude as before. My improvement having been established, I obtain

$$\vdash \Pi(a,b) \wedge \kappa_1 < b \rightarrow \exists x\, \exists y\, \exists k\, \exists j\, (x+k \equiv b \wedge y+j \equiv a \wedge a\cdot x \equiv b\cdot y + \kappa_1)\,.$$

The three equalities here imply $y\cdot x + j\cdot x \equiv x\cdot y + k\cdot y + \kappa_1$, thus $j\cdot x \equiv k\cdot y + \kappa_1$ and therefore $x\cdot j + k\cdot j \equiv k\cdot y + k\cdot j + \kappa_1$ whence

$$\vdash \Pi(a,b) \wedge \kappa_1 < b \rightarrow \exists k\, \exists j\, (b\cdot j \equiv a\cdot k + \kappa_1)$$

$$\vdash \Pi(a,b) \wedge \kappa_1 < b \rightarrow \exists j\, (b\cdot j \simeq \kappa_1\ (\textit{modulo}\ a))$$

$$\vdash \Pi(a,b) \wedge \kappa_1 < b \rightarrow \Pi(b,a)\,.$$

This completes the proof of (s_9), and now I can state the property I really shall need the predicate Π to have:

(s_{10}) $\quad \vdash \Pi(a,b) \rightarrow \exists z\, (z \simeq y_0\ (\textit{modulo}\ a) \wedge z \simeq y_1\ (\textit{modulo}\ b))\,.$

Because $\vdash \Pi(a,b) \rightarrow \exists x\, (a\cdot x \simeq \kappa_1 (\textit{modulo}\ a) \wedge \exists y (b\cdot y \simeq \kappa_1\ (\textit{modulo}\ a))$ by (s_9), and as $n\cdot x \simeq \kappa_0$ (*modulo* n) there follow

$$\vdash a\cdot x \simeq \kappa_1\ (\textit{modulo}\ b) \rightarrow a\cdot x\cdot y_1 + b\cdot y\cdot y_0 \simeq y_1\ (\textit{modulo}\ b)$$
$$\vdash b\cdot y \simeq \kappa_1\ (\text{mod}\ a) \rightarrow a\cdot x\cdot y_1 + b\cdot y\cdot y_0 \simeq y_0\ (\text{mod}\ a)$$

whence (s_{10}) follows for z as $x\cdot y\cdot y_1 + b\cdot y\cdot y_0$.

(s_{11}) $\quad \vdash \Pi(a\cdot b+\kappa_1, b)\ ,\ \vdash \Pi(b, a\cdot b+\kappa_1)$

(s_{12}) $\quad \vdash c \mid b \rightarrow \Pi(c\cdot b, a\cdot b+\kappa_1)$ $\quad$ by (s_{11}), (s_6), (s_7)

(s_{13}) $\quad \vdash c \mid b \wedge a+c \equiv d \rightarrow \Pi(a\cdot b+\kappa_1\ , d\cdot b+\kappa_1)\,.$

Because $a+c \equiv d$ implies $a\cdot b+\kappa_1+c\cdot b \equiv d\cdot b+\kappa_1$, hence $d\cdot b+\kappa_1 \simeq c\cdot b$ (*modulo* $a\cdot b+\kappa_1$). Thus (s_{13}) follows from (s_{12}) by (s_8).

(s_{14}) $\quad \vdash \forall y\, (y<x \rightarrow s(y) \mid k) \rightarrow \forall y\, (y<x \rightarrow \Pi(\sigma(y,k), \sigma(x,k)))\,.$

Because

$$\vdash \neg\, p \equiv \kappa_0 \wedge s(z)+p \equiv s(x) \rightarrow \exists y\, (p \equiv s(y) \wedge y<x)\,,$$
$$\vdash \forall y\, (y<x \rightarrow s(y) \mid k) \rightarrow (\neg\, p \equiv \kappa_0 \wedge s(z)+p \equiv s(x) \rightarrow p \mid k)\,,$$
$$\vdash z<x \rightarrow \exists p\, (\neg\, p \equiv \kappa_0 \wedge s(z)+p \equiv s(x))\,,$$
$$\vdash \forall y\, (y<x \rightarrow s(y) \mid k) \rightarrow (z<x \rightarrow \exists p\, (p \mid k \wedge s(z)+p \equiv s(x)))\,,$$
$$\vdash \forall y\, (y<x \rightarrow s(y) \mid k) \rightarrow (z<x \rightarrow \Pi(s(z)\cdot k+\kappa_1\ , s(x)\cdot k+\kappa_1))\,.$$

Here the third line is a rewriting of $z<x \rightarrow s(z)<s(x)$, and the fifth follows from the fourth by an application of (s_{13}).

All arithmetical auxiliaries having been established, I now turn to the Lemma's proof proper. From now on, also letters such as a, b, d, k shall denote (usually free) variables. I begin by defining for a term $t(x_0)$

$$\Gamma(x_1,x,d \,\|\, t(x_0)) \longleftrightarrow \neg\, x_1 \equiv \kappa_0 \wedge \Pi(x_1,d) \wedge \forall y\,(y<x \rightarrow t(y) \mid x_1)\ ;$$

observe that the variable x_0 has been replaced and does nor occur here.

(s_{15}) $\quad \vdash \forall y\,(y<x \rightarrow \neg\, t(y) \equiv \kappa_0 \wedge \Pi(t(y),d)) \rightarrow \exists x_1\, \Gamma(x_1,x,d \,\|\, t(x_0))$.

This formula $v(x)$ I shall prove by (IRA); observe that $\vdash v(\kappa_0)$ because $\vdash \Gamma(\kappa_1,\kappa_0,d \,\|\, t(x_0))$. For $\vdash v(x) \rightarrow v(s(x))$ it will suffice to show

$$\vdash \Gamma(x_1,x,d \,\|\, t(x_0)) \wedge \forall y\,(y<s(x) \rightarrow \neg\, t(y) \equiv \kappa_0 \wedge \Pi(t(y),d)) \rightarrow \Gamma(x_1 \cdot t(x), s(x), d \,\|\, t(x_0))\ .$$

Here the hypothesis $\Gamma(x_1,x,d \,\|\, t(x_0))$ implies $\vdash \neg\, x_1 \equiv \kappa_0$ and $\vdash \Pi(x_1,d)$; as $\vdash \neg\, t(x) \equiv \kappa_0 \wedge \Pi(t(x),d)$ by the second hypothesis, I find $\vdash \neg\, x_1 \cdot t(x) \equiv \kappa_0$ and, by (s_7), $\vdash \Pi(x_1 \cdot t(x), d)$. Thus it remains to show $\vdash \forall y\,(y<s(x) \rightarrow t(y) \mid x_1 \cdot t(x))$. By (q_5) this is deductively equivalent to $\forall y\,(y<x \rightarrow t(y) \mid x_1 \cdot t(x)) \wedge t(x) \mid x_1 \cdot t(x)$. The latter divisibility is trivial, and further $\vdash \forall y\,(y<x \rightarrow t(y) \mid x_1 \cdot t(x))$ follows trivially from $\vdash \forall y\,(y<x \rightarrow t(y) \mid x_1)$ which comes from $\vdash \Gamma(x_1,x,d \,\|\, t(x_0))$.

From now on, I consider the term $t(x_0) = \sigma(x_0,k)$. Replacing d by $\sigma(x,k)$, I find

(s_{16}) $\quad \vdash \forall y\,(y<x \rightarrow s(y) \mid k) \rightarrow \exists x_1\, \Gamma(x_1,x,\sigma(x,k) \,\|\, \sigma(x_0,k))$;

the hypotheses of (s_{15}) hold by (s_{14}) and since $\vdash \neg\, \sigma(p,q) \equiv \kappa_0$ (by (PQ2)).

(s_{17}) $\quad \vdash \forall y\,(y<s(x) \rightarrow s(y) \mid k) \wedge z \leq x \rightarrow \exists x_1\, \Gamma(x_1, s(z), \sigma(s(z),k) \,\|\, \sigma(x_0,k))$.

If (s_{16}) is written as $A(x) \rightarrow B(x)$ then also $\vdash A(s(x)) \rightarrow B(s(x))$, and (s_{17}) then becomes $A(s(x)) \wedge z \leq x \rightarrow B(s(z))$. But the hypotheses of (s_{17}) permit to deduce $A(s(z))$ since $\vdash z \leq x \rightarrow s(z) \leq s(x)$, $\vdash y<s(z) \rightarrow y<s(x)$.

(s_{18}) $\quad \vdash \Gamma(x_1, s(z), \sigma(s(z),k) \,\|\, \sigma(x_0,k)) \rightarrow \exists q\,(q \simeq a(\mathit{modulo}\ x_1) \wedge q \simeq b(\mathit{modulo}\ \sigma(s(z),k))$.

The hypothesis implies $\vdash \Pi(x_1, \sigma(s(z),k))$, and so I apply (s_{10}). – Let now $b(y)$ be a further term, then (s_{18}) gives in particular

(s_{19}) $\quad \vdash \Gamma(x_1,s(z), \sigma(s(z),k) \,\|\, \sigma(x_0,k)) \rightarrow \exists q\,(q \simeq a(\mathit{modulo}\ x_1) \wedge q \simeq b(s(z))\ (\mathit{modulo}\ \sigma(s(z),k))$.

(s_{20})
$$\begin{aligned}\vdash \forall y\,(y \leq z \rightarrow a \simeq b(y)\ (\mathit{modulo}\ \sigma(y,k)))\\ \wedge\ \Gamma(x_1, s(z), \sigma(s(z),k) \,\|\, \sigma(x_0,k))\\ \wedge\ q \simeq a\ (\mathit{modulo}\ x_1)\\ \wedge\ q \simeq b(s(z))\ (\mathit{modulo}\ \sigma(s(z),k))\\ \rightarrow \forall y\,(y \leq s(z) \rightarrow q \simeq b(y)\ (\mathit{modulo}\ \sigma(y,k)))\ .\end{aligned}$$

The second line implies $\vdash \forall y \ (y < s(z) \to \sigma(y,k) \mid x_1)$, and so the third gives $\vdash \forall y \ (y \leq z \to q \simeq a \ (modulo \ \sigma(y,k)))$ by (s_2). Consequently, the first line gives $\vdash \forall y \ (y \leq z \to q \simeq b(y) \ (modulo \ \sigma(y,k)))$, and so (q_5) and the fourth line imply the conclusion.

(s_{21})
$$\vdash \forall y \ (y < s(x) \to s(y) \mid k) \wedge z \leq x$$
$$\to (\ \exists q \ \forall y \ (y \leq z \to q \simeq b(y) \ (modulo \ \sigma(y,k)))$$
$$\to \exists p \ \forall y \ (y \leq s(z) \to p \simeq b(y) \ (modulo \ \sigma(y,k)))) \ ) \ .$$

This follows simply by a chain of arguments from (s_{18}), (s_{19}), (s_{20}).

(s_{22})
$$\vdash \forall y \ (y < s(x) \to s(y) \mid k) \wedge z \leq x$$
$$\to \exists p \ \forall y \ (y \leq z \to p \simeq b(y) \ (modulo \ \sigma(y,k))) \ .$$

The formula (s_{21}) can be written as $A(x) \wedge z \leq x \to (C(z) \to C(s(z)))$, and (s_{22}) then is $A(x) \wedge z \leq x \to C(z)$. But $\vdash C(\kappa_0)$ since $\vdash y \leq \kappa \to y \equiv \kappa_0$ by (b_{10}), (b_{14}). Thus induction with (IRA) proves $\vdash A(x) \wedge z \leq x \to C(z)$. In particular, there holds

(s_{23})
$$\vdash \forall y \ (y < s(x) \to s(y) \mid k)$$
$$\to \exists p_0 \forall y \ (y \leq x \to p \simeq b(y) \ (modulo \ \sigma(y,k))) \ .$$

Replacing x by $s(x)$ in (s_{16}), I find

$$\vdash \forall y \ (y < s(x) \to s(y) \mid k) \ \to \ \exists x_2 \ \Gamma(x_2, s(x), \sigma(s(x),k) \mid \sigma(x_0,k)) \ ,$$

whence, in particular, $\forall y \ (y < s(x) \to \sigma(y,k) \mid x_2)$. Applying (s_{18}), I find

$$\vdash \forall y \ (y < s(x) \to s(y) \mid k)$$
$$\to \ \exists q \ (q \simeq p_0 (mod \ x_2) \wedge q \simeq b(modulo \ \sigma(s(x),k)) \ .$$

But (s_2) now gives $\forall y \ (y < s(x) \to (q \simeq p_0 (mod \ x_2) \to q \simeq p_0 (mod \ \sigma(y,k)))$, and so with (s_{23}) I arrive at

(s_{24})
$$\vdash \forall y \ (y < s(x) \to s(y) \mid k)$$
$$\to \ \exists q \ \forall y \ (y \leq x \to (q \simeq b(y) \ (modulo \ \sigma(y,k))$$
$$\wedge \ q \simeq b(modulo \ \sigma(s(x),k))) \ ) \ .$$

From now on, I consider the term $b(y) = \beta(m,n,y)$. Then (s_{24}) improves to

(s_{25})
$$\vdash \forall y \ (y < s(x) \to (s(y) \mid k \wedge \beta(m,n,y) \leq k)) \wedge b \leq k$$
$$\to \ \exists q \ \forall y \ (y \leq x \to (\beta(q,k,y) \equiv \beta(m,n,y) \wedge \beta(q,k,s(x)) \equiv b)) \ .$$

Recall that $\vdash r \leq k \to r < \sigma(x,k)$ and $\vdash \beta(p,k,y) \equiv MOD(p, \sigma(y,k))$ by definition of σ and β. Thus by (s_1)

$$\vdash p \simeq r \ (modulo \ \sigma(y,k)) \wedge r \leq k \ \to \ r \equiv \beta(p,k,y) \ .$$

Assuming the additional hypotheses of (s_{25}), the congruences $q \simeq r$ in the conclusion of (s_{24}) can be sharpened to $q \simeq r \wedge r \leq k$, hence to $p \equiv r$.

(s_{26}) $\vdash \exists k \ w(k,x)$ where $w(k,x)$ is the formula
$$\forall y \ (y < s(x) \to (s(y) \mid k \wedge \beta(m,n,y) \leq k)) \wedge b \leq k) \ .$$

This is a formula u(x) which I prove by (IRA). Distinguishing cases for $\beta(m,n,\kappa_0)$ and b by (b_{17}), I find $\vdash \exists k_0\ (\beta(m,n,\kappa_0) \leq k_0 \wedge b \leq k_0)$, hence also $\vdash w(k_0,\kappa_0)$ and $\vdash u(\kappa_0)$ since $\vdash y < s(\kappa_0) \rightarrow s(y) \equiv \kappa_1$. A straightforward argument then shows

$$\vdash w(k,x) \wedge k_1 \equiv k \cdot s(x) \rightarrow \forall y\ (y < s(s(x)) \rightarrow s(y) \mid k_1) \wedge b \leq k_1 .$$

Now $\vdash \exists q\ (\beta(m,n,\ s(x)) \equiv k_1 \cdot q + r \wedge r \leq k_1)$ by (a_{11}), hence

$$\vdash \exists q_1\ (\beta(m,n,\ s(x)) \leq k_1 \cdot q_1) .$$

But $\vdash s(y) \mid k_1 \rightarrow s(y) \mid k_1 \cdot q_1$ and $\vdash b \leq k_1 \rightarrow b \leq k_1 \cdot q_1$; hence

$$\vdash w(k,x) \wedge k_1 \equiv k \cdot s(x) \rightarrow \exists q_1\ w(k_1 \cdot q_1, s(x)) ,$$

and this is the induction step $\vdash w(k,x) \rightarrow \exists k_2\ w(k_2,\ s(x))$. This concludes the proof of (s_{26}). From (s_{26}) and (s_{25}) there follows Lemma 1.

In a slightly different setting, a proof of Theorem 1 can be found in HILBERT–BERNAYS 34 , p.422 ff.; there the proof of the statement, called here Lemma 1, makes essential use of the principle (NISA) and, therefore, works only for PA.

3. A First Look at PR

Let Lp be an extension of L, the language of PA, by function symbols for certain primitive recursive functions. I do not specify this set and shall choose it depending on the task to be considered; in every such case it will be finite, though those who want to think globally may also wish to add symbols for *all* primitive recursive functions.

Let PR be the definitorial extension of PA by the defining equations for these terms, and with the induction schema (ISA) extended to Lp–formulas; PR is often called a system of ***primitive recursive arithmetic***. The new function symbols I shall call PR–***symbols***, and if one of my functions was defined by primitive recursion from others, then it follows from the Theorem of the last section that the recursion equations, for the terms corresponding to them, become provable in PR. The terms of Lp formed from the PR-symbols I shall call *PR–terms*. Provability shall be that from PR.

Clearly, a more precise description requires to start here not from functions but from their descriptions, for instance in the form of *schemata S*. These may be given as trees with nodes carrying functions such that

(s1) a one–node tree carrying an initial primitive recursive function (i.e. s, c_0^1, p_i^k) is a schema for that function,

(s2) if $S(f^k)$, $S(h_0^m)$,..., $S(h_{k-1}^m)$ are schemata for the indicated functions then the schema $S(f^k \circ \langle h_0^m,...,h_{k-1}^m \rangle)$ is the tree in which the given ones are put side by side and a new node on top is added which carries $f^k \circ \langle h_0^m, \dots, h_{k-1}^m \rangle$,

(s3) if $S(g^k)$, $S(r^{k+2})$ are schemata for the indicated functions then the schema $S(R(g^k, r^{k+2}))$ is the tree in which the given ones are put side by side and a new node on top is added which carries the function $f^{k+1} = R(g^k, r^{k+2})$ defined by primitive recursion from g^k, r^{k+2}.

Thus a primitive recursive function shall always be a function together with (a specified) one of its descriptions.

The PR-terms then are nothing but another sort of descriptions of this kind, only that they now are expressed in a formal language expanding L (and not in informal mathematical terminology as were the schemata). Also, since L contains already the operation symbols + and · (which I now occasionally will write with the more noticeable symbol ×), I include these into the list of basic PR-terms

(t1) $\mathcal{C}_0(x)$, $s(x)$, $x+y$, $x \times y$, $\mathcal{P}_i^n(\xi)$

of which the first and the last need defining equations, namely $\mathcal{C}_0(x) \equiv \kappa_0$ and $\mathcal{P}_i^n(\xi) \equiv x_i$ where $\xi = \langle x_0, \dots, x_{n-1} \rangle$. From the basic terms, all other PR-terms arise under superposition and the functor of primitive recursion

(t2) If $\mathcal{H}_0(\zeta), \dots, \mathcal{H}_{k-1}(\zeta)$ are PR-terms of the same arity m and if $\mathcal{G}(\xi)$ is a PR-term of arity k then $\mathcal{G}(\mathcal{H}_0(\zeta), \dots, \mathcal{H}_{k-1}(\zeta))$ is a PR-term of arity m ,

(t3) If $\mathcal{G}(\xi)$ is a PR-term of arity k and if $\mathcal{R}(\xi, x_n, x_{n+1})$ is a PR-term of arity k+2 then $\mathbf{R}[\mathcal{G}; \mathcal{R}](\xi, x_n)$ is a PR-term of arity k+1 with the defining equations

$$\mathbf{R}[\mathcal{G}; \mathcal{R}](\xi, \kappa_0) \equiv \mathcal{G}(\xi) \ ,$$
$$\mathbf{R}[\mathcal{G}; \mathcal{R}](\xi, s(x)) \equiv \mathcal{R}(\xi, x, \mathbf{R}[\mathcal{G}; \mathcal{R}](\xi, x)) \ .$$

Recall that a function f^k is said to be represented by a term $t(\xi)$ if, for every sequence α of arguments, there holds $\vdash t(\kappa \cdot \alpha) \equiv \kappa(f^k(\alpha))$. I now observe

(1) If the function g is represented by $\mathcal{G}(\xi)$ and if functions h_i, $i<k$, are represented by $\mathcal{H}_i(\zeta)$ then the superposition $g \circ \langle h_0, \dots, h_{k-1} \rangle$ is represented by $\mathcal{G}(\mathcal{H}_0(\zeta), \dots, \mathcal{H}_{k-1}(\zeta))$.

Because for any α in ω^m

$$\vdash \mathcal{H}_i(\kappa \cdot \alpha) \equiv \kappa(h_i(\alpha)) \quad \text{for every } i<m \ ,$$
$$\vdash \mathcal{G}(\mathcal{H}_0(\kappa \cdot \alpha), \dots, \mathcal{H}_{k-1}(\kappa \cdot \alpha)) \equiv \mathcal{G}(\kappa(h_0(\alpha)), \dots, \kappa(h_{k-1}(\alpha)))$$
$$\vdash \mathcal{G}(\kappa(h_0(\alpha)), \dots, \kappa(h_{k-1}(\alpha))) \equiv \kappa(g(h_0(\alpha)), \dots, h_{k-1}(\alpha)))$$
$$\kappa(g(h_0(\alpha)), \dots, h_{k-1}(\alpha)) = \kappa(g \circ \langle h_0, \dots, h_{k-1} \rangle (\alpha)) \ .$$

(2) If f is given as $f(\alpha, 0) = g(\alpha)$, $f(\alpha, n+1) = r(\alpha, n, f(\alpha, n))$ and if g and r are represented by $\mathcal{G}$ and $\mathcal{R}$ then f is represented by $\mathbf{R}[\mathcal{G}; \mathcal{R}]$.

Because $\vdash \mathbf{R}[G\,;R\,](\xi,\kappa_0) \equiv G(\xi)$ implies for any α in ω^m

$$\vdash \mathbf{R}[G\,;R\,](\kappa\cdot\alpha,\kappa_0) \equiv G(\kappa\cdot\alpha) \quad , \text{ but}$$
$$\vdash G(\kappa\cdot\alpha) \equiv \kappa(g(\alpha)) \quad , \text{ whence}$$
$$\vdash \mathbf{R}[G\,;R\,](\kappa\cdot\alpha,\kappa_0) \equiv \kappa(f(\alpha,0)) \ .$$

Also $\vdash \kappa_{n+1} \equiv s(\kappa_n)$ and $\vdash \mathbf{R}[G\,;R\,](\xi,s(x)) \equiv R(\xi,x,\mathbf{R}[G\,;R\,](\xi,x))$ imply

$$\vdash \mathbf{R}[G\,;R\,](\kappa\cdot\alpha,\kappa_{n+1}) \equiv \mathbf{R}[G\,;R\,](\kappa\cdot\alpha,s(\kappa_n))$$
$$\vdash \mathbf{R}[G\,;R\,](\kappa\cdot\alpha,s(\kappa_n)) \equiv R(\kappa\cdot\alpha,\kappa_n,\mathbf{R}[G\,;R\,](\kappa\cdot\alpha,\kappa_n)) \ .$$

So if I know that

(3) $\vdash \mathbf{R}[G\,;R](\kappa\cdot\alpha,\kappa_n) \equiv \kappa(f(\alpha,n))$

then also

$$\vdash \mathbf{R}[G\,;R\,](\kappa\cdot\alpha,\kappa_{n+1}) \equiv R(\kappa\cdot\alpha,\kappa_n,\kappa(f(\alpha,n)))$$
$$\vdash \mathbf{R}[G\,;R\,](\kappa\cdot\alpha,\kappa_{n+1}) \equiv \kappa(r(\alpha,n,f(\alpha,n)))$$
$$\vdash \mathbf{R}[G\,;R\,](\kappa\cdot\alpha,\kappa_{n+1}) \equiv \kappa(f(\alpha,n+1)) \ .$$

Consequently, (exterior) induction in ω proves (3) for every n.

Already in QA recursive functions were representable, hence they will be representable by terms in suitable extensions. But for PR there holds even the

LEMMA 2 For every k–ary primitive recursive function with a description there is a PR–term, with k variables, which represents it.

Every PR–Term, with k variables, represents a (uniquely determined) k–ary primitive recursive function.

The first statement is proved by induction on descriptions: s is represented by $s(x)$ since $\vdash s(\kappa_n) \equiv \kappa_{n+1}$ holds in PA. The constant function c_0^1 is represented by $C_0(x)$, and p_i^k is represented by $P_i^k(x_0,\ldots,x_{k-1})$. Closure under superposition and primitive recursion follows from (1) and (2). – It should be noticed that the representing term is unique once the representing terms for the initial functions s, c_0^1, p_i^k have been chosen.

The second statement is proved by induction on the definition of PR–terms. It is clear for the basic PR–terms, and closure under (t2), (t3) follows from (1) and (2).

I introduce the *elementary* functors $\Sigma_{\leq}$, $\Sigma_{<}$, $\Pi_{\leq}$, $\Pi_{<}$ which produce from (n+1)–ary PR–terms $T(\xi,x_n)$ the PR–terms $\mathbf{R}[G\,;R\,](\xi,x_n)$, where for

$$\Sigma_{\leq} \ : \ G(\xi) = T(\xi,\kappa_0) \qquad R(\xi,x_n,x_{n+1}) = x_{n+1} + T(\xi,s(x_n))$$
$$\Sigma_{<} \ : \ G(\xi) = \kappa_0 \qquad R(\xi,x_n,x_{n+1}) = x_{n+1} + T(\xi,x_n)$$

$$\Pi_{\leq} \ : \ G(\xi) = T(\xi,\kappa_0) \qquad R(\xi,x_n,x_{n+1}) = x_{n+1} \times T(\xi,s(x_n))$$

$$\Pi_{<} \ : \ G(\xi) = \kappa_1 \qquad R(\xi,x_n,x_{n+1}) = x_{n+1} \times T(\xi,x_n)$$

i.e.

$$\Sigma_{\leq}T(\xi,\kappa_0) \equiv T(\xi,\kappa_0) \qquad \Sigma_{\leq}T(\xi,s(x_n)) \equiv \Sigma_{\leq}T(\xi,x_n) + T(\xi,s(x_n)) \ ,$$

$$\Sigma_{<}T(\xi,\kappa_0) \equiv \kappa_0 \qquad \Sigma_{<}T(\xi,s(x_n)) \equiv \Sigma_{\leq}T(\xi,x_n) \ ,$$

$$\Pi_{\leq}T(\xi,\kappa_0) \equiv T(\xi,\kappa_0) \qquad \Pi_{\leq}T(\xi,s(x_n)) \equiv \Pi_{\leq}T(\xi,x_n) \times T(\xi,s(x_n)) \ ,$$

$$\Pi_{<}T(\xi,\kappa_0) \equiv \kappa_1 \qquad \Pi_{<}T(\xi,s(x_n)) \equiv \Pi_{<}T(\xi,x_n) \times T(\xi,x_n) \ .$$

The following is a list of PR-terms and their defining equations. Unless explained explicitly, they represent the elementary functions introduced in Chapter 5 with the corresponding names:

$$CSG(x) : CSG(\kappa_0) \equiv \kappa_1 \ , \ CSG(s(x)) \equiv C_0(CSG(x)) \ ,$$

$$SG(x) \ \equiv \ CSG(CSG(x)) \ ,$$

$$CS(x) \ \equiv \ CS(\kappa_0) \equiv \kappa_0 \ , \ CS(s(x)) \equiv P^2_1(CS(x), x) \ ,$$

$$x \dot{-} y \ : x \dot{-} \kappa_0 \equiv P^1_0(x) \ , \ x \dot{-} s(y) \equiv CS(x \dot{-} y)) \ ,$$

$$ID_2(x,y) \ \equiv \ SG(s(x) \dot{-} y) \times (s(y) \dot{-} x)) \ , \qquad \text{represents } \chi_{=}$$

$$CH_{<}(x,y) \ \equiv \ SG(y \dot{-} x) \ , \qquad \text{represents } \chi_{<}$$

$$PO(x,y) : PO(x,\kappa_0) \equiv \kappa_1 \ , \ PO(x, s(y)) \equiv PO(x,y) \times x \ ,$$

represents exponentiation and will be written as x^y

The definition of the next terms requires the functor $\mu_{<}$, producing from (n+1)-ary PR-terms $T(\xi,x_n)$ the PR-terms $\mu_{<}T(\xi,x_n)$ such that

$$\mu_{<}T(\xi,\kappa_0) \ \equiv \ \kappa_0 \ ,$$

$$\mu_{<}T(\xi,s(x_n)) \equiv SG(\Pi_{\leq}T(\xi,x_n)) \times s(x_n) \ + \ CSG(\Pi_{\leq}T(\xi,x_n)) \times$$
$$(\ SG(\Pi_{<}T(\xi,x_n)) \times x_n + CSG(\Pi_{<}T(\xi,x_n)) \times \mu_{<}T(\xi,x_n) \ ).$$

Now I continue with

$$QU(x,y) \ \equiv \ SG(x) \times \mu_{<}B(x,y,s(y)) \quad \text{with } B(x,y,z) = CH_{<}(y, x \cdot s(z)) \ ,$$

$$MOD(x,y) \ \equiv \ y \dot{-} x \cdot QU(x,y) \ ,$$

$$DIV(x,y) \ \equiv \ CSG(MOD(x,y)) \ ,$$

$$DIVN(x) \ \equiv \ \Sigma_{<}DIV_1(x,x) \dot{-} \kappa_2 \quad \text{with}$$
$$DIV_1(x,y) = DIV(P^2_1(x,y), P^2_0(x,y)) \ ,$$

represents the sum of proper divisors of x,

$$PRIN(x) \ \equiv \ ID_2(DIVN(x), \kappa_0) \ ,$$

represents the characteristic function of the set of primes

$$PRINO(x) : PRINO(\kappa_0) \equiv \kappa_2 \;,\; PRINO(s(x)) \equiv R(PRINO(x)) \quad \text{with}$$
$$R(y) = \mu_< B(y, PO(\kappa_2, PO(\kappa_2,y))) \;,\; B(y,z) = CH_<(y,z) \times PRIN(z),$$
$$EXP(x,y) \equiv \mu_< B(x,y,y) \text{ with } B(x,y,z) = CSG(DIV(PRINO(x)^{s(z)}, y)),$$
$$LEN(x) \equiv SG(x) \times SG(CS(x)) \times \mu_< B(x,x) \quad \text{with}$$
$$B(x,y) = \Pi_{\le} C(x,y,x-y) \;,\; C(x,y,z) = CSG(DIV(PRINO(y+z), x)) \;.$$

The defining equations here are formulas with free variables. In ordinary arithmetic, induction in ω often is used in order to prove, in the usual mathematical manner, various relationships between the functions expressed by these terms. If they are simple enough, such relationships can be expressed by free variable formulas in the language of PR, and as long as their ordinary proofs do not use more than can be expressed in that language, PR–induction can take the rôle of ω–induction in order to find proofs in PR for those formulas. – I now mention a few provabilities:

(c_0) $\vdash y \equiv \kappa_n \to \neg\, y \equiv \kappa_m$ for $n \neq m$.

Because

$\vdash y \equiv \kappa_n \wedge y \equiv \kappa_m \to \kappa_n \equiv \kappa_m$

$\vdash \neg\, \kappa_n \equiv \kappa_m \to \neg\,(y \equiv \kappa_n \wedge y \equiv \kappa_m)$

$\vdash \neg\,(y \equiv \kappa_n \wedge y \equiv \kappa_m)$ by (PC2)

$\vdash y \equiv \kappa_n \to \neg\, y \equiv \kappa_m$ propositionally .

(c_1) $\vdash \neg\, x \equiv \kappa_0 \to CSG(x) \equiv \kappa_0$

$\vdash \neg\, CSG(x) \equiv \kappa_0 \to x \equiv \kappa_0$.

Because $\vdash CSG(s(y)) \equiv \kappa_0$ by definition of CSG and C_0. So

$\vdash x \equiv s(y) \to CSG(x) \equiv \kappa_0$

$\vdash \exists y\,(x \equiv s(y)) \to \exists y\,(CSG(x) \equiv \kappa_0)$

$\vdash \exists y\,(x \equiv s(y)) \to CSG(x) \equiv \kappa_0$

$\vdash \neg\, x \equiv \kappa_0 \to \exists y\,(x \equiv s(y))$ by (PQ0)

$\vdash \neg\, x \equiv \kappa_0 \to CSG(x) \equiv \kappa_0$.

(c_2) $\vdash CSG(x) \equiv \kappa_1 \leftrightarrow x \equiv \kappa_0$.

$\vdash CSG(x) \equiv \kappa_1 \to \neg\, CSG(x) \equiv \kappa_0$ by (c_0)

$\vdash CSG(x) \equiv \kappa_1 \to x \equiv \kappa_0$ by (c_1) .

(c_3) $\vdash CSG(x) \equiv \kappa_0 \leftrightarrow \neg\, x \equiv \kappa_0$.

$\vdash x \equiv \kappa_0 \to (CSG(\kappa_0) \equiv \kappa_1 \to CSG(x) \equiv \kappa_1)$

$\vdash x \equiv \kappa_0 \to CSG(x) \equiv \kappa_1$

$\vdash \neg\, CSG(x) \equiv \kappa_1 \to \neg\, x \equiv \kappa_0$

$\vdash CSG(x) \equiv \kappa_0 \to \neg\, CSG(x) \equiv \kappa_1$ by (c_0)

$\vdash CSG(x) \equiv \kappa_0 \to \neg\, x \equiv \kappa_0$.

(c_4) $\vdash SG(x) \equiv \kappa_1 \leftrightarrow \neg\, x \equiv \kappa_0$.

Because $\vdash SG(x) \equiv \kappa_1 \leftrightarrow CSG(x) \equiv \kappa_0$ by (c_2), so (c_3) can be applied.

(c_5) $\vdash s(x) \equiv x+\kappa_1$.

This is a statement A(x) which I prove by PA–induction.

$\vdash s(\kappa_0) \equiv \kappa_1$ by (PQ71)
$\vdash \kappa_0+\kappa_1 \equiv \kappa_1$ by (a_1)
$\vdash s(\kappa_0) \equiv \kappa_0+\kappa_1$
$\vdash A(\kappa_0)$.

$\vdash s(x) \equiv x+\kappa_1 \rightarrow s(s(x)) \equiv s(x+\kappa_1)$
$\vdash s(x+\kappa_1) \equiv s(\kappa_1+x)$ by (a_3)
$\vdash s(\kappa_1+x) \equiv \kappa_1+s(x)$ by (PQ4)
$\vdash s(x+\kappa_1) \equiv s(x)+\kappa_1$ by (a_3)
$\vdash s(x) \equiv x+\kappa_1 \rightarrow s(s(x)) \equiv s(x)+\kappa_1$
$\vdash A(x) \rightarrow A(s(x))$.

(c_6) $\vdash \neg\, x \equiv \kappa_0 \rightarrow CS(x) + \kappa_1 \equiv x$.

$\vdash CS(s(y)) \equiv y$ by definition
$\vdash CS(y+\kappa_1) \equiv y$
$\vdash CS(y+\kappa_1)+\kappa_1 \equiv y+\kappa_1$
$\vdash \neg\, x \equiv \kappa_0 \rightarrow \exists y(x \equiv s(y))$
$\vdash \neg\, x \equiv \kappa_0 \rightarrow \exists y(x \equiv y+\kappa_1)$
$\vdash \neg\, x \equiv \kappa_0 \rightarrow \exists y(x \equiv y+\kappa_1 \wedge CS(y+\kappa_1)+\kappa_1 \equiv y+\kappa_1)$
$\vdash \neg\, x \equiv \kappa_0 \rightarrow \exists y(x \equiv y+\kappa_1 \wedge CS(x)+\kappa_1 \equiv x)$
$\vdash \neg\, x \equiv \kappa_0 \rightarrow CS(x)+\kappa_1 \equiv x)$.

(c_7) $\vdash y<x \rightarrow \neg\, x\dot{-}y \equiv \kappa_0 \wedge (x\dot{-}y)+y \equiv x$.

This is a statement A(x,y) which I prove by PR–induction on y.

$\vdash \kappa_0<x \rightarrow \neg\, x \equiv \kappa_0$
$\vdash x\dot{-}\kappa_0 \equiv x \wedge x+\kappa_0 \equiv x$
$\vdash A(x,\kappa_0)$.

$\vdash s(y)<x \rightarrow y<x$ by (b_8),
$\vdash A(x,y) \wedge s(y)<x \rightarrow \neg\, x\dot{-}y \equiv \kappa_0$
$\vdash A(x,y) \wedge s(y)<x \rightarrow CS(x\dot{-}y)+\kappa_1 \equiv x\dot{-}y$ by (c_6)
$\vdash A(x,y) \wedge s(y)<x \rightarrow (x\dot{-}s(y))+\kappa_1 \equiv x\dot{-}y$
$\vdash A(x,y) \wedge s(y)<x \rightarrow (x\dot{-}y)+y \equiv x \wedge (x\dot{-}s(y))+\kappa_1 \equiv x\dot{-}y$
$\vdash A(x,y) \wedge s(y)<x \rightarrow ((x\dot{-}s(y))+\kappa_1)+y \equiv x$
$\vdash A(x,y) \wedge s(y)<x \rightarrow (x\dot{-}s(y))+(\kappa_1+y) \equiv x$ by (a_0)
$\vdash A(x,y) \wedge s(y)<x \rightarrow (x\dot{-}s(y))+(y+\kappa_1) \equiv x$ by (a_3)
$\vdash A(x,y) \wedge s(y)<x \rightarrow (x\dot{-}s(y))+s(y) \equiv x$
$\vdash A(x,y) \wedge s(y)<x \rightarrow (x\dot{-}s(y) \equiv \kappa_0 \rightarrow s(y) \equiv x)$
$\vdash A(x,y) \wedge s(y)<x \rightarrow (\neg\, s(y) \equiv x \rightarrow \neg\, x\dot{-}s(y) \equiv \kappa_0)$
$\vdash s(y)<x \rightarrow \neg\, s(y) \equiv x$
$\vdash A(x,y) \wedge s(y)<x \rightarrow \neg\, x\dot{-}s(y) \equiv \kappa_0$
$\vdash A(x,y) \wedge s(y)<x \rightarrow \neg\, x\dot{-}s(y) \equiv \kappa_0 \wedge (x\dot{-}s(y))+s(y) \equiv x$

$$\vdash A(x,y) \to (s(y)<x \to \neg\, x\dot{-}s(y) \equiv \kappa_0 \wedge (x\dot{-}s(y))+s(y) \equiv x)$$
$$\vdash A(x,y) \to A(x,s(y)) .$$

(c₈) $\vdash \neg\, x\dot{-}y \equiv \kappa_0 \to y<x \wedge (x\dot{-}y)+y \equiv x$.

This is a statement $A(x,y)$ which I prove by PR-induction on y.

$$\vdash \neg\, x\dot{-}\kappa_0 \equiv \kappa_0 \to \neg\, x \equiv \kappa_0$$
$$\vdash \neg\, x\dot{-}\kappa_0 \equiv \kappa_0 \to \kappa_0<x$$
$$\vdash \neg\, x\dot{-}\kappa_0 \equiv \kappa_0 \to \kappa_0<x \wedge x+\kappa_0 \equiv x$$
$$\vdash \neg\, x\dot{-}\kappa_0 \equiv \kappa_0 \to \kappa_0<x \wedge (x\dot{-}\kappa_0)+\kappa_0 \equiv x$$
$$\vdash A(x,\kappa_0) .$$

$$\vdash \neg\, x\dot{-}s(y) \equiv \kappa_0 \to \neg\, CS(x\dot{-}y) \equiv \kappa_0$$
$$\vdash z \equiv \kappa_0 \to CS(z) \equiv \kappa_0$$
$$\vdash \neg\, CS(z) \equiv \kappa_0 \to \neg\, z \equiv \kappa_0$$
$$\vdash \neg\, x\dot{-}s(y) \equiv \kappa_0 \to \neg\, x\dot{-}y \equiv \kappa_0$$
$$\vdash A(x,y) \wedge \neg\, x\dot{-}s(y) \equiv \kappa_0 \to (x\dot{-}y)+y \equiv x$$
$$\vdash \neg\, x\dot{-}y \equiv \kappa_0 \to CS(x\dot{-}y)+\kappa_1 \equiv x\dot{-}y$$
$$\vdash \neg\, x\dot{-}s(y) \equiv \kappa_0 \to (x\dot{-}s(y))+\kappa_1 \equiv x\dot{-}y$$
$$\vdash A(x,y) \wedge \neg\, x\dot{-}s(y) \equiv \kappa_0 \to ((x\dot{-}s(y))+\kappa_1)+y \equiv x$$
$$\vdash A(x,y) \wedge \neg\, x\dot{-}s(y) \equiv \kappa_0 \to (x\dot{-}s(y))+(\kappa_1+y) \equiv x \quad \text{by } (a_0)$$
$$\vdash A(x,y) \wedge \neg\, x\dot{-}s(y) \equiv \kappa_0 \to (x\dot{-}s(y))+(y+\kappa_1) \equiv x \quad \text{by } (a_3)$$
$$\vdash A(x,y) \wedge \neg\, x\dot{-}s(y) \equiv \kappa_0 \to (x\dot{-}s(y))+s(y) \equiv x$$
$$\vdash A(x,y) \wedge \neg\, x\dot{-}s(y) \equiv \kappa_0 \to ((x\dot{-}s(y))+s(y) \equiv x) \to s(y)<x)$$
$$\vdash A(x,y) \wedge \neg\, x\dot{-}s(y) \equiv \kappa_0 \to s(y)<x)$$
$$\vdash A(x,y) \wedge \neg\, x\dot{-}s(y) \equiv \kappa_0 \to (s(y)<x \wedge (x\dot{-}s(y))+s(y) \equiv x)$$
$$\vdash A(x,y) \to (\neg\, x\dot{-}s(y) \equiv \kappa_0 \to s(y)<x \wedge (x\dot{-}s(y))+s(y) \equiv x)$$
$$\vdash A(x,y) \to A(x,s(y)) .$$

(c₉) $\vdash \neg\, x\dot{-}y \equiv \kappa_0 \leftrightarrow y<x$.

By (c₇) and (c₈) .

(c₁₀) $\vdash SG(x\dot{-}y) \equiv \kappa_1 \leftrightarrow y<x$

$$\vdash CH_<(x,y) \equiv \kappa_1 \leftrightarrow x<y .$$

By (c₄) and (c₉) .

(c₁₁) $\vdash SG(s(x)\dot{-}y) \equiv \kappa_1 \leftrightarrow y \leq x$.

By (c₁₀) and (b₆) .

(c₁₂) $\vdash SG(x+y) \equiv \kappa_0 \leftrightarrow SG(x) \equiv \kappa_0 \wedge SG(y) \equiv \kappa_0$

$$\vdash SG(x+y) \equiv \kappa_1 \leftrightarrow SG(x) \equiv \kappa_1 \vee SG(y) \equiv \kappa_1 .$$

By (c₄) and (a₆) .

(a₁₄) $\vdash x\cdot y \equiv \kappa_0 \leftrightarrow x \equiv \kappa_0 \vee y \equiv \kappa_0$.

$$\vdash \neg\, y \equiv \kappa_0 \to x \leq x\cdot y \quad \text{by } (b_{24})$$
$$\vdash \neg\, y \equiv \kappa_0 \wedge x\cdot y \equiv \kappa_0 \to x \leq \kappa_0$$
$$\vdash \neg\, y \equiv \kappa_0 \wedge x\cdot y \equiv \kappa_0 \to x \equiv \kappa_0 \quad \text{by } (b_0), (b_3) .$$

The direction from right to left follows from (PQ5), (a_9).

(c_{13}) $\vdash SG(x\cdot y) \equiv \kappa_0 \leftrightarrow SG(x) \equiv \kappa_0 \vee SG(y) \equiv \kappa_0$.
$\vdash SG(x\cdot y) \equiv \kappa_1 \leftrightarrow SG(x) \equiv \kappa_1 \wedge SG(y) \equiv \kappa_1$

(c_{14}) $\vdash ID_2(x,y) \equiv \kappa_1 \leftrightarrow x \equiv y$.

By (c_{11}) and (c_{13}) .

(a_{15}) $\vdash (x \times y) \times z \equiv x \times (y \times z)$.

(c_{15}) $\vdash x^{y+z} \equiv x^y \times x^z$.

This is a statement A(x,y,z) which I prove by PA-induction on z.

$$\begin{array}{ll}
\vdash PO(x,\ y+\kappa_0) \equiv PO(x,y) & \\
\vdash PO(x,\ y+\kappa_0) \equiv PO(x,y) \times \kappa_1 & \text{by } (a_{10}) \\
\vdash PO(x,\ y+\kappa_0) \equiv PO(x,y) \times PO(x,\kappa_0) & \\
\vdash A(x,y,\kappa_0) &
\end{array}$$

$$\begin{array}{ll}
\vdash PO(x,\ y+s(z)) \equiv PO(x,\ s(y+z)) & \\
\vdash PO(x,\ y+s(z)) \equiv PO(x,\ y+z) \times x & \\
\vdash A(x,y,z) \rightarrow PO(x,\ y+s(z)) \equiv (PO(x,y) \times PO(x,z)) \times x & \\
\vdash A(x,y,z) \rightarrow PO(x,\ y+s(z)) \equiv PO(x,y) \times (PO(x,z)) \times x) & \text{by } (a_{15}) \\
\vdash A(x,y,z) \rightarrow PO(x,\ y+s(z)) \equiv PO(x,y) \times PO(x,\ s(z)) & \\
\vdash A(x,y,z) \rightarrow A(x,y,s(z)) \ . &
\end{array}$$

(c_{16}) $\vdash SG(\Sigma_{\leq} T(\xi,\kappa_n)) \equiv \kappa_1 \leftrightarrow \mathbb{V} < SG(T(\xi,\kappa_i)) \equiv \kappa_1 \mid i \leq n >$
$\vdash SG(\Sigma_{<} T(\xi,\kappa_n)) \equiv \kappa_1 \leftrightarrow \mathbb{V} < SG(T(\xi,\kappa_i)) \equiv \kappa_1 \mid i < n >$
$\vdash SG(\Pi_{\leq} T(\xi,\kappa_n)) \equiv \kappa_1 \leftrightarrow \mathbb{A} < SG(T(\xi,\kappa_i)) \equiv \kappa_1 \mid i \leq n >$
$\vdash SG(\Pi_{<} T(\xi,\kappa_n)) \equiv \kappa_1 \leftrightarrow \mathbb{A} < SG(T(\xi,\kappa_i)) \equiv \kappa_1 \mid i < n >$.

A PR-term $T(\xi)$ shall be called *Boolean* if

(4) $\vdash T(\xi) \equiv \kappa_0 \vee T(\xi) \equiv \kappa_1$.

Clearly, $C_0(x)$ is so, but also $CSG(x)$ is Boolean because

$$\vdash x \equiv \kappa_0 \rightarrow CSG(x) \equiv CSG(\kappa_0) \equiv \kappa_1 \qquad \text{by definition .}$$

$$\begin{array}{ll}
\vdash x \equiv s(y) \rightarrow CSG(x) \equiv CSG(s(y)) & \\
\vdash x \equiv s(y) \rightarrow CSG(x) \equiv C_0(CSG(x)) & \text{by definition} \\
\vdash x \equiv s(y) \rightarrow CSG(x) \equiv \kappa_0 & \\
\vdash \exists y\ (x \equiv s(y)) \rightarrow \exists y\ (CSG(x) \equiv \kappa_0) & \\
\vdash \exists y\ (x \equiv s(y)) \rightarrow CSG(x) \equiv \kappa_0 & \\
\vdash \neg\, x \equiv \kappa_0 \rightarrow \exists y (x \equiv s(y)) & \\
\vdash \neg\, x \equiv \kappa_0 \rightarrow CSG(x) \equiv \kappa_0 \ . &
\end{array}$$

$$\vdash x \equiv \kappa_0 \vee \neg\, x \equiv \kappa_0 \rightarrow CSG(x) \equiv \kappa_1 \vee CSG(x) \equiv \kappa_0 \ .$$

Together with $T(\xi)$ also every term $T(H_0(\zeta), \ldots, H_{n-1}(\zeta))$ ist Boolean. Hence together with $CSG(x)$ also $SG(x,y)$, $DIV(x,y)$ are Boolean, and together with $SG(x,y)$ also $ID_2(x,y)$, $CH_<(x,y)$ are so, finally together with $ID_2(x,y)$ also $PRIN(x,y)$ is Boolean.

If, and only if, $T(\xi)$ is Boolean then each of the two equivalent conditions holds

$$\begin{array}{ll}(5) & \vdash \neg\, T(\xi) \equiv \kappa_0 \;\leftrightarrow\; T(\xi) \equiv \kappa_1 \\ & \vdash \neg\, T(\xi) \equiv \kappa_1 \;\leftrightarrow\; T(\xi) \equiv \kappa_0 \;.\end{array}$$

Because if $T(\xi)$ is Boolean then $\vdash \neg\, T(\xi) \equiv \kappa_0 \;\rightarrow\; T(\xi) \equiv \kappa_1$ follows from (4) propositionally, and $\vdash T(\xi) \equiv \kappa_1 \;\rightarrow\; \neg\, T(\xi) \equiv \kappa_0$ follows from (c_4). On the other hand, (5) implies

$$\vdash T(\xi) \equiv \kappa_0 \;\vee\; \neg\, T(\xi) \equiv \kappa_0 \;\rightarrow\; T(\xi) \equiv \kappa_0 \;\vee\; T(\xi) \equiv \kappa_1 \;.$$

If, and only if, $T(\xi)$ is Boolean then

$$(6) \qquad \vdash T(\xi) \equiv \kappa_1 \;\leftrightarrow\; SG(T(\xi)) \equiv \kappa_1 \;.$$

Because if $T(\xi)$ is Boolean then

$$\begin{array}{ll}\vdash T(\xi) \equiv \kappa_1 \;\leftrightarrow\; \neg\, T(\xi) \equiv \kappa_0 & \text{by (5)} \\ \vdash \neg\, T(\xi) \equiv \kappa_0 \;\leftrightarrow\; SG(T(\xi)) \equiv \kappa_1 & \text{by } (c_4)\;,\end{array}$$

and this implies (6). On the other hand,

$$\begin{array}{ll}\vdash \neg\, T(\xi) \equiv \kappa_0 \;\leftrightarrow\; SG(T(\xi)) \equiv \kappa_1 & \text{by } (c_4) \\ \vdash SG(T(\xi)) \equiv \kappa_1 \;\leftrightarrow\; T(\xi) \equiv \kappa_1 & \text{by (6)}\end{array}$$

and this implies the first equivalence of (5). – It follows from (6) that the Boolean terms $T(\xi)$ are precisely those satisfying $\vdash T(\xi) \equiv SG(T(\xi))$. If $T(\xi,x_n)$ is Boolean then

$$\begin{array}{ll}(c_{17}) & \vdash \exists x_n\,(x_n \leq y \wedge T(\xi,x_n) \equiv \kappa_1) \;\leftrightarrow\; SG(\Sigma_{\leq} T(\xi,y)) \equiv \kappa_1 \\ & \vdash \exists x_n\,(x_n < y \wedge T(\xi,x_n) \equiv \kappa_1) \;\leftrightarrow\; SG(\Sigma_{<} T(\xi,y)) \equiv \kappa_1 \\ & \vdash \forall x_n\,(x_n \leq y \wedge T(\xi,x_n) \equiv \kappa_1) \;\leftrightarrow\; SG(\Pi_{\leq} T(\xi,y)) \equiv \kappa_1 \\ & \vdash \forall x_n\,(x_n < y \wedge T(\xi,x_n) \equiv \kappa_1) \;\leftrightarrow\; SG(\Pi_{<} T(\xi,y)) \equiv \kappa_1 \;.\end{array}$$

I prove the first statement $A(\xi,y)$ by PA-induction on y.

$$\begin{array}{ll}\vdash x_n \equiv \kappa_0 \wedge T(\xi,x_n) \equiv \kappa_1 \;\leftrightarrow\; T(\xi,\kappa_0) \equiv \kappa_1 & \\ \vdash x_n \leq \kappa_0 \wedge T(\xi,x_n) \equiv \kappa_1 \;\leftrightarrow\; T(\xi,\kappa_0) \equiv \kappa_1 & \\ \vdash \exists x_n\,(x_n \leq \kappa_0 \wedge T(\xi,x_n) \equiv \kappa_1) \;\leftrightarrow\; \exists x_n\, T(\xi,\kappa_0) \equiv \kappa_1 & \\ \vdash \exists x_n\,(x_n \leq \kappa_0 \wedge T(\xi,x_n) \equiv \kappa_1) \;\leftrightarrow\; T(\xi,\kappa_0) \equiv \kappa_1 & \\ \vdash \exists x_n\,(x_n \leq \kappa_0 \wedge T(\xi,x_n) \equiv \kappa_1) \;\leftrightarrow\; SG(T(\xi,\kappa_0)) \equiv \kappa_1 & \text{by (5)} \\ \vdash \Sigma_{\leq} T(\xi,\kappa_0) \equiv T(\xi,\kappa_0) \;\rightarrow\; SG(\Sigma_{\leq} T(\xi,\kappa_0)) \equiv SG(T(\xi,\kappa_0)) & \\ \vdash SG(\Sigma_{\leq} T(\xi,\kappa_0)) \equiv SG(T(\xi,\kappa_0)) & \\ \vdash \exists x_n\,(x_n \leq \kappa_0 \wedge T(\xi,x_n) \equiv \kappa_1) \;\leftrightarrow\; SG(\Sigma_{\leq} T(\xi,\kappa_0)) \equiv \kappa_1 & \\ \vdash A(\xi,\kappa_0)\;. & \end{array}$$

$$\begin{array}{l}\vdash x_n \leq s(y) \;\wedge\; T(\xi,x_n) \equiv \kappa_1 \\ \qquad\qquad \leftrightarrow\; (x_n \leq y \;\vee\; x_n \equiv s(y)) \;\wedge\; T(\xi,x_n) \equiv \kappa_1 \\ \vdash x_n \leq s(y) \;\wedge\; T(\xi,x_n) \equiv \kappa_1 \\ \qquad\qquad \leftrightarrow\; (x_n \leq y \;\wedge\; T(\xi,x_n) \equiv \kappa_1) \;\vee\; T(\xi,s(y) \equiv \kappa_1)\end{array}$$

$$\vdash \exists x_n\,(x_n \le s(y) \wedge T(\xi,x_n) \equiv \kappa_1)$$
$$\leftrightarrow \exists x_n\,(x_n \le y \wedge T(\xi,x_n) \equiv \kappa_1) \vee \exists x_n\, T(\xi,s(y) \equiv \kappa_1)$$
$$\vdash \exists x_n\,(x_n \le s(y) \wedge T(\xi,x_n) \equiv \kappa_1)$$
$$\leftrightarrow \exists x_n\,(x_n \le y \wedge T(\xi,x_n) \equiv \kappa_1) \vee T(\xi,s(y) \equiv \kappa_1)$$
$$\vdash A(\xi,y) \rightarrow (\exists x_n\,(x_n \le s(y) \wedge T(\xi,x_n) \equiv \kappa_1)$$
$$\leftrightarrow SG(\Sigma_{\le} T(\xi,y)) \equiv \kappa_1 \vee T(\xi,s(y) \equiv \kappa_1))$$
$$\vdash A(\xi,y) \rightarrow (\exists x_n\,(x_n \le s(y) \wedge T(\xi,x_n) \equiv \kappa_1)$$
$$\leftrightarrow SG(\Sigma_{\le} T(\xi,y)) \equiv \kappa_1 \vee SG(T(\xi,s(y))) \equiv \kappa_1 \text{ by (5)}$$
$$\vdash A(\xi,y) \rightarrow (\exists x_n\,(x_n \le s(y) \wedge T(\xi,x_n) \equiv \kappa_1)$$
$$\leftrightarrow SG(\Sigma_{\le} T(\xi,y) + T(\xi,s(y))) \equiv \kappa_1 \quad \text{by } (c_{12})$$
$$\vdash A(\xi,y) \rightarrow (\exists x_n\,(x_n \le s(y) \wedge T(\xi,x_n) \equiv \kappa_1)$$
$$\leftrightarrow SG(\Sigma_{\le} T(\xi,s(y))) \equiv \kappa_1$$
$$\vdash A(\xi,y) \rightarrow A(\xi,s(y)) \;.$$

The proof of the other statements are analogous, and in the case of universal quantifiers (c_{13}) will take the place of (c_{12}).

To every Boolean PR-term $BO(\xi)$, I assign the *associated* predicate $\mathrm{BO}(\xi)$ with the defining formula

(5) $\quad \mathrm{BO}(\xi) \leftrightarrow BO(\xi) \equiv \kappa_1$.

In particular, there are the associated predicates $\mathbf{DIV}(x,y)$ and $\mathbf{PRIN}(x)$ for divisibility and for being a prime number. It follows from (c_{14}) and (c_{10}) that the predicates associated to $ID_2(x_0,x_1)$ and $CH_<(x_0,x_1)$ are provably equivalent to the predicates $x_0 \equiv x_1$ and $x_0 < x_1$.

Assume for the moment that PA is consistent. If $BO(\xi)$ is a Boolean PR-term and if f^k is the (unique) function represented by $BO(\xi)$, then f^k can only take the values 0 and 1. Because $\vdash BO(\kappa\cdot\alpha) \equiv \kappa_0 \;\vee\; BO(\kappa\cdot\alpha) \equiv \kappa_1$ by (4) for every $\alpha\in\omega^k$. So $\vdash \kappa_n \equiv \kappa_0 \vee \kappa_n \equiv \kappa_1$ if $\vdash BO(\kappa\cdot\alpha) \equiv \kappa_n$ in consequence of $f^k(\alpha) = n$. But if $n>1$ then $\vdash \neg\,\kappa_n \equiv \kappa_0 \wedge \neg\,\kappa_n \equiv \kappa_1$ already by (PC2), and so PA would be inconsistent. – Observe, though, that for the Boolean PR-terms listed above, e.g. $DIV(x,y)$ and $PRIN(x,y)$, the functions represented by them are seen as 0-1-valued by exterior mathematical arguments, without formal use of the consistency of PA.

Assume now that the Boolean PR-term $BO(\xi)$ represents a 0-1-valued function $f^k(\alpha)$. Then f^k is the characteristic function of a relation R in ω^k, and there holds that

(6) $\quad \mathrm{BO}(\xi)$ represents the relation R.

Because if $\alpha\in R$ then $\kappa(f(\alpha)) = \kappa_1$, hence $\vdash BO(\kappa\cdot\alpha) \equiv \kappa_1$ and $\vdash \mathrm{BO}(\kappa\cdot\alpha)$. And if not $\alpha\in R$ then $\kappa(f(\alpha)) = \kappa_0$, hence $\vdash BO(\kappa\cdot\alpha) \equiv \kappa_0$; but $\vdash \neg\,\kappa_0 \equiv \kappa_1$, hence $\vdash \neg\, BO(\kappa\cdot\alpha) \equiv \kappa_1$ and so $\vdash \neg\,\mathrm{BO}(\kappa\cdot\alpha)$.

Conversely, if the Boolean PR-term $BO(\xi)$ is such that (6) holds, then it represents the characteristic function χ of R. Because if $\chi(\alpha) = 1$, hence $\alpha\in R$, then $\vdash \mathrm{BO}(\kappa\cdot\alpha)$ whence $\vdash BO(\kappa\cdot\alpha) \equiv \kappa_1$. And if $\chi(\alpha) = 0$, hence not

$\alpha \in R$, then $\vdash \neg BO(\kappa \cdot \alpha)$ whence $\vdash \neg BO(\kappa\cdot\alpha) \equiv \kappa_1$. But $\vdash BO(\kappa\cdot\alpha) \equiv \kappa_0 \vee BO(\kappa\cdot\alpha) \equiv \kappa_1$, hence $\vdash \neg BO(\kappa\cdot\alpha)\equiv\kappa_1 \to BO(\kappa\cdot\alpha)\equiv\kappa_0$ and therefore $\vdash BO(\kappa\cdot\alpha)\equiv\kappa_0$.

To any Δ_0-formula $v(\xi)$ of PR, where $\xi = <x_0, \ldots, x_{n-1}>$ comprises the variables free in v, I assign an *associated* Boolean term $V[v](\xi)$, namely for

$$v : T_0(\xi) \equiv T_1(\xi) \quad \text{set} \quad V[v] = ID_2(T_0(\xi), T_1(\xi))$$

$$\neg u(\xi) \quad \text{set} \quad V[v](\xi) = CSG(V[u](\xi))$$

$$(u \wedge w)(\xi) \quad \text{set} \quad V[v](\xi) = V[u](\xi) \times V[w](\xi)$$

$$(u \vee w)(\xi) \quad \text{set} \quad V[v](\xi) = SG(V[u](\xi) + V[w](\xi))$$

$$(u \to w)(\xi) \quad \text{set} \quad V[v](\xi) = SG(CSG V[u](\xi)) + V[w](\xi))$$

$$\exists x_n < T(\xi)\, u(\xi, x_n) \quad \text{set} \quad V[v](\xi, y) = SG(\Sigma_< V[u](\xi, T(\xi)))$$

$$\forall x_n < T(\xi)\, u(\xi, x_n) \quad \text{set} \quad V[v](\xi, y) = SG(\Pi_< V[u](\xi, T(\xi))) \;.$$

LEMMA 3 For every Δ_0-formula $v(\xi)$ of PR :

$$\vdash v(\xi) \leftrightarrow V[v](\xi) \equiv \kappa_1 .$$

For the inital case this is $\vdash T_0(\xi) \equiv T_1(\xi) \leftrightarrow ID_2(T_0(\xi), T_1(\xi))$ which follows from (c_{14}). If

$$\vdash u(\xi) \leftrightarrow V[u](\xi) \equiv \kappa_1$$
$$\vdash w(\xi) \leftrightarrow V[w](\xi) \equiv \kappa_1$$

then
$$\vdash (u(\xi) \to w(\xi)) \leftrightarrow (V[u](\xi) \equiv \kappa_1 \to V[w](\xi) \equiv \kappa_1)$$
$$\vdash (u(\xi) \to w(\xi)) \leftrightarrow (\neg V[u](\xi) \equiv \kappa_1 \vee V[w](\xi) \equiv \kappa_1) .$$

Now $\vdash \neg V[u](\xi) \equiv \kappa_1 \leftrightarrow V[u](\xi) \equiv \kappa_0$ as $V[u](\xi)$ is Boolean, hence

$$\vdash (u(\xi) \to w(\xi)) \leftrightarrow (V[u](\xi) \equiv \kappa_0 \vee V[w](\xi) \equiv \kappa_1)$$
$$\vdash (u(\xi) \to w(\xi)) \leftrightarrow (CSG\,V[u](\xi) \equiv \kappa_1 \vee V[w](\xi) \equiv \kappa_1) \text{ by } (c_2)$$
$$\vdash (u(\xi) \to w(\xi)) \leftrightarrow SG(CSG\,V[u](\xi) + V[w](\xi)) \equiv \kappa_1$$
$$\vdash (u(\xi) \to w(\xi)) \leftrightarrow V[u \to w](\xi) \equiv \kappa_1 .$$

The case of the other propositional connectives is analogous. Further, for v as $\exists x_n < T(\xi)\, u(\xi, x_n)$

$$\exists x_n < T(\xi)\, u(\xi, x_n) \leftrightarrow \exists x_n < T(\xi)\, V[u](\xi, x_n) \equiv \kappa_1$$
$$\exists x_n < T(\xi)\, u(\xi, x_n) \leftrightarrow SG(\Sigma_< V[u](\xi, T(\xi)) \equiv \kappa_1 \text{ by } (c_{17})$$
$$\exists x_n < T(\xi)\, u(\xi, x_n) \leftrightarrow SG(V[v](\xi)) \equiv \kappa_1 ,$$

and the case of v as $\forall x_n < T(\xi)\, u(\xi, x_n)$ is analogous.

From this point on now, I will abstain occasionally from writing out certain PR-proofs in full detail, and appeal instead to the reader's conviction that he can do this himself. Clearly, a proof of an arithmetical statement, employing tools from, say, analysis or topology, will not immediately, if at all, translate into a proof expressed in PA or PR. Also, arithmetical proofs making use of higher order notions, such as sequences or maps between sets, will not translate without additional preparations.

But the attentive reader will have noticed that I *did* prove two series of simple arithmetical statements, first (a1)–(a7) in Chapter 1.16 and then (a8)–(a17) in this book's Chapter 5, in a rather painstaking (and partly unusual) manner, employing induction (sometimes in the form of the smallest number principle) where the familiar treatment would have proceeded in a shorter and more 'conceptual' manner (e.g. in the cases of (a5), (a7), (a16), (a17)). The point of this approach, becoming obvious now, was that *these* proofs indeed do immediately translate into PR–proofs.

In particular, the mathematical arguments establishing (a2) and (a9) translate directly into PR–proofs of

$$\vdash \ \mathrm{PRIN}(y) \wedge \mathrm{DIV}(y, x\cdot z) \ \rightarrow \ \mathrm{DIV}(y,x) \vee \mathrm{DIV}(y,z) \ ,$$

$$\vdash \ \mathrm{PRIN}(x) \wedge \mathrm{PRIN}(y) \wedge \neg x \equiv y \wedge \mathrm{DIV}(x^{v}, z\cdot y^{w}) \ \rightarrow \ \mathrm{DIV}(x^{v}, z) \ .$$

The mathematical arguments establishing (a10)–(a17) translate directly into PR-proofs of the free variable formulas

$$\text{(p1)} \ \vdash \ \neg\, \kappa_0 \equiv x \ \rightarrow \ x \equiv \Pi_{<} T(x,\ \mathit{LEN}(x)) \ \text{ for } \ T(x,z) = \mathit{PRINO}(z)^{\mathit{EXP}(z,x)}$$

$$\text{(p2)} \ \vdash \ F(x) < F(s(x)) \wedge \kappa_0 < G(x) \wedge T(z) = \mathit{PRINO}(F(z))^{G(z)}$$
$$\wedge\ x \equiv \Pi_{<} T(r) \wedge \mathrm{PRIN}(y)$$
$$\rightarrow \ (\mathrm{DIV}(y,x) \ \leftrightarrow \ \exists q{<}r\ y \equiv \mathit{PRINO}(F(q))\ )$$
$$\wedge\ (\exists q{<}r\ y \equiv \mathit{PRINO}(F(q)) \ \rightarrow \ G(s) \equiv \mathit{EXP}(F(r), x)\ ).$$

During the applications of recursive functions, many of them were introduced by course of values recursion according to the principles (W) and (WR) of Chapter 5. It now is remarkable that also the formulas, expressing such course of values recursions, become provable as *free variable* formulas. I first consider the situation described in Theorem 5.1, and use the notation employed in its proof; assume that there are PR–terms G and R representing the functions g^k and r^{k+2}. I introduce a PR–term Q for the function q^{k+1} with the recursion equations

$$Q(\xi, \kappa_0) \ \equiv \ \kappa_2^{\ s(G(\xi))}$$
$$Q(\xi,\ s(x)) \ \equiv \ Q(\xi, x)\cdot \mathit{PRINO}(s(x))^{s(R(\xi, x,\ Q(\xi, x)))}$$

and shall show that the PR–term F defined as

$$F(\xi,\ s(x)) \ = \ \mathit{CS}(\mathit{EXP}(x,\ Q(\xi, x))$$

satisfies the course of values formulas

(w1) $\vdash F(\xi, \kappa_0) \equiv G(\xi)$,

(w2) $\vdash F(\xi, s(x)) \equiv R(\xi, x, Q(\xi,x))$,

(w3) $\vdash Q(\xi,x) \equiv \Pi_{\leq} T(\xi,x)$ for $T(\xi,y) = PRINO(y)^{s(F(\xi,y))}$.

Now the mathematical proof from Chapter 5 can be imitated as well. PA-induction shows

$$\vdash \forall z\ (\mathrm{DIV}(PRINO(z), Q(\xi, x)) \rightarrow z \leq x)$$

whence

$$\vdash EXP(s(x), Q(\xi, s(x)) \equiv s(R(\xi, x, Q(\xi,x)) .$$

Thus (w1) and (w2) can be proved making use of (p1), (p2) above. PA-induction also shows

$$\vdash \forall y\ (y \leq x \rightarrow EXP(y, Q(\xi,y)) \equiv EXP(y, Q(\xi,x))\ )$$

whence

(w4) $\vdash \forall y\ (y \leq x \rightarrow F(\xi,y) \equiv CS(EXP(y, Q(\xi,x))\ )$

and from that I conclude upon (w3).

Course of values recursion most often makes use of regression functions. Here I can restrict myself to the following situation:

LEMMA 4 Let $FXP(y,x)$, $U(x)$, $V(x)$, $W(u,v)$, $L(x)$ be PR-terms and assume that

(r1) $\vdash FXP(y,x) < x$.

Then I can find a PR-term $F(x)$ such that $\vdash F(\kappa_0) \equiv \kappa_0$ and

(r2) $\vdash \neg\kappa_0 \equiv x \rightarrow F(x) \equiv U(x) + V(x) \times \Pi_{<} T(x, L(x)>$

for $T(x,v) = W(v, F(FXP(v,x))$.

I define $R(x,z)$ to be

$$U(s(x)) + V(s(x)) \times \Pi_{<} T_0(s(x),z, L(s(s(x)))) \text{ for}$$

$$T_0(x,z,v) = W(v, CS(EXP(FXP(v,x), z))) ,$$

and from R I define the PR-term Q as above. Now $\vdash F(s(x)) \equiv R(x, Q(x))$ by (w2), and by definition of $R(x,z)$ this gives

$$\vdash F(s(x)) \equiv U(s(x)) + V(s(x)) \times \Pi_{<} T_0(s(x), Q(x), L(s(s(x)))) .$$

Since $\vdash FXP(v,s(x)) < s(x)$ by (r1), hence $\vdash FXP(v,s(x)) \leq x$, there follows from (w4)

$$\vdash F(FXP(v,s(x))) \equiv CS(EXP(FXP(v,s(x)), Q(x))) .$$

Now $T_0(s(x),Q(x),v) = W(v, CS(EXP(FXP(v,s(x)))$ implies

$$\vdash T_0(s(x), Q(x), v) \equiv W(v,\ F(FXP(v, s(x))))$$
$$\vdash T_0(s(x), Q(x), v) \equiv T(s(x), v)$$
$$\vdash \Pi_< T_0(s(x),\ Q(x),\ L(s(s(x))) \equiv \Pi_< T(s(x),\ L(s(s(x)))) .$$

Thus

$$\vdash F(s(x)) \equiv U(s(x)) + V(s(x)) \times \Pi_< T(s(x),\ L(s(s(x)))) .$$

Replacing x by $CS(x)$ and making use of $\vdash \kappa_0 < x \rightarrow s(CS(x)) \equiv x$, this implies (r2) .

In view of the preceding discussion, this Lemma is no surprise, and it will be applied mainly to Boolean terms F which represent the characteristic functions f of relations. The next Lemma shows that statements, about terms defined by course of values recursion, can also be proved by course of values induction:

LEMMA 5 Let F be defined from $FXP(y,x)$, $U(x)$, $V(x)$, $W(u,v)$, $L(x)$ by Lemma 4 under the hypotheses $\vdash L(x) \leq x$ and (r1) and

(r3) $\vdash \neg\kappa_0 \equiv x \rightarrow F(x) \equiv U(x) + V(x) \times \Pi_< T(x,\ L(x))$

for $T(x,v) = F(FXP(v,x))$.

Let D be a further PR-term and assume that U, V and D are Boolean and that

(a) $\vdash U(x) + V(x)$ is Boolean (i.e. U and V are disjoint sets),

(b) $\vdash U(x) \leq D(x)$ (i.e. U is a subset of D),

(c) $\vdash V(x) \times \Pi_< T_1(x,\ L(x)) \leq D(x)$ for $T_1(x,v) = D(FXP(v,x))$

(i.e. D is closed under the course of values conditions of F).

Then $\vdash F(x) \leq D(x)$ (and, in particular, F is Boolean).

The schema (ISO) of order induction, which remains valid in the definitorial extension PR of PA, here gives

$$\vdash (\forall x\ (\forall v\ (v<x \rightarrow F(v) \leq D(v))) \rightarrow F(x) \leq D(x)) \rightarrow F(x) \leq D(x)$$

such that only the premiss of that implication has to be verified. This may be done under the further hypothesis $\kappa_0 < x$ since $\vdash F(\kappa_0) \leq D(\kappa_0)$ follows from $\vdash F(\kappa_0) \equiv \kappa_0$. Now $\vdash U(x) \equiv \kappa_1 \rightarrow D(x) \equiv \kappa_1$ by (b), and $\vdash U(x) \equiv \kappa_1 \rightarrow F(x) \equiv \kappa_1$ by (a) and (r3). Hence $\vdash U(x) \equiv \kappa_1 \rightarrow F(x) \leq D(x)$. But $\vdash U(x) \equiv \kappa_0 \vee U(x) \equiv \kappa_1$ since U is Boolean, and so it remains to be shown

(r4) $\vdash \kappa_0 < x \wedge U(x) \equiv \kappa_0 \wedge (\forall v\ (v<x \rightarrow F(v) \leq D(v))) \rightarrow F(x) \leq D(x)$.

It follows from (r1) that

$$\vdash (\forall v\ (v<x \rightarrow F(v) \leq D(v))) \rightarrow \forall y\ (v<L(x) \rightarrow T(x,v) \leq T_1(x,v))$$

whence also

$$\vdash (\forall v\ (v<x \to F(v) \leq D(v))) \ \to\ \Pi_{<} T(x,\ L(x)) \ \leq\ \Pi_{<} T_1(x,\ L(x)).$$

Thus (c) gives

$$\vdash (\forall v\ (v<x \to F(v) \leq D(v))) \ \to\ V(x) \times \Pi_{<} T_1(x,\ L(x)) \ \leq\ D(x),$$

and together with (r3) this establishes (r4).

Chapter 11. Unprovability of Consistency

1. Introduction: Gödel's Theorem and Provability Conditions

In Chapter 2, it was shown that the additive fragment of arithmetic with order is consistent; in doing so a certain part of arithmetic itself was employed which made use of induction and recursion. In any case, not more of arithmetic was used than is expressed in the axioms and schemata of PA.

As discussed in this Book's Introduction, consistency proofs by the means of arithmetic were the aims of Hilbert's program (in the narrower sense), and so the next question to investigate would be the consistency of PA. It will be the content of this Chapter to show that the consistency of PA *cannot* be proven by means of PA itself.

The background to this insight comes from GÖDEL's Incompleteness Theorem 4.3, the hypotheses of which hold for PA. Its first part (B1a) states that there is a sentence w with

(BF) $\vdash w \leftrightarrow \neg \mathrm{BEW}(\gamma_w)$.

and that then

(G) if PA is consistent, then *not* $\vdash w$.

Of course, this was established by mathematical arguments, securing the representability of (primitive) recursive functions. Assume for the moment that this mathematical argumentation could be recaptured within PA itself in the form

(GF) $\vdash \mathrm{CON}_{PA} \rightarrow \neg \mathrm{BEW}(\gamma_w)$

for some sentence CON_{PA} expressing that PA is consistent. Then I could conclude that

THEOREM 1 If PA is consistent then *not* $\vdash \mathrm{CON}_{PA}$.

For $\vdash \mathrm{CON}_{PA}$ and (GF) would imply $\vdash \neg \mathrm{BEW}(\gamma_w)$, hence $\vdash w$ by (BF), and this would contradict (G).

Theorem 1 is GÖDEL's 31 *Theorem on Unprovability* (sometimes called his 2nd Incompleteness Theorem); a first complete proof of (GF) was published in HILBERT-BERNAYS 39.

There are various possibilities to formulate consistency, and here I shall use one of its classical forms saying that, *if* $\neg v$ is provable *then* v is *not* provable. In order to express this in PA, I may assume that there is a defined term $NE(x)$, in the extension PR of PA, such that $NE(\kappa(\ulcorner v \urcorner)) = \kappa(\ulcorner \neg v \urcorner)$ (recall that γ_v is $\kappa(\ulcorner v \urcorner)$) and then define CON_{PA} as

$$\forall x\,(BEW(NE(x)) \to \neg BEW(x)) .$$

For sentences v I also shall use the abbreviations

$\Box v$ for the sentence $BEW(\gamma_v) = BEW(\kappa(\ulcorner v \urcorner))$,

$\Box\neg v$ for the sentence $BEW(\kappa(\ulcorner \neg v \urcorner)) = BEW(NE(\kappa(\ulcorner v \urcorner)))$

such that $\Box\Box w$ is $BEW(\kappa(\ulcorner \Box v \urcorner))$) and $\Box\neg\Box w$ is $BEW(NE(\kappa(\ulcorner \Box v \urcorner)))$.

For the proof of (GF), HILBERT–BERNAYS 39 isolated three so-called *provability conditions* and showed that, once these had been secured, the provability (GF) was not difficult to conclude. The third of these provability conditions involved references to recursive terms, and it was LÖB 55 who observed that a condition simpler to state was (I) sufficient to derive (GF) and (II) a consequence of the HILBERT–BERNAYS conditions. The provability conditions according to LÖB then are for sentences v and u:

(BL1) if $\vdash v$ then $\vdash \Box v$,

(BL2) $\vdash \Box(v \to u) \to (\Box v \to \Box u)$,

(BL3) $\vdash \Box v \to \Box\Box v$,

and from them indeed (GF) is an easy consequence. Because let w be the sentence satisfying (BF) from GÖDEL's (1st) incompleteness theorem. Since

$$\vdash CON_{PA} \to (BEW(NE(x)) \to \neg BEW(x))$$

I can substitute $\kappa(\ulcorner \Box w \urcorner)$ for x such that

$$\vdash CON_{PA} \to (BEW(NE(\kappa(\ulcorner \Box w \urcorner))) \to \neg BEW(\kappa(\ulcorner \Box w \urcorner)))$$

i.e.

(0) $\vdash CON_{PA} \to (\Box\neg\Box w \to \neg\Box\Box w)$.

From that point on, I only need to use propositional arguments and the provability conditions. First, (BF) may be written as $\vdash w \leftrightarrow \neg\Box w$. Hence in particular

(1) $\vdash w \to \neg\Box w$

$\vdash \Box(w \to \neg\Box w)$ by (BL1)

(2) $\vdash \Box w \to \Box\neg\Box w$ by (BL2) .

Now

$$\vdash (\Box\neg\Box w \to \neg\Box\Box w) \to ((\Box w \to \Box\neg\Box w) \to (\Box w \to \neg\Box\Box w)) ,$$

propositionally, whence (2) gives

$$\vdash (\Box\neg\Box w \to \neg\Box\Box w) \to (\Box w \to \neg\Box\Box w)$$

whence (0) gives

(3) $\vdash \mathrm{CON}_{PA} \to (\Box w \to \neg\Box\Box w)$.

Also

$\Box w \to \Box\Box w$ by (BL3)

$\neg\Box\Box w \to \neg\Box w$

and so again (3) gives

$\vdash \mathrm{CON}_{PA} \to (\Box w \to \neg\Box w)$.

But (BF) also implies

(4) $\vdash \neg\Box w \to w$

whence again

$\vdash \mathrm{CON}_{PA} \to (\Box w \to w)$

or

(5) $\vdash \mathrm{CON}_{PA} \wedge \Box w \to w$.

On the other hand, (4) trivially gives

(6) $\vdash \mathrm{CON}_{PA} \wedge \neg\Box w \to w$,

whence (5) and (6) by tertium non datur give

$\vdash \mathrm{CON}_{PA} \to w$,

and now the desired (GF) follows by (1).

So it remains to establish (BL1)-(BL3) in order have proven Theorem 1. As for (BL1), this follows from the definitions as observed in (B0) of the proof of Theorem 4.3. The remaining sections of this Chapter are devoted to the proofs of (BL2) and (BL3) for the notion of provability as used so far in this book.

LÖB found his form of the provability conditions when deriving from them what is known now as LÖB's Theorem:

THEOREM 2 For every sentence v: if $\vdash \Box v \to v$ then $\vdash v$.

For a proof let $u(x_0)$ be the formula $\mathrm{BEW}(x_0) \to v$ and let a be a fixpoint of u, i.e. $\vdash a \leftrightarrow (\mathrm{BEW}(\gamma_a) \to v)$. Thus

0. $\vdash a \to (\Box a \to v)$
1. $\vdash \Box(a \to (\Box a \to v))$ by (BL1) ,
2. $\vdash \Box a \to (\Box\Box a \to \Box v)$ by (BL2) twice ,
3. $\vdash (\Box a \to \Box\Box a) \to (\Box a \to \Box v)$ propositionally ,
4. $\vdash \Box a \to \Box v$ from 3, (BL3) and modus ponens ,
5. $\vdash \Box a \to v$ since $\vdash \Box v \to v$ by hypothesis ,
6. $\vdash a$ since $\vdash (\Box a \to v) \to a$ for the fixpoint a ,

7. $\vdash \Box a$ by (BL1) ,
8. $\vdash v$ from 7, 5 by modus ponens .

COROLLARY 1 If *not* $\vdash v$ then *not* $\vdash \neg \mathrm{BEW}(\gamma_v)$.

Because *not* $\vdash v$ implies *not* $\vdash \Box v \rightarrow v$ by the theorem. Since ex absurdo quodlibet assures $\vdash \neg p \rightarrow (p \rightarrow q)$, there follows *not* $\vdash \neg \Box v$.

As a consequence, I now obtain Theorem 1 not only for the particular form of CON_{PA} chosen above, but for any sentence CON_{PA} whose underivability expresses consistency - e.g. a sentence $\neg\, \kappa_0 \equiv \kappa_1$.

The proofs of both Theorem 1 and 2 make use of the Fixpoint Lemma from Chapter 4, and JEROSLOV 73 did use different fixpoints to obtain several extensions of Theorem 1. The conditions (BL1)-(Bl3), written with the symbol $\Box$, appear also as rules of proof in systems of propositional modal logic; there is a voluminuous literature about this connection, some of which is reported in BOOLOS 79 and 93 and in SMORYNSKI 85.

Both Theorems 1 and 2 depend on the fact that the provability predicate BEW(x) has the properties (BL1)-(BL3); it was introduced in Chapter 4 as $\exists u$ DED(x,u) where DED(x,u) was representing the relation *ded*, but not further specified. The importance of this particular choice of BEW(x) for my Theorems is illustrated by the following example from FEFERMAN 60 (cf. also KREISEL 65 and HILBERT-BERNAYS 70) which extends the idea used in ROSSER's fixpoint construction in Theorem 4.5.

Assume that there is a notion of u being a deduction of a formula v, and assume that every deduction u determines a natural number $|d|$ as its *complexity* (say its length, or the length of a maximal path in a deduction tree, or its code under an arithmetization); let me also say that u is a 0-deduction. Define now that u is 1-deduction of v if

$$\text{u is a 0-deduction of v and for all 0-deductions } u_1 \text{ of } w_1 \text{ and } u_2 \text{ of } w_2: \\ \text{if } |u_1| \lesssim |u| \text{ and } |u_2| \lesssim |u| \text{ then } w_1 \neq \neg w_2 .$$

Clearly, 1-provability implies 0-provability, and if 0-provability is known to be consistent then 1-provability coincides with it. But in any case now, 1-provability *is* consistent. Because if d and e would be 1-deductions of v and of $\neg v$ then either $|d| \lesssim |e|$ or $|e| \lesssim |d|$, and so $u_1 = d$, $u_2 = e$, $w_1 = v$, $w_2 = \neg v$ in the first case violate that e is a 1-deduction and in the second case that d is a 1-deduction.

In the case that 0-provability is the usual one, arithmetized by the relation *ded*, I can consider right away the arithmetization ded_1 of 1-provability, taking the the numbers u as their own complexities. As it follows from the definition of codes of deductions that *ded*(v,u) implies $v \lesssim u$, the relation $ded_1(v,u)$ is defined as

$$ded(v,u) \text{ and for all } u_1 \lesssim u,\ u_2 \lesssim u,\ w_1 \lesssim u,\ w_2 \lesssim u :$$
$$(ded(w_1,u_1) \text{ and } ded(w_1,u_1) \text{ implies } not\ w_1 = ne(w_2)) .$$

So together with *ded* also ded_1 will be (elementary) recursive and will be represented by a formula $\mathrm{DED}_1(x,u)$. The above argument, that 1-provability is consistent, translates into the statement

$$\text{if } ded_1(ne(v), d)) \text{ then } not\ ded_1(v,e)$$

hence

$$\vdash \mathrm{DED}_1(NE(\kappa_v), \kappa_d) \rightarrow \neg \mathrm{DED}_1(\kappa_v, \kappa_e) .$$

Assuming certain facts about the formula $\mathrm{DED}(x,u)$ (which all will be established during the developments of next Section), this statement about representability can be improved to the provability of the free variable formula

$$\vdash \mathrm{DED}_1(NE(x), y) \rightarrow \neg \mathrm{DED}_1(x, z)$$

i.e.

$$\vdash (\mathrm{BEW}_1(NE(x)) \rightarrow \neg\mathrm{BEW}_1(x))$$

for $\mathrm{BEW}_1(x) = \exists u\ \mathrm{DED}_1(x,u)$. Thus PA does actually does prove its own consistency with respect to the provability predicate $\mathrm{BEW}_1(x)$.

Plan of the Remaining Sections of this Chapter

In the provability condition (BL2),

$$\vdash \mathrm{BEW}(\kappa(\ulcorner v \rightarrow u \urcorner)) \rightarrow (\mathrm{BEW}(\kappa(\ulcorner v \urcorner)) \rightarrow \mathrm{BEW}(\kappa(\ulcorner u \urcorner))) ,$$

there occurs the arithmetical code $\kappa(\ulcorner v \rightarrow u \urcorner)$ of $v \rightarrow u$. The arithmetization in Chapter 7 provided a primitive recursive (actually elementary) function *imp* which computes this code from $\kappa(\ulcorner v \urcorner)$ and $\kappa(\ulcorner u \urcorner)$. In Section 3, I shall present the *free variable formalization* of (the arithmetization of) syntax which depends (a) on the use of PR-terms and (b) on the availability of PA-induction to prove formulas involving them. This has the effect that statements expressed by formulas, which so far were only representing the arithmetizing functions of syntax, now become PR-provable for free variables. In particular, if the PR-term $\mathit{IMP}(x,y)$ represents the function *imp* then for the formula $\mathrm{DED}(x,u)$ from $\mathrm{BEW}(x) = \exists u\ \mathrm{DED}(x,u)$ it will be shown in Theorem 3 that

$$\vdash \mathrm{DED}(\mathit{IMP}(x,y),u) \wedge \mathrm{DED}(y,v) \rightarrow \exists w\, \mathrm{DED}(y,w)$$

whence

$$\vdash \mathrm{BEW}(\mathit{IMP}(x,y)) \rightarrow (\mathrm{BEW}(x) \rightarrow \mathrm{BEW}(y))$$

from where (BL2) immediately follows.

In order to prove (BL3), I shall specify $\mathrm{DED}(u,x)$ as $\mathit{DED}(u,x) \equiv \kappa_1$ where $\mathit{DED}(u,x)$ is a PR-term. This has the effect that $v = \mathrm{BEW}(\kappa_a)$ becomes a Σ_1-PR-*sentence*, and for such sentences v Theorem 4 will say that, indeed,

(*) $\vdash$ v $\to$ BEW(κ(⌜v⌝)) .

In particular, if v is BEW(κ(⌜w⌝)) then this becomes

$\vdash$ BEW(κ(⌜w⌝)) $\to$ BEW(κ(⌜BEW(κ(⌜w⌝))⌝))

which proves (BL3) (with w in place of v).

The statement (*) concerns sentences, and it will be established as the special case of an analogous statement about Σ_1–PR–*formulas*. Now provability is preserved under replacements of variables; hence if v in (*) contains variables and such replacement is performed, then κ(⌜v⌝) in BEW(–) still remains unchanged, leading to various implications which, all keeping the same right side BEW(κ(⌜v⌝)), cannot possibly be provable. For a formula v, therefore, the place of κ(⌜v⌝) in (*) must be taken by a term GG(v),

(§) $\vdash$ v $\to$ BEW(GG(v)) ,

which contains the same variables as v and which, if the variables of v are replaced by constants transforming v into a sentence u, under this same replacement becomes κ(⌜u⌝). Thus GG(v) will be a term which is a formal description, or for short *an internal form*, of the uniformity present in the construction of κ(⌜v⌝) and κ(⌜u⌝).

The Σ_1–PR–formulas for which (§) is to be established are defined as those which are provably equivalent to formulas of the form $\exists$x $T(\zeta,x) \equiv \kappa_1$ where $T(\zeta,x)$ is a PR–term. So it will suffice to show that

$\vdash$ ($\exists$x $T(\zeta,x) \equiv \kappa_1$) $\to$ BEW(GG($\exists$x $T(\zeta,x) \equiv \kappa_1$))

of which it will be seen rather easily (at the start of Section 4) that it can be reduced to the statement of Lemma 6, viz.

(%) $\vdash$ ($T(\zeta,x) \equiv z$) $\to$ BEW(GG($T(\zeta,x) \equiv z$)) ,

which in Lemma 8 will be proved by induction on the construction of the PR–term $T(\zeta,x)$. So (at least) for the internal forms occurring in (%) their transformation under replacements of variables must become tractable

To this end, I construct Section 3 an internal copy of the arithmetization from Chapter 7. This means that I shall introduce representing PR–terms for the functions arithmetizing the syntax of my language L_F. In particular, I assign to

every predicate symbol p of L_F a term $p^{\S}(\xi)$ which is the free variable formalization of the function arithmetizing $p(\xi)$,

every L_F–formula v a term v^*, called its *internalization*, which has the same free variables, in which $p^{\S}$ stands in place of the p, and which the place of the logical operations is taken by the terms representing their arithmetizations,

every L_F–formula v a formula $v^{\ddagger}$, called its *codification*, by replacing in v^* every variable z by the constant γ_z;

it then can be shown that $v^{\ddagger}$ is provably equal to the code $\gamma_v = \kappa(\ulcorner v \urcorner)$ of v. For these internal formulas then, internal replacement maps can be defined mirroring the usual function rep, and the properties of internal replacements can be be expressed by provable identities employing replacement terms *REPT*(-,-,-) and *REPF*(-,-,-). Finally, the internal forms *GG*(v) will be defined as particular replacement terms, and then their behaviour under replacements can be expressed.

References

G. Boolos: The Unprovability of Consistency. Cambridge 1979

G. Boolos: The Logic of Provability. Cambridge 1993

S. Fefermann: Arithmetization of Metamathematics in a General Setting. Fund.Math. 49 (1960) 35–92

K. Gödel: Über formal unentscheidbare Sätze der Principia Mathematica und verwandter Systeme I. Monatshefte Math.Phys. 38 (1931) 173–198

D. Hilbert, P. Bernays: Grundlagen der Mathematik II. Berlin 1939. Zweite Auflage. Berlin 1970

R.G. Jeroslov: Redundancies in the Hilbert–Bernays derivability conditions for Gödel's second incompleteness theorem. J.Symb.Logic 38 (1973) 359–367

G. Kreisel: Mathematical Logic. in: T.L.Saaty: Lectures on Modern Mathematics, vol. 3, New York 1965, pp. 95–195

M.H. Löb: Solution of a Problem of Leon Henkin. J.Symb.Logic 20 (1955) 115–118

C. Smorynski: Self Reference and Modal Logic. Berlin 1985

2. Free Variable Formalization of Syntax

It is the aim of this section to construct PR-terms, representing the primitive recursive functions used to arithmetize syntax in Chapter 7, and to express some characteristic properties of those arithmetizing functions by PR-provable free variable formulas built from those terms. For arithmetizing functions I shall use the notations from Chapter 7, and for PR-terms the notations introduced in Section 3 of the last Chapter. Also, it will be convenient to have the κ_n introduced by explicit definition as $s(\kappa_{n-1})$.

PR-Terms for Terms

The language L, whose arithmetization is to be formalized, is that of PA. So there are four function symbols f_i, namely s, κ_0, $+$, $\cdot$, which, for later reference, I collect into a set FF. I choose I_a to be the set $\{2,3,4,12\}$; thus its characteristic function ina is represented by

$$INA(z) = CS(ID_2(z,\kappa_2) + ID_2(z,\kappa_3) + ID_2(z,\kappa_4) + ID_2(z,\kappa_{12}))\ .$$

The arithmetizing functions for s and κ_0 I write as S and K_0. Thus K_0 is a constant function whose only value is $\ulcorner\kappa_0\urcorner = 512 = 2^c$ with $c = 3^2$, and S is the function with $S(a) = 134217728 \cdot 3^a = 2^c \cdot 3^a$ with $c = 3^3$. It follows from $\kappa_{n+1} = s(\kappa_n)$ that $\ulcorner\kappa_{n+1}\urcorner = \ulcorner s(\kappa_n)\urcorner = S(\ulcorner\kappa_n\urcorner)$.

Corresponding to the notations of Chapter 7 I shall use the abbreviations

$$E_i(y) \quad \text{for} \quad EXP(\kappa_i, y),$$
$$D_i(y) \quad \text{for} \quad EXP(\kappa_i, E_0(y)),$$
$$ART_a(y) \quad \text{for} \quad E_0(E_1(E_0(y))).$$

The set var of arithmetized variables consists of the values of $va(n) = 2^{2^{2+n}}$; making use of the PR-term PO for exponentiation I define a term

$$VA(z) = \kappa_2^{\kappa_2^{\kappa_2+z}}$$

which represents va. To the formula $\mathrm{VAR}(y) = \exists z{<}y\ y \equiv VA(z)$ belongs a term which I denote as $VAR(y)$:

$$VAR(y) \equiv \Sigma_{<} T(y,y) \qquad \text{with} \quad T(y,z) \equiv ID_2(y,\ VA(z))\ .$$

While L has numerous constant *terms* (i.e. terms without variables), there is only the one *constant* κ_0, and so the set con consists of $\ulcorner\kappa_0\urcorner$ only; it is represented by the term $CON(y) = ID_2(y, \kappa(\ulcorner\kappa_0\urcorner))$. Since PO represents exponentiation there holds

$$\vdash \kappa(\ulcorner\kappa_0\urcorner) = \kappa(2^{3^2}) \equiv \kappa_2^{\kappa_3^{\kappa_2}} .$$

Thus $\vdash \neg\ \kappa_i \equiv \kappa_j$ for $i \neq j$ implies that $\vdash \neg\ VA(\kappa_0) \equiv \kappa(\ulcorner\kappa_0\urcorner)$ whence PA-induction gives $\vdash \neg\ VA(z) \equiv \kappa(\ulcorner\kappa_0\urcorner)$. Hence $\vdash \mathrm{VAR}(z) \rightarrow \neg\,\mathrm{CON}(z)$.

Lemma 10.4 provides a term $TERM(x)$ with the recursion equation

$$\text{(ta)} \quad TERM(x) \equiv VAR(x) + CON(x) + VT(x) \times \Pi_{<} T(x, ART_a(x)) \quad \text{for} \quad T(x,v) = TERM(EXP(v+\kappa_1, x))$$

where $VT(x)$ is the Boolean term $V[v]$ for the formula $v(x)$:

$$\text{(tb)} \quad D_0(x) \equiv \kappa_0 \wedge IA(D_1(x)) \equiv \kappa_1 \wedge LEN(E_0(x)) \equiv \kappa_2 \wedge \neg\ ART_a(x) \equiv \kappa_0 \wedge ART_a(x) + \kappa_1 \equiv LEN(x)) ,$$

and $TERM(x)$ represents the characteristic function of *term*. Employing the same $VT(x)$, I define a term $CTERM(x)$ with a condition (tac) obtained from (ta) by omitting $VAR(x)$ and changing the phrase $TERM$ to $CTERM$. Thus $\vdash CTERM(x) \rightarrow TERM(x)$, amd $CTERM(x)$ represents the codes of constant terms.

PR-Terms for Formulas

The language L here has only one predicate symbol $\equiv$ whose index I choose as 4 ; hence its arithmetizing function is $F_=(a,b) = 2^{725} \cdot 3^a \cdot 5^b$. So $INB(z) = ID_2(z, \kappa_4)$ represents the characteristic function of the set I_b; I abbreviate $E_0(D_2(x))$ as $ART_b(x)$. Lemma 10.4 provides a term $ATOM(x)$, representing *atom*, with the recursion equation $ATOM(x) \equiv V[v]$ for the formula $v(x)$:

$$D_0(x) \equiv \kappa_0 \wedge D_1(x) \equiv \kappa_0 \wedge INB(D_2(x)) \equiv \kappa_1 \wedge LEN(E_0(x)) \equiv \kappa_3 \wedge ART_b(x) + \kappa_1 \equiv LEN(x) \wedge \Pi_{<} T(x, ART_b(x)) \equiv \kappa_1$$

for $T(x,v) = TERM(EXP(v+\kappa_1, x))$. Terms $ET(x,y)$, $VEL(x,y)$, $IMP(x,y)$, $NE(x)$, $QUNIV(x,y)$, $QEXI(x,y)$ representing the functions *et*, *vel*, *imp*, *ne*, *quniv*, *qexi* I define directly as

$$\kappa_2^{\kappa_7} \cdot \kappa_3^{x} \cdot \kappa_5^{y} , \quad \kappa_2^{\kappa_7^{\kappa_2}} \cdot \kappa_3^{x} \cdot \kappa_5^{y} , \quad \kappa_2^{\kappa_7^{\kappa_3}} \cdot \kappa_3^{x} \cdot \kappa_5^{y} , \quad \kappa_2^{\kappa_7^{\kappa_4}} \cdot \kappa_3^{x} \cdot \kappa_5^{y}$$

$$\kappa_2^{\kappa_{11}^{x}} \cdot \kappa_3^{y} , \quad \kappa_2^{\kappa_{13}^{x}} \cdot \kappa_3^{y} .$$

Lemma 10.4 provides a term $FORM(x)$, representing *form*, with the recursion equation $FORM(x) \equiv V[v]$ for the formula $v(x)$:

$$\mathrm{ATOM}(x) \vee (D_0(x) \equiv \kappa_0 \wedge D_1(x) \equiv \kappa_0 \wedge D_2(x) \equiv \kappa_0 \wedge (a(x) \vee b(x) \vee c(x) \vee d(x)))$$

where

$$a(x):\quad (D_3(x) \equiv \kappa_1 \vee D_3(x)) \equiv \kappa_2 \vee D_3(x) \equiv \kappa_3) \wedge LEN(E_0(x)) \equiv \kappa_4$$
$$\wedge\, LEN(x) \equiv \kappa_3 \wedge FORM(E_1(x)) \equiv \kappa_1 \wedge FORM(E_2(x)) \equiv \kappa_1,$$
$$b(x):\quad D_3(x) \equiv \kappa_4 \wedge LEN(E_0(x)) \equiv \kappa_4 \wedge LEN(x) \equiv \kappa_2 \wedge FORM(E_1(x)) \equiv \kappa_1,$$
$$c(x):\quad D_3(x) \equiv \kappa_0 \wedge LEN(E_0(x)) \equiv \kappa_5 \wedge LEN(x) \equiv \kappa_2 \wedge \mathrm{VAR}(D_4(x))$$
$$\wedge\, FORM(E_1(x)) \equiv \kappa_1,$$
$$d(x):\quad D_3(x) \equiv \kappa_0 \wedge D_4(x) \equiv \kappa_0 \wedge LEN(E_0(x)) \equiv \kappa_6 \wedge LEN(x) \equiv \kappa_2$$
$$\wedge\, \mathrm{VAR}(D_5(x)) \wedge FORM(E_1(x)) \equiv \kappa_1.$$

As an illustration, let me show that $\vdash \mathrm{FORM}(QUNIV(x,y)) \rightarrow \mathrm{VAR}(x))$. Because $\vdash D_i(QUNIV(x,y)) \equiv \kappa_0$ for $i \neq 4$ by arithmetic, and since $k \neq 0$ implies $\vdash \neg\kappa_k \equiv \kappa_0$ there follows $\vdash \neg a(z) \wedge \neg b(z) \wedge \neg d(z)$ for $QUNIV(x,y)$ as z. Hence $\vdash \mathrm{FORM}(QUNIV(x,y)) \rightarrow c(QUNIV(x,y))$, but $\vdash c(z) \rightarrow \mathrm{VAR}(D_4(z))$ and $\vdash D_4(QUNIV(x,y)) \equiv x$.

As for syntactical relations, Lemma 10.4 provides a term $OCCT(x,y)$ with the recursion equation $OCCT(x) \equiv V[v]$ for the formula $v(x)$:

$$\mathrm{VAR}(x) \wedge \mathrm{TERM}(y) \wedge (\neg x \equiv y \rightarrow \kappa_1 < LEN(y) \wedge \Sigma_< T_0(x,y,LEN(y)) \equiv \kappa_1)$$

where $T_0(x,y,v) = OCCT(y, EXP(v+\kappa_1, y))$. – It follows by an application of Lemma 10.5 that

(v1) $\quad \vdash \mathrm{VAR}(x) \wedge \mathrm{CTERM}(y) \rightarrow \neg\mathrm{OCCT}(x,y)$.

Further, Lemma 10.4 provides a term $OCCF(x,y)$ with the recursion equation $OCCF(x) \equiv V[v]$ for the formula $v(x)$:

$$\mathrm{VAR}(x) \wedge \mathrm{FORM}(y) \wedge ((\mathrm{ATOM}(y) \wedge \Sigma_< T_0(x,y,LEN(y)) \equiv \kappa_1)$$
$$\vee\, (\neg\mathrm{ATOM}(y) \wedge \Sigma_< T_1(x,y,LEN(y)) \equiv \kappa_1))$$

where $T_0(x,y,v)$ is as above and $T_1(x,y,v) = OCCF(y, EXP(v+\kappa_1, y))$. Further, Lemma 10.4 provides a term $OCCFREE(x,y)$ with the recursion equation $OCCFREE(x) \equiv V[v]$ for the formula $v(x)$:

$$\mathrm{OCCF}(x,y) \wedge (\mathrm{ATOM}(y) \vee (\neg\mathrm{ATOM}(y) \wedge \neg D_4(y) \equiv x \wedge \neg D_5(y) \equiv x$$
$$\wedge\, \Sigma_< T_2(x,y,LEN(y))) \equiv \kappa_1))$$

where $T_2(x,y,v) = OCCFREE(y, EXP(v+\kappa_1, y))$. Finally, Lemma 10.4 provides a term $FREEFOR(x,y)$ with the recursion equation $FREEFOR(x) \equiv V[v]$ for the formula $v(x)$:

$$\mathrm{VAR}(x) \wedge \mathrm{TERM}(y) \wedge \mathrm{FORM}(z)$$
$$\wedge\, (\mathrm{ATOM}(z) \vee (\neg\mathrm{ATOM}(z) \wedge (a_0(x) \vee b_0(x) \vee c_1(x) \vee c_2(x))))$$

where

$a_0(x)$: $(D_3(z) \equiv \kappa_1 \vee D_3(z) \equiv \kappa_2 \vee D_3(z) \equiv \kappa_3)$
$\wedge\ FREEFOR(x,y,\ E_1(z)) \equiv \kappa_1 \wedge FREEFOR(x,y,\ E_2(z)) \equiv \kappa_1$

$b_0(x)$: $D_3(z) \equiv \kappa_4 \wedge FREEFOR(x,y,\ E_1(z)) \equiv \kappa_1$

$c_i(x)$: $\neg D_{i+3}(z) \equiv \kappa_0 \ \wedge\ d_i(x)$

$d_i(x)$: $x \equiv D_{i+3}(z) \equiv x \vee (FREEFOR(x,y,\ E_1(z)) \equiv \kappa_0$
$\wedge (\neg \mathrm{OCCFREE}(x,\ E_1(z)) \vee \neg \mathrm{OCCT}(x,\ y)))$.

Observe that the only non-inductive reference to y here occurs in the OCCT-formulas at the end of $d_i(x)$. Now clause $c(x)$ in the definition of $FORM(x)$ implies $\vdash \mathrm{FORM}(z) \wedge \neg D_4(z) \equiv \kappa_0 \rightarrow \mathrm{VAR}(D_4(z))$, and (v1) above implies $\vdash \mathrm{CTERM}(y) \rightarrow \neg \mathrm{OCCT}(D_4(z), y)$; hence $\vdash \mathrm{CTERM}(y) \rightarrow c_1(x)$. As the same holds for $c_2(x)$, I can apply Lemma 10.5 and obtain

(v2) $\mathrm{VAR}(x) \wedge \mathrm{CTERM}(y) \wedge \mathrm{FORM}(z) \rightarrow FREEFOR(x,y,z) \equiv \kappa_1$.

The values $rept(b,a,n)$ of the function $rept$ were defined as a if $b \epsilon var$ and $n = b$, as $F_f(rept(b,a,-) \cdot \alpha)$ if $b \epsilon var$ and $n \epsilon term$ and $n = F_f(\alpha)$, and as n otherwise. Making use of Lemma 10.4, I define a term $REPT(x,y,z)$ with the recursion equation:

$$REPT(x,y,z) \equiv V[v] \times y + V[u] \times z + V[w] \times \kappa_2^{E_0(z)} \times \Pi_{<} T(x,y,z,\ ART_a(z)) \qquad \text{(rt)}$$

where

$v(x,z) = \mathrm{VAR}(x) \wedge x \equiv z$,

$u(x,z) = \neg \mathrm{VAR}(x) \vee \neg \mathrm{TERM}(z) \vee \mathrm{CTERM}(z) \vee ART_a(z) \equiv \kappa_0$
$\vee (\mathrm{VAR}(x) \wedge \mathrm{VAR}(z) \wedge \neg x \equiv z)$,

$w(x,z) = \mathrm{VAR}(x) \wedge \mathrm{TERM}(z) \wedge \neg \mathrm{CTERM}(z) \wedge \neg ART_a(z) \equiv \kappa_0$,

$T(x,y,z,v) = PRINO(v+\kappa_1)^{REPT(x,\,y,\ EXP(v+\kappa_1,\ z)}$.

Further, Lemma 10.4 provides a term $REPF(x,y,z)$, representing $repf$, with the recursion equation

$$\begin{aligned} REPF(x,y,z) \equiv\ & (ATOM(z) \times \kappa_2^{E_0(z)} \times \Pi_{<} T_1(x,y,z,\ ART_b(z))) \\ & + V[u] \times \kappa_2^{E_0(z)} \times \Pi_{<} T_2(x,y,z,\ ART_b(z)) \\ & + V[\neg ATOM(z) \vee \neg u] \times z \end{aligned} \qquad \text{(rf)}$$

where

$u(x,z) = \mathrm{FORM}(z) \wedge \neg \mathrm{ATOM}(z) \wedge \neg D_4(z) \equiv x \wedge \neg D_5(z) \equiv x$

$T_1(x,y,z,v) = PRINO(v+\kappa_1)^{REPT(x,y,\ EXP(v+\kappa_1,\ z)}$

$T_2(x,y,z,v) = PRINO(v+\kappa_1)^{REPF(x,y,\ EXP(v+\kappa_1,\ z)}$.

There follows immediately from the definitions

$$\vdash \ REPT(x,y,IMP(z,z')) \equiv IMP(REPT(x,y,z), REPT(x,y,z'))$$

and related free variable provabilities for *ET*, *VEL* and *NE*. Also

$$\text{(v3)} \quad \vdash \ z \equiv z' \ \rightarrow \ REPT(x,y,z) \equiv REPT(x,y,z') \\ \wedge \ REPF(x,y,z) \equiv REPF(x,y,z') \ .$$

Observe further that $\vdash \ REPF(x,y, \ QUNIV(x,z)) \equiv QUNIV(x,z)$, because $\vdash \neg \, \mathrm{ATOM}(QUNIV(x,z))$ and because $\vdash \ D_4(QUNIV(x,z)) \equiv x$ implies also $\vdash \neg \, u(x, \ QUNIV(x,z))$. Finally, two replacements of distinct variables by constant terms commute:

$$\text{(v4)} \quad \vdash \ \mathrm{VAR}(x_0) \wedge \mathrm{VAR}(x_1) \wedge \neg \, x_0 \equiv x_1 \wedge \mathrm{CTERM}(y_0) \wedge \mathrm{CTERM}(y_1) \wedge \mathrm{FORM}(z) \\ \rightarrow \quad REPF(x_1,y_1,REPF(x_0,y_0,z)) \equiv REPF(x_0,y_0,REPF(x_1,y_1,z)) .$$

The proof proceeds by showing first the analogous statement for *REPT* and the hypothesis TERM(z), making use of Lemma 10.5 to express the induction on terms. From this I obtain (v4) under the hypothesis ATOM(z), and then proceed by Lemma 10.5, this time expressing induction on formulas.

PR-Terms for Deductions and the Proof Predicate

The term $AXIOM(x)$, representing the set *axiom* of axioms for logic, I define as $V[\mathrm{FORM}(x) \wedge (a_0(x) \vee a_1(x) \vee \ldots)]$ where the $a_i(x)$ correspond to the finitely many axiom schemata. The formula corresponding to the schema $v \rightarrow (w \rightarrow v)$ is

$$E_0(x) \equiv \kappa_7^{\kappa_3} \wedge E_0(E_2(x)) \equiv \kappa_7^{\kappa_3} \wedge E_1(x) \equiv E_2(E_2(x)) \ ,$$

and the formula corresponding to the schema $\forall v \, (u \rightarrow w) \rightarrow (u \rightarrow \forall v w)$, with v not free in u, is

$$E_0(x) \equiv \kappa_7^{\kappa_3} \ \wedge \ E_0(E_1(E_1(x))) \equiv \kappa_7^{\kappa_3} \ \wedge \ E_0(E_2(x)) \equiv \kappa_7^{\kappa_3} \\ \wedge \ \neg D_4(E_1(x)) \equiv \kappa_0 \ \wedge \ D_4(E_1(x)) \equiv D_4(E_2(E_2(x))) \\ \wedge \ E_1(E_1(E_1(x))) \equiv E_1(E_2(x)) \ \wedge \ E_2(E_1(E_1(x))) \equiv E_1(E_2(E_2(x))) \\ \wedge \ \neg \, OCCFREE(D_4(E_1(x)), E_1(E_2(x))) \ .$$

Let s(p) be the schema $v = \forall v \ w \rightarrow \mathrm{rep}(v,t \mid w)$ with $\zeta(v,t)$ free for w. The formula v(x,y):

$$E_0(x) \equiv \kappa_7^{\kappa_3} \wedge FREEFOR(D_4(E_1(x)), y, E_1(E_1(x))) \\ \wedge \ E_2(x) \equiv REPF(D_4(E_1(x)), y, E_1(E_1(x)))$$

evaluates to 1 at $<\ulcorner v \urcorner, \ulcorner t \urcorner>$ if, and only if, v with t satisfies s(i). But then t is uniquely determined by v and $\ulcorner t \urcorner < \ulcorner v \urcorner$. Consequently, the formula $a_p(x) = \exists y < x \ v(x,y)$ represents the codes of formulas satisfying s(p). –

The term representing the codes of formulas $\mathrm{rep}(v,t \mid w) \rightarrow \exists vw$ is defined analogously.

In order to include the equality axioms into $\mathit{AXIOM}(x)$, I need a PR-term which represents the arithmetizations of formulas $x \equiv y$; since $F_{=}(a,b) = 2^{725} \cdot 3^{a} \cdot 5^{b}$, I define the desired term as

$$\equiv^{\S}(x,y) = \kappa_2^{\kappa_{725}} \cdot \kappa_3^{x} \cdot \kappa_5^{y} \quad .$$

But equality axioms involve function symbols, and so I also need terms which represent *their* arithmetizations. For κ_0 I have already $\kappa(\ulcorner \kappa_0 \urcorner)$ which I now write $\kappa_0^{\S}$. Since s, $+$ and $\cdot$ have the indices 2, 4 and 12, the terms representing their arithmetizations will be defined as

$$s^{\S}(x) = \kappa_2^{\kappa_9} \cdot \kappa_3^{x} \ , \quad +^{\S}(x,y) = \kappa_2^{\kappa_{81}} \cdot \kappa_3^{x} \cdot \kappa_5^{y} \ , \quad \times^{\S}(x,y) = \kappa_2^{\kappa_{531441}} \cdot \kappa_3^{x} \cdot \kappa_5^{y} \ .$$

Thus the equality axiom $u \equiv v \rightarrow u+w \equiv v+w$ is represented by a further formula $a_{\bullet}(x)$, namely

$$\begin{aligned} E_0(x) \equiv \kappa_7^{\kappa_3} \wedge E_0E_1(x) \equiv \kappa_{725} \wedge E_0E_2(x) \equiv \kappa_{725} & \\ \wedge\, E_0E_1E_2(x) \equiv \kappa_{81} \wedge E_0E_2E_2(x) \equiv \kappa_{81} & \\ \wedge\, E_1E_1(x) \equiv E_1E_1E_2(x) \wedge E_2E_1(x) \equiv E_1E_2E_2(x) & \\ \wedge\, E_2E_1E_2(x) \equiv E_2E_2E_2(x) & \\ \wedge\, \mathrm{VAR}(E_1E_1(x)) \wedge \mathrm{VAR}(E_2E_1(x)) \wedge \mathrm{VAR}(E_2E_1E_2(x)) & . \end{aligned}$$

The other equality axioms are dealt with analogously, and thus the definition of $\mathit{AXIOM}(x)$ will be completed.

The term $\mathit{PA}(x)$, representing the specific axioms PA, I define analogously as $V[\mathrm{FORM}(x) \wedge (b_1(x) \vee \ldots \vee b_7(x))]$ where the $b_i(x)$ correspond to the schemata (PQi) for $0<i<7$ and to (ISA). For instance, (PQ1) corresponds to $b_1(x)$:

$$\begin{aligned} E_0(x) \equiv \kappa_7^{\kappa_3} \wedge E_0E_1(x) \equiv \kappa_{725} \wedge E_0E_2(x) \equiv \kappa_{725} & \\ \wedge\, E_0E_1E_1(x) \equiv \kappa_9 \wedge E_0E_2E_1(x) \equiv \kappa_9 & \\ \wedge\, E_1E_2(x) \equiv E_1E_1E_1(x) \wedge E_2E_2(x) \equiv E_1E_2E_1(x) & \\ \wedge\, \mathrm{VAR}(E_1E_2(x)) \wedge \mathrm{VAR}(E_2E_2(x)) & \ , \end{aligned}$$

and (ISA) corresponds to $b_7(x)$:

$$\begin{aligned} & \mathrm{FORM}(x) \wedge E_0(x) \equiv \kappa_7^{\kappa_3} \wedge E_0(E_1(x)) \equiv \kappa_7 \wedge \neg D_4(E_2(E_1(x))) \equiv \kappa_0 \\ & \quad \wedge D_4(E_2(E_1(x))) \equiv D_4(E_2(x)) \wedge E_1(E_2(E_1(x))) \equiv \kappa_7^{\kappa_3} \\ & \quad \wedge E_1(E_1(E_2(E_1(x)))) \equiv E_1(E_2(x)) \wedge \mathrm{OCCFREE}(D_4(E_2(x)), E_1(E_2(x))) \\ & \quad \wedge E_1(E_1(x)) \equiv \mathit{REPT}(D_4(E_2(x)), \kappa_0^{\S}, E_1(E_2(x))) \\ & \quad \wedge \mathrm{FREEFOR}(D_4(E_2(x)), s^{\S}(D_4(E_2(x))), E_1(E_2(x))) \\ & \quad \wedge E_2(E_1(E_2(E_1(x)))) \equiv \mathit{REPT}(D_4(E_2(x)), s^{\S}(D_4(E_2(x))), E_1(E_2(x))) \ . \end{aligned}$$

I next introduce the abbreviation $ECAR(x) = EXP(CS(LEV(x)), x)$. Lemma 10.4 provides the PR-term $DEDU(x)$, representing the set *dedu*, with the recursion equation $DEDU(u) \equiv V[v]$ for the formula v(u):

$$\neg LEN(u) \equiv \kappa_0$$
$$\wedge ((LEN(u) \equiv \kappa_1 \wedge (\mathrm{AXIOM}(ECAR(u)) \vee \mathrm{PA}(ECAR(u))))$$
$$\quad \vee (\neg LEN(u) \equiv \kappa_1) \wedge DEDU(CDR(u)) \equiv \kappa_1$$
$$\quad\quad \wedge (\mathrm{AXIOM}(ECAR(u)) \vee \mathrm{PA}(ECAR(u))$$
$$\quad\quad\quad \vee \exists z_1 < LEN(u)\ \exists z < z_1\ (EXP(z_1,u) \equiv IMP(EXP(z,u), ECAR(u)))$$
$$\quad\quad\quad \vee \exists z_1 < LEN(u)\ \exists z < z_1\ (EXP(z_1,u) \equiv IMP(EXP(z,u), ECAR(u)))$$
$$\quad\quad \vee \exists z_1 < LEN(u)\ \exists z < u\ (\mathrm{VAR}(z) \wedge QUNIV(z, EXP(z_1,u)) \equiv ECAR(u)))).$$

From $DEDU(u)$ I define the term $DED(x,u) = DEDU(u) \times ID_2(x, ECAR(u))$, and $DED(x,u)$ then represents the relation *ded*. From $DED(x,u)$ I define the provability formula

$$\mathrm{BEW}(x) \quad = \quad \exists u\, DED(x,u) \equiv \kappa_1 .$$

At this stage I have established, for PA, once more the result that *ded* is represented by a formula. Formerly, I used various facts about recursive relations and functions, and then I found in Chapter 9 that recursive relations were representable. This was a long way, but the results did hold not only for PA but already for PB, PC and PQ. Here now I have apparently proceeded more directly and have, for recursive relations such as *ded*, explicitly constructed representing formulas which belong to PR-terms. Yet the road was not much shorter if I take into account the long preparation in Chapter 10 and the fact that not all PA-provabilities were presented down to the last details.

But my aim, of course, was not to give a second proof of the earlier result. Rather, I shall now show two properties of the provability predicate BEW(u):

THEOREM 3 (B2) $\vdash \mathrm{BEW}(x) \wedge \mathrm{VAR}(y) \wedge \mathrm{CTERM}(z) \rightarrow \mathrm{BEW}(REPF(y,z,x))$,

(B3) $\vdash \mathrm{BEW}(IMP(x,y)) \rightarrow (\mathrm{BEW}(x) \rightarrow \mathrm{BEW}(y))$.

As for (B2), it follows from logic that it suffices to prove this implication not for $\mathrm{BEW}(x) = \exists u\ \mathrm{DED}(x,u)$ but for $\mathrm{DED}(x,u)$ for a sufficiently new u; let $B(u,x,y,z)$ stand for $\mathrm{DED}(x,u) \wedge \mathrm{VAR}(y) \wedge \mathrm{CTERM}(z)$. If $a_p(x)$ is that part of the definition of $AXIOM(x)$ which describes the schema s(p), then (v2) implies

$$\vdash \mathrm{VAR}(y) \wedge \mathrm{CTERM}(z) \wedge \mathrm{FORM}(x)$$
$$\wedge\ x_1 \equiv \mathit{IMP}(\mathit{QUNIV}(y,x), \mathit{REPF}(y,z,x)) \ \rightarrow\ a_p(x_1)$$

Now $\vdash \mathrm{DED}(x,u) \rightarrow \mathrm{DEDU}(u) \wedge x \equiv \mathit{ECAR}(u)$. But

$$\vdash \mathrm{DEDU}(u) \wedge u \equiv \mathit{ECAR}(u) \wedge \mathrm{VAR}(y)$$
$$\wedge\ u_0 \equiv u \times \mathit{PRINO}(\mathit{LEN}(u))^{\mathit{QUNIV}(y,x)} \ \rightarrow\ \mathrm{DEDU}(u_0).$$

Further

$$\vdash \mathrm{B}(u,x,y,z) \wedge \mathrm{DEDU}(u_0)$$
$$\wedge\ u_1 \equiv u_0 \times \mathit{PRINO}(\mathit{LEN}(u)+\kappa_1)^{V[\,a_p(x_1)\,]} \ \rightarrow\ \mathrm{DEDU}(u_1)$$

$$\vdash \mathrm{B}(u,x,y,z) \wedge \mathrm{DEDU}(u_1)$$
$$\wedge\ u_2 \equiv u_1 \times \mathit{PRINO}(\mathit{LEN}(u)+\kappa_2)^{\mathit{REPF}(y,z,x)} \ \rightarrow\ \mathrm{DEDU}(u_2).$$

Hence

$$\vdash \mathrm{B}(u,x,y,z) \ \rightarrow\ \mathrm{DED}(\mathit{REPF}(y,z,x),\, u_2)\ ,$$
$$\vdash \mathrm{B}(u,x,y,z) \ \rightarrow\ \mathrm{BEW}(\mathit{REPF}(y,z,x))\ .$$

This proves (B2).

Before continuing with (B1), I introduce the PR-terms

$$\mathit{LEV}(x) \equiv \mathit{CS}(\mathit{LEN}(x))\ ,$$
$$\mathit{CAR}(x) \equiv \mathit{CSGID}_2(\mathit{LEN}(x),\kappa_0) \times \mathit{PRINO}(\mathit{LEV}(x))^{\mathit{EXP}(\mathit{LEV}(x),\,x)}\ ,$$
$$\mathit{CDR}(x) \equiv \mathit{CSGID}_2(\mathit{LEN}(x),\kappa_0) \times \mathit{QU}(\mathit{CAR}(x),\, x)\ ,$$

and Lemma 10.4 I obtain the term $\mathit{CAT}(x,y)$ with the recursion equation

$$\vdash\ \mathit{CAT}(x,y) \equiv V[\,\mathit{LEN}(y) \equiv \kappa_0] \times x$$
$$+\ V[\neg\, \mathit{LEN}(y) \equiv \kappa_0] \times \mathit{CAT}(x,\, \mathit{CDR}(y))$$
$$\times\ \mathit{PRINO}(\mathit{LEV}(x)+\mathit{LEV}(y))^{\mathit{ECAR}(y)}.$$

An easy PA-induction shows

$$\vdash \mathit{LEN}(\mathit{CAT}(x,y)) \equiv \mathit{LEN}(x) + \mathit{LEN}(y)$$
$$\vdash z < \mathit{LEN}(x) \ \rightarrow\ \mathit{EXP}(z,x) \equiv \mathit{EXP}(z,\, \mathit{CAT}(x,y))$$
$$\vdash z < \mathit{LEN}(y) \ \rightarrow\ \mathit{EXP}(z,y) \equiv \mathit{EXP}(z+\mathit{LEN}(x),\, \mathit{CAT}(x,y))$$
$$\vdash \mathit{LEN}(x) \leq z < \mathit{LEN}(\mathit{CAT}(x,y)) \ \rightarrow\ \mathit{EXP}(z \dot{-} \mathit{LEN}(x),y) \equiv \mathit{EXP}(z,\, \mathit{CAT}(x,y))$$
$$\vdash \kappa_0 < \mathit{LEN}(y) \ \rightarrow\ \mathit{ECAR}(\mathit{CAT}(x,y)) \equiv \mathit{ECAR}(y)\ .$$

Returning to the formula $\mathrm{DEDU}(u)$, there follows by induction

$$\vdash\ \mathrm{DEDU}(u) \ \rightarrow\ (\mathrm{DEDU}(v) \ \rightarrow\ \mathrm{DEDU}(\mathit{CAT}(u,v)))\ .$$

I next observe that

$$\text{(d0)}\quad \vdash \mathrm{DED}(\mathit{IMP}(x,y),u) \wedge \mathrm{DED}(x,v) \wedge w \equiv \mathit{CAT}(u,v) \times \mathit{PRINO}(\mathit{LEN}(\mathit{CAT}(u,v)))^{y} \rightarrow \mathrm{DED}(y,w)$$

because

$$\vdash w \equiv \mathit{CAT}(u,v) \times \mathit{PRINO}(\mathit{LEN}(\mathit{CAT}(u,v)))^{y} \rightarrow y \equiv \mathit{ECAR}(w)$$

$$\vdash \mathrm{DED}(x,v) \rightarrow x \equiv \mathit{ECAR}(v) \equiv \mathit{EXP}(\mathit{LEV}(v),v)$$

$$\equiv \mathit{EXP}(\mathit{LEV}(u) + \mathit{LEN}(v),\ \mathit{CAT}(u,v))$$

$$\equiv \mathit{EXP}(\mathit{LEV}(u) + \mathit{LEN}(v),\ w)$$

$$\vdash \mathrm{DED}(\mathit{IMP}(x,y),u) \rightarrow \mathit{IMP}(x,y) \equiv \mathit{EXP}(\mathit{LEV}(u),u) \equiv \mathit{EXP}(\mathit{LEV}(u),w)$$

$$\vdash \quad \mathit{EXP}(\mathit{LEV}(u),w) \equiv \mathit{IMP}(\mathit{EXP}(\mathit{LEV}(u) + \mathit{LEN}(v),\ w)\ ,\ \mathit{ECAR}(w))$$

$$\vdash \exists z_1 < \mathit{LEN}(w) \wedge \mathit{LEV}(u) < z_1$$

$$\wedge\ \mathit{EXP}(\mathit{LEV}(u),w) \equiv \mathit{IMP}(\mathit{EXP}(z_1,w),\mathit{ECAR}(w)))$$

$$\vdash \exists z_1 < \mathit{LEN}(w)\ \exists z < z_1\ (\mathit{EXP}(z,w) \equiv \mathit{IMP}(\mathit{EXP}(z_1,w),\mathit{ECAR}(w)))\ .$$

By definition of DEDU(w) this implies $\vdash$ DEDU(w). Finally, I obtain from (d0) by logic

$$\vdash \mathrm{DED}(\mathit{IMP}(x,y),u) \wedge \mathrm{DED}(x,v) \rightarrow \exists w\,\mathrm{DED}(y,w)$$

$$\vdash \exists u\ \mathrm{DED}(\mathit{IMP}(x,y),u) \wedge \exists v\,\mathrm{DED}(x,v) \rightarrow \exists w\,\mathrm{DED}(y,w)\ .$$

This concludes the proof of Theorem 3.

At the beginning of this section, I gave the name FF to the set of the four function symbols f_i of PA. It is straightforward to generalize all that has been developed here to the case that FF contains further function symbols, defined by PR-terms, provided that (1) it then contains, for every such f_i, also the function symbols used in the defining equations of f_i, (2) the set I_a of the indices i of these function symbols in the extended signature Δ_a remains recursive, and (3) there are at least a many unused PR-symbols *not* in FF than there are in FF. Condition (2) is trivially satisfied if FF is finite; if FF is infinite and if f_i is to represent a primitive recursive function f then the index $i(f_i)$ can be defined by recursion on the schemata S(f) which describes f. In any case, indices i should be such that *exp*(0,i) is the arity of f_i and f. Under these circumstances, the definition (af) of arithmetizations works also for the new f_i and the definition of *TERM*(x) remains unchanged. That condition (3) is required can be seen from the appearance of the new function symbols $f_i^{\,s}$ in the definition of *AXIOM*(x) for the case of equality axioms. Further, *PA*(x) then will have to be extended by the formalized defining equations (not the recursion equations!) of the f_i. In the same way, also defined predicate symbols such as $<$ may be included into an arithmetized and formalized extension of PA.

3. Internalization

In Chapter 7 syntax was arithmetized by primitive recursive functions, and in Chapter 9 it was shown that, already for fragments of arithmetic, primitive recursive functions and relations were representable by formulas of their respective languages. This was sufficient in order to obtain the theorems on undecidability and incompleteness. During the last section I have formalized the arithmetization of syntax by constructing such representing formulas in the language L_P of PR, affording the luxury to express in Theorem 3 (B2) and (B3) some properties of BEW(x) by PR-provable free variable formulas built from PR-terms. It was rather accidentally, namely in order to write down the formalized equality axioms, used in the definition of BEW(x) via *AXIOM*(x), that I actually did mention the PR-terms $\equiv^{\mathfrak{s}}(x,y)$, $\kappa_0^{\mathfrak{s}}$, $s^{\mathfrak{s}}(x)$, $+^{\mathfrak{s}}(x,y)$, $\times^{\mathfrak{s}}(x,y)$ which formalize the original terms and predicate symbols of the language L of PA.

In the present section, I shall study such formalizing terms more systematically. I shall begin by setting up

The Internal Copy

of the *arithmetization* of L in PR.

Being primitive recursive, the arithmetizing functions F_f for L_F can be represented by terms themselves. Repeating what I did accidentally in the last section, I now state explicitly that, to every operation symbol f in FF, I assign a *new* PR-symbol $f^{\mathfrak{s}}$ of the same arity, *not* in FF:

If f (and F_f) have arity 0 then $n = \ulcorner f \urcorner$ is the only value of F_f, and $f^{\mathfrak{s}}$ is a constant term such that $\vdash f^{\mathfrak{s}} \equiv \kappa_n$; in particular, $\kappa_0^{\mathfrak{s}} \equiv \kappa(\kappa_0)$.

If f (and F_f) have positive arity k, and if $i(f)$ is the index of f in Δ_a, then $f^{\mathfrak{s}}$ has the defining equation

$$\text{(ff)} \quad \vdash f^{\mathfrak{s}}(\xi) \equiv \kappa_2^{\kappa_3^{\kappa(i(f))}} \times PRINO(\kappa_1)^{x_0} \times \ldots \times PRINO(\kappa_k)^{x_{k-1}}$$

for $\xi = \langle x_0, \ldots, x_{k-1} \rangle$. [Should FF have been enlarged by further symbols f for certain PR-terms, then the defining equation of $f^{\mathfrak{s}}(\xi)$ would be connected to that of $f(\xi)$ only by the number $i(f)$.]

That $f^{\mathfrak{s}}(\xi)$ then represents F_f is seen as follows. The arithmetic functions occurring on the right in the definition

$$F_f(\alpha) = 2^{3^{i(f)}} \cdot prino(1)^{\alpha(0)} \cdots prino(k)^{\alpha(k-1)}$$

of F_f are represented by the corresponding terms. Hence the three terms

$$\kappa(F_f(\alpha)) = \kappa\,(\,2^{3^{i(f)}} \cdot prino(1)^{\alpha(0)} \cdots prino(k)^{\alpha(k-1)}\,)$$

$$\kappa(2^{3^{i(f)}}) \times \kappa\,(\,prino(1)^{\alpha(0)}) \times \ldots \times \kappa(prino(k)^{\alpha(k-1)}\,)$$

$$f^{\$}(\kappa \cdot \alpha) = \kappa_2^{\kappa_3^{\kappa(i(f))}} \times PRINO(\kappa_1)^{\kappa(\alpha(0))} \times \ldots \times PRINO(\kappa_k)^{\kappa(\alpha(k-1))}$$

are provably equal. – [It should be noticed that the definition (ff) depends on the particular definition (af) of my basic arithmetization. Would a different arithmetization have been chosen in Chapter 7 (either with other distinguishing prime factors than 2, or by an altogether different coding than that through prime factorization exponents) then a different definition (af*) of F_f would have been mirrored in a different definition (ff*).]

It is important to notice that, for any sequence $\tau = \langle t_0, \ldots, t_{k-1} \rangle$ of terms

(tf) $\qquad \vdash \mathrm{TERM}(t_0) \wedge \ldots \wedge \mathrm{TERM}(t_{k-1}) \rightarrow \mathrm{TERM}(f^{\$}(\tau))$,

(tfc) $\qquad \vdash \mathrm{CTERM}(t_0) \wedge \ldots \wedge \mathrm{CTERM}(t_{k-1}) \rightarrow \mathrm{CTERM}(f^{\$}(\tau))$.

To see this, recall the definition (ta) of $TERM(x)$ and notice that the index $i(f)$ and the arity k are related by $k = exp(0, i(f))$. Hence there holds $\vdash \kappa_k \equiv EXP(\kappa_0, \kappa(i(f))$ and $\vdash \kappa_k \equiv ART_a(f^{\$}(\xi))$, $\vdash \kappa_k + \kappa_1 \equiv LEN(f^{\$}(\xi))$. Also, $k \in I_a$ implies $\vdash IA(\kappa_k) \equiv \kappa_1$. . Thus $\vdash v(f^{\$}(\xi))$ for the formula $v(x)$ with $VT(x) = V[v]$ in (tb), and so there follows

$$\vdash \Pi_{<} T(f^{\$}(\xi), \kappa_k) \equiv \kappa_1 \rightarrow TERM(f^{\$}(\xi)) \equiv \kappa_1$$

for $T(f^{\$}(\xi), y) = TERM(EXP(y + \kappa_1, f^{\$}(\xi)))$. The definition (ff) implies

$$\vdash EXP(\kappa_{i+1}, f^{\$}(\xi)) \equiv x_i$$

for $i \lesssim k-1$. Also, for any $T(\xi, y)$, induction in ω shows for $n > 0$

$$\vdash \Pi_{<} T(\xi, \kappa_{n+1}) \equiv \Pi_{<} T(\xi, \kappa_n) \times T(\xi, \kappa_n) \equiv T(\xi, \kappa_0) \times \ldots \times T(\xi, \kappa_n) ,$$

hence

$$\vdash \Pi_{<} T(f^{\$}(\xi), \kappa_k) \equiv TERM(EXP(\kappa_0 + \kappa_1, f^{\$}(\xi))) \times \ldots$$
$$\times\, TERM(EXP(\kappa_{k-1} + \kappa_1, f^{\$}(\xi))) .$$

Since $\vdash \kappa_i + \kappa_1 \equiv \kappa_{i+1}$, this simplifies to

$$\vdash \Pi_{<} T(f^{\$}(\xi), \kappa_k) \equiv TERM(x_0) \times \ldots \times TERM(x_{k-1})$$

whence

$$\vdash \mathrm{TERM}(x_0) \wedge \ldots \wedge \mathrm{TERM}(x_{k-1}) \rightarrow \mathrm{TERM}(f^{\$}(\xi)) .$$

But here the variables x_i may be replaced by terms t_i. The case of (tfc) is analogous.

To every predicate symbol p, introduced for L, I assign a *new* function symbol $p^{\$}$ of the same arity, *not* in FF, and choose the term $p^{\$}(\xi)$ as the free variable formalization of the function F_p arithmetizing the atomic formula $p(\xi)$. Thus if p has the index i and the arity k then

(af) $\quad \vdash p^{\$}(\xi) \equiv \kappa_2^{\kappa_5^{\kappa_i}} \times PRINO(\kappa_1)^{x_0} \times \ldots \times PRINO(\kappa_k)^{x_{k-1}}$, with $\xi = <x_0, \ldots, x_{k-1}>$, is the defining equation of $p^\S$. Again, $p^\S(\xi)$ represents F_p, and for $\tau = <t_0, \ldots, t_{k-1}>$ there holds

(ta) $\vdash \mathrm{TERM}(t_0) \wedge \ldots \wedge \mathrm{TERM}(t_{k-1}) \rightarrow \mathrm{ATOM}(p^\S(\tau))$.

The new function symbols $f^\S$ I call *internal*. As the function symbol s is distinct from the internal $s^\S$, no $\kappa_n = s(\kappa_{n-1})$ for $n>0$ is value of an internal function symbol. In particular, no $\gamma_x = \kappa(\ulcorner x \urcorner)$ is so.

I define the algebra T^* of *internal terms* as that generated, by the internal function symbols within $T(L_P)$, from the variables x, the constant $\kappa_0{}^\S$ *and* the constants γ_x. Being an algebra of terms, T^* has the the property of unique readability with respect to its operations $f^\S$ and its generators.

The new function symbols $p^\S$ I call *internal predicate* symbols, and T_L-terms $p^\S(\tau)$, with sequences τ of internal terms, I call *internal atomic formulas.* Being terms, the atomic internal formulas are also uniquely readable.

I define the algebra L^* of internal *formulas* as that generated from the internal atomic formulas by

> the operations induced by the 2-ary P_R-terms $ET(y_0,y_1)$, $VEL(y_0,y_1)$, $IMP(y_0,y_1)$, $NE(y)$, and instead of $ET(a_0,a_1)$, $VEL(a_0,a_1)$, $IMP(a_0,a_1)$, $NE(a)$ I write $a_0 \wedge^\S a_1$, $a_0 \vee^\S a_1$, $a_0 \rightarrow^\S a_1$, $\neg^\S a$,
>
> the operations sending an internal formula a, which does not contain γ_x, into $\forall^\S \gamma_x a_x = QUNIV(\gamma_x, a_x)$ and $\exists^\S \gamma_x a_x = QEXI(\gamma_x, a_x)$ where a_x is the result of replacing x in a by γ_x.

To every L_F-term h, I assign an internal term h^*, called its *internalization*, by setting $x^* = x$ for variables, $\kappa_0{}^* = \kappa_0{}^\S$, and $f(\tau)^* = f^\S(\tau^*)$ for composite terms. The internalization map thus is an isomorphism of the algebra $T(L)$ of L-terms into T^* which is the identity for variables. Observe that h and its internalization h^* contain the same variables and that h^* does *not* contain any constants γ_x.

To every L-formula v, I assign an internal formula v^*, called its *internalization*, by setting $p(\tau)^* = p^\S(\tau^*)$ for atomic formulas, and $(v \wedge w)^* = v^* \wedge^\S w^*$, $(v \vee w)^* = v^* \vee^\S w^*$, $(v \rightarrow w)^* = v^* \rightarrow^\S w^*$, $(\neg v)^* = \neg^\S v^*$, $(Qx\, v)^* = Q^\S \gamma_x v^*{}_x$. The internalization map thus is an isomorphism of the algebra of straight L-formulas onto the algebra L^*. Observe that the free variables of v are the variables occurring in v^*.

To every L_F-term h, I assign an internal term $h^\ddagger$, called its *codification*, by replacing in h^* every variable z by the constant γ_z. Thus $x^\ddagger = \gamma_x$, and the map is a homomorphism of $T(L)$ onto the subalgebra $T^\ddagger$ of T^* consisting of constant terms.

To every L_F-formula v, I assign an internal formula $v^\ddagger$, called its *codification*, by replacing in v^* every variable z by the constant γ_z. Also this map is a homomorphism from the algebra Fm of L–formulas onto an algebra $L^\ddagger$ of constant terms in L^*.

Then there holds

(ti) $\vdash \mathrm{TERM}(h^\ddagger)$ and $\vdash \mathrm{TERM}(v^\ddagger)$ for every $h^\ddagger$ and every $v^\ddagger$.

Observe first that $\vdash \mathrm{VAR}(\gamma_x)$ for every variable x. Because if $x = x_n$ then

$$\vdash \gamma_x = \kappa(\ulcorner x_n \urcorner) = \kappa(2^{2^{2+n}}) \equiv \kappa_2^{\kappa_2^{\kappa_2+\kappa_n}} = \mathit{VA}(\kappa_n) .$$

Thus $\vdash \kappa_n < \kappa_2^{\kappa_2^{\kappa_2+\kappa_n}}$ implies $\vdash \kappa_n < \gamma_x \wedge \gamma_x \equiv \mathit{VA}(\kappa_n)$ whence $\vdash \exists z < \gamma_x \wedge \gamma_x \equiv \mathit{VA}(z)$, i.e. $\vdash \mathrm{VAR}(\gamma_x)$. Next, $\vdash \mathrm{VAR}(\gamma_x) \to \mathrm{TERM}(\gamma_x)$ gives $\vdash \mathrm{TERM}(\gamma_x)$. Thus if $\gamma \cdot \xi$ is $<\kappa(\ulcorner x_0 \urcorner), \ldots, \kappa(\ulcorner x_{n-1} \urcorner)$ for $\xi = <x_0, \ldots, x_{k-1}>$ then (tf) implies $\vdash \mathrm{TERM}(f^\$(\gamma \cdot \xi))$; induction on terms h then gives $\vdash \mathrm{TERM}(h^\ddagger)$. But now (ta) implies $\vdash \mathrm{FORM}(p(\tau)^\ddagger)$ for internal atomic formulas, and from that induction on formulas shows $\vdash \mathrm{FORM}(v^\ddagger)$ for arbitrary $v^\ddagger$.

Internalization and Codification

The operations $f^\$$ represent the operations F_f which compute $g_T(h)$, and $x^\ddagger$ actually is γ_x. So the constant term $h^\ddagger$ will evaluate to γ_h, and in Lemma 1 I shall show that this is provably so: $\vdash h^\ddagger \equiv \gamma_h = \kappa(\ulcorner h \urcorner)$. The operations $p^\$$ represent the operations F_p which compute $g_F(p(\tau))$, and the operations $\wedge^\$, \vee^\$, \to^\$, \neg^\$, \forall^\$ \gamma_x, \exists^\$ \gamma_x$ represent the operations which then compute $g_F(v)$. So the constant term $v^\ddagger$ will evaluate to γ_h, and in Lemma 1 I will show that this is provably so: $\vdash v^\ddagger \equiv \gamma_v = \kappa(\ulcorner v \urcorner)$. The terms in $T^\ddagger$ and $Fm^\ddagger$ are, therefore, tools to internally express in PR the mathematical computation of codes.

In the following I shall use repeatedly expressions $c(\zeta)$, referring to c as a term h or a formula v *together* with a sequence ζ of pairwise distinct variables which contains the variables occurring in h, respectively occurring free in v. Also, given $c(\zeta)$ and a sequence τ, of the same length as ζ and consisting of terms, I write $c(\tau)$ for $\mathrm{rep}(\eta \mid c)$ with $\eta(\zeta_i) = \tau_i$ for every member ζ_i of ζ. If ζ' is decomposed into a sequence ζ and a single variable x, then I shall write $c(\zeta, x)$ and $c(x, \zeta)$ in place of $c(\zeta')$. – An object c and its internalization c^* contain the same (free) variables; thus if $c(\zeta)$ is defined then so is $c^*(\zeta)$, and $c^\ddagger = c^*(\gamma \cdot \zeta)$.

For every term $h(\zeta)$ with ζ of length m, I define a primitive recursive function $\Phi(h, \zeta)$ of arity m by:

$$\Phi(\kappa_0, \zeta) = c_0^m$$
$$\Phi(x, \zeta) = p_i^m \qquad \text{where } x = \zeta_i,$$
$$\Phi(f(\tau), \zeta) = F_f \circ <\Phi(\tau_0, \zeta), \ldots, \Phi(\tau_{n-1}, \zeta)> .$$

Induction on terms shows that

(xt) $\quad g_T(h) = \Phi(h, \zeta)(g_T \cdot \zeta)$.

This holds for $h = x_i$ and $h = \kappa_0$. Assume that (xt) holds for the τ_i in $h = f(\tau)$ and observe that if ζ contains the variables in $f(\tau)$ then it also contains the variables of every τ_i. Then it follows from the definition of g_T that

$$\begin{aligned} g_T(f(\tau)) &= F_f(g_T(\tau_0), \dots, g_T(\tau_{n-1})) \\ &= F_f(\Phi(\tau_0, \zeta)(g_T \cdot \zeta), \dots, \Phi(\tau_{n-1}, \zeta)(g_T \cdot \zeta)) \\ &= F_f \circ \langle \Phi(\tau_0, \zeta), \dots, \Phi(\tau_{n-1}, \zeta) \rangle (g_T \cdot \zeta) . \end{aligned}$$

For every formula $v(\zeta)$ with ζ of length m, I define a primitive recursive function $\Phi(v, \zeta)$ of arity m, first by

$$\begin{aligned} &\Phi(p(\tau), \zeta) = F_p(\Phi(\tau_0, \zeta), \dots, \Phi(\tau_{n-1}, \zeta)) , && \Phi(\neg v, \zeta) = ne(\Phi(v, \zeta)), \\ &\Phi(v \wedge w, \zeta) = et(\Phi(v, \zeta), \Phi(w, \zeta)) , && \Phi(v \vee w, \zeta) = vel(\Phi(v, \zeta), \Phi(w, \zeta)), \\ &\Phi(v \to w, \zeta) = imp(\Phi(v, \zeta), \Phi(w, \zeta)) . \end{aligned}$$

Given $(Qxv)(\zeta)$; if x is in ζ then I set $\Phi(\forall xv, \zeta) = quniv(\ulcorner x \urcorner, \Phi(v, \zeta))$ and $\Phi(\exists xv, \zeta) = qexiv(\ulcorner x \urcorner, \Phi(v, \zeta))$. If x is not in ζ but occurs free in v then let ζ' be ζ prolonged by x, i.e. $\zeta'_m = x$. Then $\Phi(v, \zeta')$ is defined as an (m+1)-ary function, and if $\Phi(v, \zeta')_{\ulcorner x \urcorner}$ denotes the m-ary function obtained by keeping the last argument constant to $\ulcorner x \urcorner$, then I set

$$\Phi(\forall xv, \zeta) = quniv(\ulcorner x \urcorner, \Phi(v, \zeta')_{\ulcorner x \urcorner}), \quad \Phi(\exists xv, \zeta) = qexi(\ulcorner x \urcorner, \Phi(v, \zeta')_{\ulcorner x \urcorner}) .$$

Induction on formulas shows again

(xf) $\quad g_F(v) = \Phi(v, \zeta)(g_T \cdot \zeta)$

since e.g.

$$\begin{aligned} g_F(\forall xv) &= quniv(\ulcorner x \urcorner, g_F(v)) \\ &= quniv(\ulcorner x \urcorner, \Phi(v, \zeta')(g_T \cdot \zeta')) \\ &= quniv(\ulcorner x \urcorner, \Phi(v, \zeta')_{\ulcorner x \urcorner}(g_T \cdot \zeta)) \\ &= quniv(\ulcorner x \urcorner, \Phi(v, \zeta')_{\ulcorner x \urcorner})(g_T \cdot \zeta) \\ &= \Phi(\forall xv, \zeta)(g_T \cdot \zeta) . \end{aligned}$$

I now use internalizations to see that

(yt) $\quad h^*$ represents $\Phi(h, \zeta)$.

This holds for $h = \zeta_i$ since then $h^* = \zeta_i$, and the variable ζ_i represents p^m_i. Assume that (yt) holds for the τ_i in $h = f(\tau)$. Let α be in ω^m, with m the length of ζ, and let β be the sequence of values $\Phi(\tau_i, \zeta)(\alpha) = \beta_i$. Then

$$\vdash \tau_i^*(\kappa \cdot \alpha) \equiv \kappa_{\beta(i)}$$

by inductive hypothesis, hence also

$$\vdash f^{\$}(\kappa \cdot \beta) \equiv f^{\$}(\tau_0^*(\kappa \cdot \alpha) , \dots , \tau_{n-1}^*(\kappa \cdot \alpha))$$

by equality logic. Since $t(\kappa \cdot \alpha) = rep(\zeta; \kappa \cdot \alpha; t)$ and since $rep(\zeta; \kappa \cdot \alpha; -)$ is homomorphic, the right side here is $f^{\$}(\tau^*)(\kappa \cdot \alpha)$ whence

$$\vdash f^{\mathfrak{s}}(\kappa\cdot\beta) \equiv (f^{\mathfrak{s}}(\tau^*))(\kappa\cdot\alpha)\ ,$$

$$\vdash f^{\mathfrak{s}}(\kappa\cdot\beta) \equiv (f(\tau))^*(\kappa\cdot\alpha)\ .$$

since $f(\tau)^* = f^{\mathfrak{s}}(\tau^*)$. On the other hand,

$$\Phi(f(\tau),\ \zeta)(\alpha) = \mathrm{F}_f \circ <\Phi(\tau_0,\zeta), \ldots, \Phi(\tau_{n-1},\zeta)>(\alpha) = \mathrm{F}_f(\beta)$$

by definition of $\Phi(f(\tau),\ \zeta)$. Hence

$$\text{if } \Phi(f(\tau),\zeta)(\alpha) = \mathrm{n} \ \text{ then } \ \mathrm{F}_f(\beta) = \mathrm{n}\ ,$$

$$\text{if } \Phi(f(\tau),\zeta)(\alpha) = \mathrm{n} \ \text{ then } \vdash f^{\mathfrak{s}}(\kappa\cdot\beta) \equiv \kappa_{\mathrm{n}}$$

since $f^{\mathfrak{s}}$ represents F_f . Thus

$$\text{if } \Phi(f(\tau),\zeta)(\alpha) = \mathrm{n} \ \text{ then } \vdash (f(\tau))^*(\kappa\cdot\alpha) \equiv \kappa_{\mathrm{n}}\ .$$

(yf) $\quad$ v^* represents $\Phi(\mathrm{v},\ \zeta)$.

Consider $\mathrm{v} = p(\tau)$; let α be in ω^{m} and let β be defined as above. Then again $\vdash \tau_i^*(\kappa\cdot\alpha) \equiv \kappa_{\beta(i)}$ and

$$\vdash p^{\mathfrak{s}}(\kappa\cdot\beta) \equiv p^{\mathfrak{s}}(\tau_0^*(\kappa\cdot\alpha)\ , \ldots , \tau_{n-1}^*(\kappa\cdot\alpha)) \equiv (p(\tau))^*(\kappa\cdot\alpha)\ .$$

Also $\Phi(p(\tau),\ \zeta)(\alpha) = \mathrm{F}_p(\beta)$, and since $p^{\mathfrak{s}}(\xi)$ represents F_p it follows from $\Phi(p(\tau),\zeta)(\alpha) = \mathrm{n}$ that $\vdash p^{\mathfrak{s}}(\kappa\cdot\beta) \equiv \kappa_{\mathrm{n}}$. Hence

$$\text{if } \Phi(p(\tau),\zeta)(\alpha) = \mathrm{n} \ \text{ then } \vdash (p(\tau))^*(\kappa\cdot\alpha) \equiv \kappa_{\mathrm{n}}\ .$$

Consider next $\mathrm{v}\wedge\mathrm{w}$. Then $\Phi(\mathrm{v},\zeta)(\alpha) = \mathrm{n}$, $\Phi(\mathrm{w},\zeta)(\alpha) = \mathrm{m}$ imply

$$\vdash \mathrm{v}^*(\kappa\cdot\alpha) \equiv \kappa_{\mathrm{n}} \quad \text{and} \quad \vdash \mathrm{w}^*(\kappa\cdot\alpha) \equiv \kappa_{\mathrm{m}}\ .$$

Since

$$\Phi(\mathrm{v}\wedge\mathrm{w},\zeta)(\alpha) = et(\Phi(\mathrm{v},\zeta)(\alpha), \Phi(\mathrm{w},\zeta)(\alpha)) = et(\mathrm{n},\ \mathrm{m}) = \mathrm{r}$$

implies $\vdash ET(\kappa_{\mathrm{n}},\ \kappa_{\mathrm{m}}) \equiv \kappa_{\mathrm{r}}$, there follows

$$\vdash (\mathrm{v}\wedge\mathrm{w})^*(\kappa\cdot\alpha) = ET(\mathrm{v}^*,\ \mathrm{w}^*)(\kappa\cdot\alpha)) = ET(\mathrm{v}^*(\kappa\cdot\alpha),\ \mathrm{w}^*(\kappa\cdot\alpha)) \equiv \kappa_{\mathrm{r}}\ .$$

Consider finally $\forall\mathrm{x}\mathrm{v}$ and ζ. Let ζ' be the prolongation of ζ with x; let α have the length of ζ, and define α' as α prolonged with $\ulcorner\mathrm{x}\urcorner$. If n is the value of $\Phi(\mathrm{v},\zeta')(\alpha')$ then $\vdash \mathrm{v}^*(\kappa\cdot\alpha') \equiv \kappa_{\mathrm{n}}$ by inductive hypothesis. Thus

$$\begin{aligned}\Phi(\forall\mathrm{x}\mathrm{v},\ \zeta)(\alpha) &= quniv(\ulcorner\mathrm{x}\urcorner,\ \Phi(\mathrm{v},\zeta')_{\ulcorner\mathrm{x}\urcorner})(\alpha) \\ &= quniv(\ulcorner\mathrm{x}\urcorner,\ \Phi(\mathrm{v},\zeta')_{\ulcorner\mathrm{x}\urcorner}(\alpha)) \\ &= quniv(\ulcorner\mathrm{x}\urcorner,\ \Phi(\mathrm{v},\zeta')(\alpha')) \\ &= quniv(\ulcorner\mathrm{x}\urcorner,\ \mathrm{n})\end{aligned}$$

whence

$$\begin{aligned}\vdash \kappa(\Phi(\forall\mathrm{x}\mathrm{v},\ \zeta)(\alpha)) &\equiv \kappa(quniv(\ulcorner\mathrm{x}\urcorner,\ \mathrm{n})) \\ &\equiv QUNIV(\gamma_{\mathrm{x}},\ \kappa_{\mathrm{n}}) \\ &\equiv QUNIV(\gamma_{\mathrm{x}},\ \mathrm{v}^*(\kappa\cdot\alpha')) \\ &= QUNIV(\gamma_{\mathrm{x}},\ \mathrm{v}^*_{\mathrm{x}}(\kappa\cdot\alpha)) \\ &= QUNIV(\gamma_{\mathrm{x}},\ \mathrm{v}^*_{\mathrm{x}})(\kappa\cdot\alpha)\end{aligned}$$

$$= \forall^{\S} \gamma_x v^*{}_x(\kappa \cdot \alpha)$$
$$= (\forall x v)^*(\kappa \cdot \alpha) \ .$$

LEMMA 1 For every L_F-term $h(\zeta)$: $\vdash \gamma_h \equiv h^*(\gamma \cdot \zeta) = h^\ddagger$,
For every L_F-formula $v(\zeta)$: $\vdash \gamma_v \equiv v^*(\gamma \cdot \zeta) = v^\ddagger$.

Consider $h(\zeta)$. Then (xt) implies $g_T(h) = \Phi(h, \zeta)(g_T \cdot \zeta)$, and (yt) implies $\vdash \kappa(g_T(h)) \equiv h^*(\kappa \cdot g_T \cdot \zeta)$. The case of formulas is analogous.

Internalization of Constants

Concerning the internalizations $\kappa_n{}^*$, I did already observe that for $n > 0$ they cannot be of the form $\kappa_{\varphi(n)}$. However, *provably* equal to such terms they are: the internalization $\kappa_0{}^*$ of κ_0 is, by definition, $\kappa_0{}^{\S} = \kappa(\ulcorner \kappa_0 \urcorner)$, and induction in ω shows that

$$\vdash \kappa_n{}^* \equiv \kappa(\ulcorner \kappa_n \urcorner) \ .$$

For assume this to hold for n. As $\kappa_{n+1} = s(\kappa_n)$ by definition, also $\kappa_{n+1}{}^* = s^{\S}(\kappa_n{}^*)$ by definition of $(s(\kappa_n))^*$, and so $\vdash \kappa_{n+1}{}^* \equiv s^{\S}(\kappa(\ulcorner \kappa_n \urcorner))$ by inductive hypothesis. But $\kappa_{n+1} = s(\kappa_n)$ also implies $\ulcorner \kappa_{n+1} \urcorner = \ulcorner s(\kappa_n) \urcorner = S(\ulcorner \kappa_n \urcorner)$ by definition of the arithmetization S of s, hence $\vdash s^{\S}(\kappa(\ulcorner \kappa_n \urcorner)) \equiv \kappa(\ulcorner \kappa_{n+1} \urcorner)$ since $s^{\S}(x)$ represents S. Thus $\vdash \kappa_{n+1}{}^* \equiv \kappa(\ulcorner \kappa_{n+1} \urcorner)$. – Obviously, $\kappa_n{}^\ddagger = \kappa_n{}^*$ since $\kappa_n{}^*$ does not contain variables. Viewing $\kappa(\ulcorner \kappa_n \urcorner)$ as depending on n, I use the abbreviation

$$\hat{\gamma}_n = \gamma(\kappa_n) = \kappa(\ulcorner \kappa_n \urcorner)$$

whence $\hat{\gamma}_0 = \kappa_0{}^*$, and so the above may be rewritten as

$$(\hat{\gamma}) \qquad \vdash \kappa_n{}^* \equiv \hat{\gamma}_n \quad \text{and} \quad \vdash s^{\S}(\hat{\gamma}_n) \equiv \hat{\gamma}_{n+1} \ .$$

The constant terms κ_n mirror ω in L; the constant terms $\hat{\gamma}_n$ mirror ω in T*, and they also mirror the arithmetizations $\ulcorner \kappa_n \urcorner$ in T*. It remains to express the connection between the κ_n and the $\kappa_n{}^*$.

To this end, I introduce a PR-term $NN(z)$ with the recursion equations

$$NN(\kappa_0) \equiv \kappa(\ulcorner \kappa_0 \urcorner) \ , \quad NN(s(z)) \equiv s^{\S}(NN(z)) \ .$$

The arithmetical function nn which counts the arithmetizations $\ulcorner \kappa_n \urcorner$, i.e. $nn(n) = \ulcorner \kappa_n \urcorner$, is primitive recursive since it can be defined by $nn(0) = \ulcorner \kappa_0 \urcorner$, $nn(s(n)) = S(nn(n))$. Induction in ω shows that NN represents nn, i.e. for every n: $\vdash NN(\kappa_n) \equiv \kappa(\ulcorner \kappa_n \urcorner) = \hat{\gamma}_n$. Because $\vdash NN(\kappa_{n+1})) = NN(s(\kappa_n)) \equiv s^{\S}(NN(\kappa_n)) \equiv s^{\S}(\kappa(\ulcorner \kappa_n \urcorner)) \equiv \kappa(\ulcorner \kappa_{n+1} \urcorner)$. Thus I have shown the

LEMMA 2a $\vdash NN(\kappa_n) \equiv \kappa_n{}^* = \hat{\gamma}_n$, saying that $NN(z)$ represents the constant L–terms κ_n by their internalizations. – Taking in (tfc) the term t as $NN(z)$, I obtain

$$\vdash \mathrm{CTERM}(NN(z)) \rightarrow \mathrm{CTERM}(s^{\natural}(NN(z))) ,$$

hence

$$\vdash \mathrm{CTERM}(NN(z)) \rightarrow \mathrm{CTERM}(NN(s(z))) .$$

Since also $\vdash \mathrm{CTERM}(\kappa(\ulcorner\kappa_0\urcorner))$, this is the schema for a PA–induction resulting in

LEMMA 2b $\quad \vdash \mathrm{CTERM}(NN(z))$.

Hence a fortiori also $\vdash \mathrm{TERM}(NN(z))$.

Certain internal terms, namely the codifications $h^{\ddagger}$, were *provably* terms in the sense that $\vdash \mathrm{TERM}(h^{\ddagger})$; they did not contain L–variables. On the other hand, $NN(z)$ is not an internal term but is also provably a term. Further, $NN(z)$ shares with internal terms the property that it does not use function symbols from FF.

Internal Forms

I define the algebra T_N^* of *quasi–internal* or *qi–terms* as that generated, by the internal function symbols, from the variables x, the constants $\kappa_0^{\natural}$ and γ_x, *and* from the terms $NN(z)$; again, all of these are constants in this algebra. I define the algebra T_C^* of *closed qi–terms* or cqi–terms in the same manner, but omitting the variables x from the generators. Being algebras of terms, T_N^* and T_C^* have the the property of unique readability with respect to their operations $f^{\natural}$ and their generators.

I define the algebras L_N^* and L_C^* of *qi* and *cqi–formulas* in the same way as the algebra L^* of internal formulas, but now generated from internal atomic formulas which only employ qi–terms, respectively cqi–terms. Again, the algebras L_N^* and L_C^* have the property of unique readability.

Clearly, a qi–term term or qi–formula is closed if the only variables occurring in it occur in subterms $NN(z)$. It follows immediately from (tf) and (ta) that $\vdash \mathrm{TERM}(r)$ for every cqi–term r and $\vdash \mathrm{FORM}(a)$ for every cqi–formula a .

The replacement function rep is defined for every language L; it acts upon formulas and, in the initial case of atomic formulas, employs homomorphic transformations of terms. I now shall define an *internal* replacement function, calling rept* the transformation of terms from T_N^* and repf* the transformation of formulas from L_N^*. Further, it will be important to restrict myself to replacements, applied to cqi–terms and formulas, of constants γ_x by cqi–terms h. In view of unique readability, I can define rept* (as to be expected) by

$$\text{rept}^*(\gamma_x, h \mid \gamma_x) = h$$
$$\text{rept}^*(\gamma_x, h \mid \gamma_z) = \gamma_z \quad \text{for } x \neq z$$
$$\text{rept}^*(\gamma_x, h \mid \kappa_0^{\mathfrak{s}}) = \kappa_0^{\mathfrak{s}}$$
$$\text{rept}^*(\gamma_x, h \mid NN(z)) = NN(z)$$

for the generators and $\text{rept}^*(\gamma_x, h, f^{\mathfrak{s}}(\tau)) \equiv f^{\mathfrak{s}}(\text{rept}^*(\gamma_x, h, \tau))$ for composite terms. I define repf* by

$$\text{repf}^*(\gamma_x, h, p^{\mathfrak{s}}(\tau)) \equiv p^{\mathfrak{s}}(\text{repf}^*(\gamma_x, h, \tau)) ,$$

for atomic cqi-formulas, and then

$$\begin{aligned}
\text{repf}^*(\gamma_x, h, a \wedge^{\mathfrak{s}} b) &= \text{repf}^*(\gamma_x, h, a) \wedge^{\mathfrak{s}} \text{repf}^*(\gamma_x, h, b) ,\\
\text{repf}^*(\gamma_x, h, a \vee^{\mathfrak{s}} b) &= \text{repf}^*(\gamma_x, h, a) \vee^{\mathfrak{s}} \text{repf}^*(\gamma_x, h, b) ,\\
\text{repf}^*(\gamma_x, h, a \to^{\mathfrak{s}} b) &= \text{repf}^*(\gamma_x, h, a^*) \to^{\mathfrak{s}} \text{repf}^*(\gamma_x, h, b^*) ,\\
\text{repf}^*(\gamma_x, h, \neg^{\mathfrak{s}} a) &= \neg^{\mathfrak{s}} \text{repf}^*(\gamma_x, h, a) ,\\
\text{repf}^*(\gamma_x, h, Q^{\mathfrak{s}} \gamma_z a_z) &= Q^{\mathfrak{s}} \gamma_z \text{repf}^*(\gamma_x, h, a_z) \quad \text{for } x \neq z ,\\
\text{repf}^*(\gamma_x, h, Q^{\mathfrak{s}} \gamma_x a_x) &= Q^{\mathfrak{s}} \gamma_x a_x
\end{aligned}$$

for composite cqi-formulas. [Observe here that, in order to treat $Q^{\mathfrak{s}} \gamma_z a_z$, I do not need to know the original term a, but only the term a_z.]

The replacement function, useful as it will be for some purposes, is limited in certain other respects. For instance, if $v(\zeta)$ is a formula then there is no direct way to conclude from $\vdash \gamma_v \equiv v^*(\gamma \cdot \zeta)$ that $\text{repf}^*(\gamma_x, h, \gamma_v)$ isprovably equal to $\text{repf}^*(\gamma_x, h, v^*(\gamma \cdot \zeta))$. Provabilities, however, can be established for the terms built with *REPT* and *REPF*, and of these I shall first develop those paralleling the definitions of rept* and repf*.

Recall the definition (rt) of $REPT(x,y,z)$; I shall use the particular formulas v, u and w employed in it. It follows from $\vdash \text{VAR}(\gamma_x)$ that $\vdash v(\gamma_x, \gamma_x)$ whence

(r0) $\qquad \vdash REPT(\gamma_x, y, \gamma_x) \equiv y$.

Also, $x \neq z$ implies $\ulcorner x \urcorner \neq \ulcorner z \urcorner$, hence $\vdash \neg\, \gamma_x \equiv \gamma_z$ since $\vdash \neg\, \kappa_n \equiv \kappa_m$ for $n \neq m$. Thus $\vdash u(\gamma_x, \gamma_z)$ whence

(r1) $\qquad \vdash REPT(\gamma_x, y, \gamma_z) \equiv \gamma_z \quad$ for $x \neq z$.

Now $\vdash ART_a(\kappa_0^{\mathfrak{s}}) \equiv \kappa_0$ implies $\vdash u(\gamma_x, \kappa_0^{\mathfrak{s}})$, and $\vdash \text{CTERM}(NN(z))$ implies $\vdash u(\gamma_x, NN(z))$; hence

(r2) $\qquad \vdash REPT(\gamma_x, y, \kappa_0^{\mathfrak{s}}) \equiv \kappa_0^{\mathfrak{s}}$,

(r3) $\qquad \vdash REPT(\gamma_x, y, NN(z)) \equiv NN(z)$.

Finally, the definition of w, together with the definition (ff) of internalizations, implies for every cqi-term $f^{\mathfrak{s}}(\tau)$

(r4) $\qquad \vdash REPT(\gamma_x, y, f^{\mathfrak{s}}(\tau)) \equiv f^{\mathfrak{s}}(REPT(\gamma_x, y, \tau))$.

In the same way, the definition (rf) of *REPF* and (ap) imply (r5) $\vdash REPF(\gamma_x, y, p^\$(\tau)) \equiv p^\$(REPT(\gamma_x, y, \tau))$,

for atomic cqi-formulas, and then

(r6)
$$\begin{aligned}
&\vdash REPF(\gamma_x, y, a \wedge^\$ b) \equiv REPF(\gamma_x, y, a) \wedge^\$ REPF(\gamma_x, y, b)\ ,\\
&\vdash REPF(\gamma_x, y, a \vee^\$ b) \equiv REPF(\gamma_x, y, a) \vee^\$ REPF(\gamma_x, y, b)\ ,\\
&\vdash REPF(\gamma_x, y, a \to^\$ b) \equiv REPF(\gamma_x, y, a^*) \to^\$ REPF(\gamma_x, y, b^*)\ ,\\
&\vdash REPF(\gamma_x, y, \neg^\$ a) \equiv \neg^\$ REPF(\gamma_x, y, a)\ ,\\
&\vdash REPF(\gamma_x, y, Q^\$ \gamma_z a_z) \equiv Q^\$ \gamma_z REPF(\gamma_x, y, a_z)\quad , x \neq z\ ,\\
&\vdash REPF(\gamma_x, y, Q^\$ \gamma_x a_x) \equiv Q^\$ \gamma_x a_x
\end{aligned}$$

for composite cqi-formulas.

I now connect the replacement terms with the replacement functions:

LEMMA 3 For all cqi-terms t:

$$\vdash REPT(\gamma_x, h, t) \equiv \mathrm{rept}^*(\gamma_x, h \mid t)\ .$$

For all cqi-formulas a and all qi-terms h not containing any constant γ_z :

$$\vdash REPF(\gamma_x, h, a) \equiv \mathrm{repf}^*(\gamma_x, h \mid a)\ .$$

For the generating cqi-terms this follows from the definition of rept* together with (r0)–(r3). All other cqi terms have the form $f^\$(\tau)$ with a sequence τ of cqi-terms; hence the statement follows with (r4) by induction on cqi-terms. By the same reasoning, it follows with (r5) for atomic cqi-formulas. For composite cqi-formulas I apply induction. So if $x \neq z$ and

$$\vdash REPF(\gamma_x, h, a_z) \equiv \mathrm{repf}^*(\gamma_x, h, a_z)$$

then

$$\vdash Q^\$ \gamma_z REPF(\gamma_x, h, a_z) \equiv Q^\$ \gamma_z\ \mathrm{repf}^*(\gamma_x, h, a_z)$$

hence with (r6)

$$\vdash REPF(\gamma_x, h, Q^\$ \gamma_z a_z) \equiv \mathrm{repf}^*(\gamma_x, h, Q^\$ \gamma_z a_z)\ .$$

I now come to the definition of internal forms. Consider an L_F-formula v, and let v* be its internalization. Let $\zeta = \langle x_0, \ldots, x_{n-1}\rangle$ be a sequence of variables containing those which are free in v (i.e. in v*). So both $v^\ddagger = v^*(\gamma \cdot \zeta)$ and $\mathrm{repf}^*(\gamma_{x(n-1)}, NN(x_{n-1}) \mid v^*(\gamma \cdot \zeta))$ are cqi-formulas. Thus

$$\begin{aligned}
\vdash\ & REPF(\gamma_{x(n-2)}, NN(x_{n-2}), \mathrm{repf}^*(\gamma_{x(n-1)}, NN(x_{n-1}) \mid v^*(\gamma \cdot \zeta)))\\
&\equiv \mathrm{repf}^*(\gamma_{x(n-2)}, NN(x_{n-2}) \mid \mathrm{repf}^*(\gamma_{x(n-1)}, NN(x_{n-1}) \mid v^*(\gamma \cdot \zeta)))\ ,
\end{aligned}$$

by Lemma 3, and since also

$$\vdash REPF(\gamma_{x(n-1)}, NN(x_{n-1}), v(\gamma \cdot \zeta)) \equiv \mathrm{repf}^*(\gamma_{x(n-1)}, NN(x_{n-1}) \mid v^*(\gamma \cdot \zeta))$$

and

$$\vdash \; REPF(\gamma_{x(n-2)}, NN(x_{n-2}), REPF(\gamma_{x(n-1)}, NN(x_{n-1}) \mid v^*(\gamma \cdot \zeta)))$$
$$\equiv \; REPF(\gamma_{x(n-2)}, NN(x_{n-2}), \mathrm{repf}^*(\gamma_{x(n-1)}, NN(x_{n-1}) \mid v^*(\gamma \cdot \zeta)))$$

by equality logic, there follows

$$\vdash \; REPF(\gamma_{x(n-2)}, NN(x_{n-2}), REPF(\gamma_{x(n-1)}, NN(x_{n-1}), v^*(\gamma \cdot \zeta)))$$
$$\equiv \; \mathrm{repf}^*(\gamma_{x(n-2)}, NN(x_{n-2}) \mid \mathrm{repf}^*(\gamma_{x(n-1)}, NN(x_{n-1}) \mid v^*(\gamma \cdot \zeta))) \; .$$

Induction then shows that

$$\vdash \; REPF(\gamma_{x(0)}, NN(x_0), REPF(\gamma_{x(1)}, NN(x_1), \ldots$$
$$REPF(\gamma_{x(n-2)}, NN(x_{n-2}), REPF(\gamma_{x(n-1)}, NN(x_{n-1}) \mid v^*(\gamma \cdot \zeta))) \ldots))$$
$$\equiv \; \mathrm{repf}^*(\gamma_{x(0)}, NN(x_0) \mid \mathrm{repf}^*(\gamma_{x(1)}, NN(x_1) \mid \; \ldots$$
$$\mathrm{repf}^*(\gamma_{x(n-2)}, NN(x_{n-2}) \mid \mathrm{repf}^*(\gamma_{x(n-1)}, NN(x_{n-1}) \mid v^*(\gamma \cdot \zeta))) \ldots)) \; ;$$

These two terms I abbreviate as

$$REPF(\gamma \cdot \zeta \; ; NN \cdot \zeta \; ; v^*(\gamma \cdot \zeta)) \quad \text{and} \quad \mathrm{repf}^*(\gamma \cdot \zeta \; ; NN \cdot \zeta \; ; v^*(\gamma \cdot \zeta))$$

and call $REPF(\gamma \cdot \zeta \; ; NN \cdot \zeta \; ; v^*(\gamma \cdot \zeta))$ an *internal form* for $v(\gamma \cdot \zeta)$. So I have shown part (a) of

LEMMA 4 Let $\zeta = \langle x_0, \ldots, x_{n-1} \rangle$ be a sequence of variables containing those which are free in v (i.e. in v^*), let π be a permutation of n, and let ζ' be a second sequence such as ζ. Then

(a) $\vdash \; REPF(\gamma \cdot \zeta \; ; NN \cdot \zeta \; ; v^*(\gamma \cdot \zeta)) \equiv \mathrm{repf}^*(\gamma \cdot \zeta \; ; NN \cdot \zeta \; ; v^*(\gamma \cdot \zeta))$

(b) $\vdash \; REPF(\gamma \cdot \zeta \; ; NN \cdot \zeta \; ; v^*(\gamma \cdot \zeta)) \equiv REPF(\gamma \cdot \zeta \cdot \pi \; ; NN \cdot \zeta \cdot \pi \; ; v^*(\gamma \cdot \zeta))$

(c) $\vdash \; REPF(\gamma \cdot \zeta \; ; NN \cdot \zeta \; ; v^*(\gamma \cdot \zeta)) \equiv REPF(\gamma \cdot \zeta' \; ; NN \cdot \zeta' \; ; v^*(\gamma \cdot \zeta'))$

(d) $\vdash \; REPF(\gamma \cdot \zeta \; ; NN \cdot \zeta \; ; v^*(\gamma \cdot \zeta)) \equiv REPF(\gamma \cdot \zeta \; ; NN \cdot \zeta \; ; \gamma_v)$

As for (b), observe that the $NN(x_j)$ do not contain any $\gamma_{x(i)}$; hence

$$\mathrm{repf}^*(\gamma_{x(0)}, NN(x_0) \mid \; \ldots \; \mathrm{repf}^*(\gamma_{x(n-1)}, NN(x_{n-1}) \mid v^*(\gamma \cdot \zeta))$$
$$= \; \mathrm{repf}^*(\gamma_{x(\pi(0))}, NN(x_{\pi(0)}) \mid \; \ldots \; \mathrm{repf}^*(\gamma_{x(\pi(n-1))}, NN(x_{\pi(n-1)} \mid v^*(\gamma \cdot \zeta)) \; ,$$

i.e.

$$\mathrm{repf}^*(\gamma \cdot \zeta \; ; NN \cdot \zeta \; ; v^*(\gamma \cdot \zeta)) = \mathrm{repf}^*(\gamma \cdot \zeta \cdot \pi \; ; NN \cdot \zeta \cdot \pi \; ; v^*(\gamma \cdot \zeta)) \; ,$$

and this together with (a) gives (b) [observe that I also might have used (v4) since $\vdash$ CTERM($NN(x_j)$)]. Further, since ζ and ζ' both contain the free variables of v^* (and in v) there holds $v^*(\gamma \cdot \zeta) = v^*(\gamma \cdot \zeta')$ and

$$\mathrm{repf}^*(\gamma \cdot \zeta \; ; NN \cdot \zeta \; ; v^*(\gamma \cdot \zeta)) = \mathrm{repf}^*(\gamma \cdot \zeta' \; ; NN \cdot \zeta' \; ; v^*(\gamma \cdot \zeta')) \; ,$$

hence also (c). Finally, (d) follows by equality logic from Lemma 1. – It should be noticed that statements analogous to those above also hold for L_F-terms $t(\zeta)$.

In the remainder of this section I collect some technical results which will be needed in the proofs of Lemma 6 and 7 in the following section. First, it is an immediate consequence of Lemma 3 that, for an atomic formula $p(\mathrm{x},\mathrm{z})$,

$$\vdash REPF(\gamma_x,\gamma_z\ ;\ NN(x),NN(z)\ ;\ p^{\$}(\gamma_x,\gamma_z)) \equiv p^{\$}(NN(x),NN(z))\ .$$

For L_F-terms $t(\zeta)$, $r(\zeta)$ and an atomic formula $p(t(\zeta), r(\zeta))$ I find

$$\text{(X)}\quad \vdash REPF(\gamma\cdot\zeta\ ;\ NN\cdot\zeta\ ;\ p^{\$}(\gamma_t,\gamma_r)) \equiv p^{\$}(t^*(NN\cdot\zeta), r^*(NN\cdot\zeta))\ .$$

Because the left term by (r5) is provably equal to

$$p^{\$}(REPT(\gamma\cdot\zeta\ ;NN\cdot\zeta\ ;\ \gamma_t), REPT(\gamma\cdot\zeta\ ;\ NN\cdot\zeta\ ;\ \gamma_r))\ ,$$

and by Lemma 1 this is provably equal to

$$p^{\$}(REPT(\gamma\cdot\zeta\ ;\ NN\cdot\zeta\ ;t^*(\gamma\cdot\zeta)), REPT(\gamma\cdot\zeta\ ;\ NN\cdot\zeta\ ;r^*(\gamma\cdot\zeta)))\ .$$

which, by the statement analogous to (a) of Lemma 4, is provably equal to

$$p^{\$}(\mathrm{repf}^*(\gamma\cdot\zeta\ ;\ NN\cdot\zeta\ ;\ t^*(\gamma\cdot\zeta)), \mathrm{repf}^*(\gamma\cdot\zeta\ ;\ NN\cdot\zeta\ ;\ r^*(\gamma\cdot\zeta)))$$
$$= p^{\$}(t^*(NN\cdot\zeta), r^*(NN\cdot\zeta))\ .$$

In the following, replacement terms and internal replacements will be connected with simple usual replacements. If z is not in ζ then I denote by $\gamma\cdot\zeta,\gamma_z$ the prolongation of $\gamma\cdot\zeta$ with γ_z, and by $NN\cdot\zeta,NN(z)$ the prolongation of $NN\cdot\zeta$ with $NN(z)$.

LEMMA 5 Let z be different from x and nor occur in ζ of $t(\zeta)$.

(a) $\vdash \mathrm{rep}(z,\kappa_n\,|\,REPF(\gamma\cdot\zeta,\gamma_z\ ;\ NN\cdot\zeta,NN(z)\ ;\ p^{\$}(\gamma_t,\gamma_z)))$
$$\equiv REPF(\gamma\cdot\zeta\ ;\ NN\cdot\zeta\ ;\ p^{\$}(\gamma_t,\hat{\gamma}_n))$$

(b) $\vdash \mathrm{rep}(z,\kappa_n\,|\,REPF(\gamma_z\ ;NN(z)\ ;\ p^{\$}(\gamma_n,\gamma_z))) \equiv p^{\$}(\gamma_n,\hat{\gamma}_n)$

(c) $\vdash \mathrm{rep}(z,x\,|\,REPF(\gamma\cdot\zeta,\gamma_z\ ;\ NN\cdot\zeta,NN(z)\ ;\ p^{\$}(\gamma_t,\gamma_z)))$
$$\equiv REPF(\gamma\cdot\zeta,\gamma_x\ ;\ NN\cdot\zeta,NN(x)\ ;\ p^{\$}(\gamma_t,\gamma_x))$$

(d) $\vdash \mathrm{rep}(z,s(x)\,|\,REPF(\gamma_x,\gamma_z\ ;\ NN(x),NN(z)\ ;\ p^{\$}(\gamma_{s(x)},\gamma_z)))$
$$\equiv REPF(\gamma_x\ ;\ NN(x)\ ;\ p^{\$}(\gamma_{s(x)},\gamma_{s(x)}))\,.$$

Assume now that y occurs in ζ, and let ζ', $t'(\zeta')$ arise from ζ, $t(\zeta)$ under the homomorphism sending y into z .

(e) $\vdash \mathrm{rep}(y,z\,|\,REPF(\gamma\cdot\zeta,\gamma_z\ ;\ NN\cdot\zeta,NN(z)\ ;\ p^{\$}(\gamma_t,\gamma_z)))$
$$\equiv REPF(\gamma\cdot\zeta'\ ;\ NN\cdot\zeta'\ ;\ p^{\$}(\gamma_{t'},\gamma_z))$$

I show (a) by computing

$\vdash$ rep(z, κ_n | $REPF(\gamma\cdot\zeta, \gamma_z\ ;\ NN\cdot\zeta, NN(z)\ ;\ p^{\S}(\gamma_t, \gamma_z)))$

$\equiv$ rep(z, κ_n | $p^{\S}(t^*(NN\cdot\zeta), NN(z))$ by (X)

$= p^{\S}(t^*(NN\cdot\zeta), NN(\kappa_n))$ because t and t* contain the same variables whence z is not in t*, hence not in $t^*(NN\cdot\zeta)$

$\equiv p^{\S}(t^*(NN\cdot\zeta), \hat{\gamma}_n)$ by Lemma 2

$\equiv REPF(\gamma\cdot\zeta\ ;\ NN\cdot\zeta\ ;\ p^{\S}(\gamma_t, \hat{\gamma}_n))$ by (X) .

Clearly, (b) is a special case of (a). I show (c) by computing

$\vdash$ rep(z, x | $REPF(\gamma\cdot\zeta, \gamma_z\ ;\ NN\cdot\zeta, NN(z)\ ;\ p^{\S}(\gamma_t, \gamma_z)))$

$\equiv$ rep(z, x | $p^{\S}(t^*(NN\cdot\zeta), NN(z))$ by (X)

$= p^{\S}(t^*(NN\cdot\zeta), NN(x))$

$\equiv REPF(\gamma\cdot\zeta, \gamma_x\ ;\ NN\cdot\zeta, NN(x)\ ;\ p^{\S}(\gamma_t, \gamma_x))$ by (X) .

I show (d) by computing

$\vdash$ rep(z, s(x) | $REPF(\gamma_x, \gamma_z\ ;\ NN(x), NN(z)\ ;\ p^{\S}(\gamma_{s(x)}, \gamma_z)))$

$\equiv$ rep(z, s(x) | $p^{\S}(s^{\S}(NN(x)), NN(z))$ by (X) ,

$= p^{\S}(s^{\S}(NN(x)), NN(s(x)))$

$\equiv p^{\S}(s^{\S}(NN(x)), s^{\S}(NN(x))$ since $NN(s(x)) \equiv s^{\S}(NN(x))$

$\equiv REPF(\gamma_x\ ;\ NN(x)\ ;\ p^{\S}(\gamma_{s(x)}, \gamma_{s(x)}))$ by (X) .

I show (e) by computing

$\vdash$ rep(y, z | $REPF(\gamma\cdot\zeta, \gamma_z\ ;\ NN\cdot\zeta, NN(z)\ ;\ p^{\S}(\gamma_t, \gamma_z)))$

$\equiv$ rep(y,z | $p^{\S}(t^*(NN\cdot\zeta), NN(z))$ by (X)

$= p^{\S}(t'^*(NN\cdot\zeta'), NN(z))$

$\equiv REPF(\gamma\cdot\zeta'\ ;\ NN\cdot\zeta'\ ;\ p^{\S}(\gamma_{t'}, \gamma_z))$ by (X) .

4. Sentences Which Imply Their Provability

The provability condition (BL1), that if $\vdash$ v then $\vdash$ BEW(γ_v), followed simply from the fact that BEW(x) represents the set *bew* of all ⌜v⌝ such that v has a proof. This is a mathematical argument about formulas, numbers and relations between numbers, and so there arises the question whether a formalized version

$$v \rightarrow \text{BEW}(\gamma_v)\ .$$

of (BL1) holds for sentences v. But a semantical argument shows that, in general, it is not true, much less provable. Because I have shown in Chapter 4 that, as the axioms of PA are incomplete, if there is a model A of PA then I can find a sentence v which is true in A but is not provable from PA. While I have said so far that the arithmetization did take place "in ω", I now may assume that it has taken place with values in the model A. Then

If v is true in A and if v $\rightarrow$ BEW(γ_v) is provable, hence true in the model A, then BEW(γ_v) is true in A.

If BEW(γ_v) is true in A then BEW(x) = $\exists$y DED(x,y) implies there is an m in A such that DED(γ_v, m) is true in A.

Having a model, PA is consistent whence representability first implies formal definability and then semantical definability. Thus DED(x,y), representing the relation *ded*, also semantically defines *ded*.

Hence the truth of DED(γ_v, m) implies that $<\gamma_v, m>$ is in *ded* and so m must be the number of a proof of v.

Thus v would be provable after all, and this is a contradiction. – I recall from Chapter 4 that the sentence v there is itself of the form $\neg$BEW(κ_m) with a suitably chosen m; hence v is provably equivalent to $\forall$y DED(κ_m, y).

A sentence of the form

$\exists$y t(y) $\equiv \kappa_1$ where t(y) is a PR-term

shall be called a *proper* Σ_1-PR-*sentence*, and a sentence provably equivalent to a proper Σ_1-PR-sentence shall be called a Σ_1-PR-*sentence*. The provability predicate BEW(x) was chosen is Section 3 such that BEW(γ_w) is indeed a Σ_1-PR-sentence. So the provability condition (BL3) will follow from the

THEOREM 4 For every Σ_1-PR-sentence: $\vdash$ v $\rightarrow$ BEW(γ_v).

established first in HILBERT–BERNAYS 39. This theorem I will obtain as a special case of a more general theorem referring to formulas instead of sentences. But for formulas v the constant γ_v must be replaced by a suitable term containing the same variables as v, and here now the internal forms introduced in Section 4 will be required.

I define for sentences v the term *GG*(v) to be γ_v. If v is a formula containing free variables then let ξ be the canonical sequence of these (ordered by growing indices); I assign to v its internal form

$$GG(v) = REPF(\gamma \cdot \xi \,;\, NN \cdot \xi \,;\, \gamma_v)$$

which has the same free variables as v. I introduce the abbreviation

$$\mathrm{BB}(v) = \mathrm{BEW}(\mathit{GG}(v))$$

which is a formula containing the same free variables as v.

Formulas $v(\xi)$ provably equivalent to formulas

$$\exists y\ t(\xi,y) \equiv \kappa_1 \qquad \text{where } t(\xi,y) \text{ is a PR-term}$$

I call Σ_1-PR-*formulas*, and so Theorem 4 becomes a special case of

THEOREM 4A For every Σ_1-PR-formula: $\vdash v \rightarrow \mathrm{BB}(v)$.

Before entering its proof, I want to introduce some variants of $\mathit{GG}(v)$ and $\mathrm{BB}(v)$. While the particular choice of their definitions was determined by the wish for uniqueness, I now define, for any formula v and any sequence ζ which comprises the canonical sequence ξ (which will be empty if v is a sentence)

$$\mathit{GG}_\zeta(v) = \mathit{REPF}(\gamma\cdot\zeta\,;\,\mathit{NN}\cdot\zeta\,;\,\gamma_v)$$

$$\mathit{GGJ}_\zeta(v) = \mathit{REPF}(\gamma\cdot\zeta\,;\,\mathit{NN}\cdot\zeta\,;\,v^*(\gamma\cdot\zeta))$$

$$\mathit{GGH}_\zeta(v) = \mathrm{repf}^*(\gamma\cdot\zeta\,;\,\mathit{NN}\cdot\zeta\,;\,v^*(\gamma\cdot\zeta)) \ .$$

It follows from Lemma 4 that $\vdash \mathit{GG}_\zeta(v) \equiv \mathit{GGJ}_\zeta(v)$, $\vdash \mathit{GG}_\zeta(v) \equiv \mathit{GGH}_\zeta(v)$, and if v is a formula with variables then also $\vdash \mathit{GG}(v) \equiv \mathit{GG}_\zeta(v)$. Also, if v is a sentence then the term $v^*(\zeta)$ does not contain variables, hence $v^*(\gamma\cdot\zeta)$ does not contain any γ_x, and so $\mathrm{repf}^*(\gamma\cdot\zeta\,;\,\mathit{NN}\cdot\zeta\,;\,v^*(\gamma\cdot\zeta))$ is $v^*(\gamma\cdot\zeta)$. In that case, therefore, $\vdash \mathit{GGH}_\zeta(v) \equiv \gamma_v$ by Lemma 1 and so again $\vdash \mathit{GGH}(v) \equiv \mathit{GG}(v)$. Consequently, for any formula v, be it with variables or a sentence, the four PR-terms $\mathit{GG}(v)$, $\mathit{GG}_\zeta(v)$, $\mathit{GGJ}_\zeta(v)$, $\mathit{GGH}_\zeta(v)$ are provably equal to each other. Setting

$$\mathrm{BB}_\zeta(v) = \mathrm{BEW}(\mathit{GG}_\zeta(v))$$

$$\mathrm{BBJ}_\zeta(v) = \mathrm{BEW}(\mathit{GGJ}_\zeta(v)) \quad , \quad \mathrm{BBH}_\zeta(v) = \mathrm{BEW}(\mathit{GGH}_\zeta(v)) \ ,$$

it also follows that $\mathrm{BB}(v)$, $\mathrm{BB}_\zeta(v)$, $\mathrm{BBJ}_\zeta(v)$, $\mathrm{BBH}_\zeta(v)$ are pairwise provably equivalent. So I may use in my following considerations whichever is convenient of these four formulas.

The main part of the proof of Theorem 4A will consist of the particular case

LEMMA 6 For every PR-term t and every variable z:
if not $z\,\epsilon\,\mathrm{occ}(t)$ then

$$\vdash t \equiv z \rightarrow \mathrm{BB}(t \equiv z).$$

Once this has been secured, and if also

(c) if $\vdash v \rightarrow w$ then $\vdash BB(v) \rightarrow BB(w)$,

(h) if not $z \epsilon occ(t)$ and $\vdash t \equiv z \rightarrow BB(t \equiv z)$ then $\vdash t \equiv \kappa_n \rightarrow BB(t \equiv \kappa_n)$

have been secured, then I conclude the proof of Theorem 4A as follows. Let the Σ_1-PR-formula $v(\xi)$ be provably equivalent to $\exists y\, t(\xi,y) \equiv \kappa_1$. Then (c) implies that $BB(v)$ is provably equivalent to $BB(\exists y\, t(\xi,y) \equiv \kappa_1)$. Hence it suffices to show

(1) $\vdash \exists y\, t(\xi,y) \equiv \kappa_1 \rightarrow BB(\exists y\, t(\xi,y) \equiv \kappa_1)$.

Let z be chosen outside of occ(t); then Lemma 6 implies

$$\vdash t(\xi,y) \equiv z \rightarrow BB(t(\xi,y) \equiv z)$$

whence by (h)

$$\vdash t(\xi,y) \equiv \kappa_1 \rightarrow BB(t(\xi,y) \equiv \kappa_1) .$$

But $\vdash t(\xi,y) \equiv \kappa_1 \rightarrow \exists y\, t(\xi,y) \equiv \kappa_1$ by logic. So (c) implies

$$\vdash BB(t(\xi,y) \equiv \kappa_1) \rightarrow BB(\exists y\, t(\xi,y) \equiv \kappa_1) ,$$

whence

$$\vdash t(\xi,y) \equiv \kappa_1 \rightarrow BB(\exists y\, t(\xi,y) \equiv \kappa_1) .$$

Here y is not free in the right formula since u and BB(u) always have the same free variables. Thus I can conclude by logic upon (1).

Before attacking the proof of Lemma 6, I shall show the

LEMMA 7 (a) If $\vdash v$ then $\vdash BB(v)$,

(b) $\vdash BB(v \rightarrow w) \rightarrow (BB(v) \rightarrow BB(w))$,

(c) If $\vdash v \rightarrow w$ then $\vdash BB(v) \rightarrow BB(w)$,

(d) If $\vdash u \rightarrow (v \rightarrow w)$ then $\vdash BB(u) \rightarrow (BB(v) \rightarrow BB(w))$,

(e) $\vdash BB(r \equiv s) \rightarrow BB(s \equiv r)$,

(f) $\vdash BB(r \equiv s) \rightarrow (BB(s \equiv t) \rightarrow BB(r \equiv t))$,

(g) If $\vdash t_0 \equiv t_1$ and $\vdash t_0 \equiv z \rightarrow BB(t_0 \equiv z)$
then $\vdash t_1 \equiv z \rightarrow BB(t_1 \equiv z)$,

(h) If not $z \epsilon occ(t)$ and $\vdash t \equiv z \rightarrow BB(t \equiv z)$
then $\vdash t \equiv \kappa_n \rightarrow BB(t \equiv \kappa_n)$.

In order to prove (a), assume $\vdash v(\zeta)$ whence $\vdash BEW(\gamma_v)$ by (BL1). It follows from $\vdash VAR(\gamma_{x(i)})$ and $\vdash CTERM(NN(x_i))$ for the x_i in ζ that I use Theorem 3 (B2) to obtain $\vdash BEW(REPF(\gamma_{x(n-1)}, NN(x_{n-1}), \gamma_v))$. Repeating this n times, I arrive at

$$\vdash BEW(REPF(\gamma_{x(o)}, NN(x_o), \dots, REPF(\gamma_{x(n-1)}, NN(x_{n-1}), \gamma_v) \dots)),$$

i.e. $\vdash \mathrm{BB}_\zeta(v)$. Applying (BL1), there follows $\vdash \mathrm{BBH}_\zeta(v)$ and therefore $\vdash \mathrm{BB}(v)$. Concerning (b), observe that it follows from $\vdash \gamma(v \to w) \equiv \mathit{IMP}(\gamma_v, \gamma_w)$ and the definition of *REPF* that

$$\vdash \mathit{REPF}(\gamma\cdot\zeta\,;\mathit{NN}\cdot\zeta\,;\gamma(v\to w))$$
$$\equiv \mathit{IMP}(\mathit{REPF}(\gamma\cdot\zeta\,;\mathit{NN}\cdot\zeta\,;\gamma_v), \mathit{REPF}(\gamma\cdot\zeta\,;\mathit{NN}\cdot\zeta\,;\gamma_w)$$

whence

$$\vdash \mathit{GG}_\zeta(v \to w) \equiv \mathit{IMP}(\mathit{GG}_\zeta(v), \mathit{GG}_\zeta(w))\ .$$

Thus (B2) implies (b). Next, (c) follows from (a) and (b), and (d) follows from (c) and (b). Further, (e) follows from (c) applied to the equality axiom $r \equiv s \to s \equiv r$, and in the same manner (f) follows from (d). As for (g) observe that $\vdash t_0 \equiv t_1$ implies $\vdash t_0 \equiv z \to t_1 \equiv z$ whence $\vdash \mathrm{BB}(t_0 \equiv z) \to \mathrm{BB}(t_1 \equiv z)$ by (c); now the second hypothesis in (g) implies $\vdash t_0 \equiv z \to \mathrm{BB}(t_1 \equiv z)$, and then $\vdash t_1 \equiv z \to t_0 \equiv z$ implies the conclusion of (g).

In order to prove (h), it will suffice to show that $\mathrm{rep}(z,\kappa_n \mid -)$ maps $\mathrm{BB}(t \equiv z)$ into a formula provably equivalent to $\mathrm{BB}(t \equiv \kappa_n)$. To begin with, I may assume that $\mathrm{BEW}(x)$ does not contain variables from $t \equiv z$ – otherwise I could transform it into a formula $\mathrm{BEW}'(y)$ with this property such that, for any term h, $\mathrm{BEW}(h)$ and $\mathrm{BEW}'(h)$ are provably equivalent. Now $\mathrm{BBJ}_\zeta(t \equiv z) = \mathrm{sub}(x, \mathit{GGJ}_\zeta(t \equiv z) \mid \mathrm{BEW}(x)) = \mathrm{rep}(x, \mathit{GGJ}_\zeta(t \equiv z) \mid \mathrm{BEW}(x))$ and its image under $\mathrm{rep}(z,\kappa_n \mid -)$ is

$$\text{(x)}\quad \mathrm{rep}(z,\kappa_n \mid \mathrm{rep}(x, \mathit{GGJ}_\zeta(t \equiv z) \mid \mathrm{BEW}(x)))$$
$$= \mathrm{rep}(x, \mathrm{rep}(z,\kappa_n \mid \mathit{GGJ}_\zeta(t \equiv z)) \mid \mathrm{BEW}(x))\ .$$

Here the replacement term on the right side is

$$\mathrm{rep}(z,\kappa_n \mid \mathit{GGJ}_\zeta(t \equiv z)) = \mathrm{rep}(z,\kappa_n \mid \mathit{REPF}(\gamma\cdot\xi\,;\mathit{NN}\cdot\xi\,;\equiv^{s}(\gamma_t, \gamma_z)))\ ,$$

and by Lemma 5(a) this is provably equal to

$$\mathit{GGJ}_\zeta(t \equiv \kappa_n) = \mathit{REPF}(\gamma\cdot\xi\,;\mathit{NN}\cdot\xi\,;\equiv^{s}(\gamma_t, \hat{\gamma}_n))\ .$$

Hence $\vdash \mathrm{rep}(z,\kappa_n \mid \mathit{GGJ}_\zeta(t \equiv z)) \equiv \mathit{GGJ}_\zeta(t \equiv \kappa_n)$, and so by equality logic the right side of (x) is provably equivalent to $\mathrm{rep}(x, \mathit{GGJ}_\zeta(t \equiv \kappa_n) \mid \mathrm{BEW}(x)))$ which is $\mathrm{BBJ}_\zeta(t \equiv \kappa_n)$. This concludes the proof of the Lemma.

It remains to prove Lemma 6. If t is one of the PA–terms κ_n, x and $s(x)$ then it will suffice to show

$$\text{(2)}\quad \vdash t \equiv z \to (\mathit{GG}(t \equiv t) \equiv \mathit{GG}(t \equiv z))\ .$$

Because then $\vdash t \equiv z \to (\mathrm{BB}(t \equiv t) \to \mathrm{BB}(t \equiv z))$ by equality logic, further $\vdash \mathrm{BB}(t \equiv t)$ by Lemma 7(a), and so Lemma 6 follows propositionally. In order see (2), equality logic shows that it suffices prove

$$\vdash \mathrm{rep}(z,t \mid \mathit{GG}(t \equiv z)) \equiv \mathit{GG}(t \equiv t)\ .$$

Taking again GGJ_ζ in place of GG, for $t = \kappa_n$ this is Lemma 5(a), for $t = z$

it is trivial, for $t = x$ and $x \neq z$ it is Lemma 5(c). For $t = s(z)$ it holds by ex falso quodlibet, and for $t = s(x)$ with $x \neq z$ it is Lemma 5(d).

Since $\vdash P_i^k(x_0, \ldots, x_{k-1}) \equiv x_i$ and $C_0(x) \equiv \kappa_0$ by PA-induction, it follows by equality logic that Lemma 6 also holds for the PR-terms $P_i^k(x_0, \ldots, x_{k-1})$ and $C_0(x)$.

While so far I could afford to talk loosely about PR-terms, now that I want to prove Lemma 6 for all of them, I need a more precise definition. Corresponding to the generation of primitive recursive functions, I define as initial PR-terms the terms x, $s(x)$, $P_i^k(x_0, \ldots, x_{k-1})$, $C_0(x)$; I define as PR-terms the initial ones, further all terms which are provably equal to a superposition of PR-terms, and finally all terms $f^{k+1}(\xi, x)$ for which there are PR-terms $g^k(\xi)$, $r^{k+2}(\xi, x, y)$ and the recursion equations for $f^{k+1}(\xi, x)$ from $g^k(\xi)$, $r^{k+2}(\xi, x, y)$ are provable. In particular, the latter is the case for the PA-terms $x+y$ and $x \cdot y$, and thus also they are PR-terms.

I shall say that a PR-term t is *good* if Lemma 6 holds for t and for every variable z not in occ(t). It follows from Lemma 7(g) that a term is good if it is provably equal to good term. I now observe that, if $t(\zeta)$ is good and if $t'(\zeta')$ arises by renaming the free variables in ζ into those of in ζ', then also $t'(\zeta')$ is good.

For a proof, it suffices to consider the case that only one variable y in ζ is changed into a variable u. Again I may assume that BEW(x) does not contain z or variables from ζ and ζ'. Thus $\mathrm{rep}(y,u \mid BBJ_\zeta(t \equiv z))$ is

$$(y) \quad \mathrm{rep}(y,u \mid \mathrm{rep}(x, GGJ_\zeta(t \equiv z) \mid BEW(x))) = \mathrm{rep}(x, \mathrm{rep}(y,u \mid GGJ_\zeta(t \equiv z)) \mid BEW(x)) .$$

For the replacement term on the right, Lemma 5(e) shows

$$\vdash \mathrm{rep}(y,u \mid GGJ_\zeta(t \equiv z)) \equiv GG_{\zeta'}(t' \equiv z) ,$$

So by equality logic $\mathrm{rep}(y,u \mid BBJ_\zeta(t \equiv z))$ becomes provably equivalent to $BBJ_{\zeta'}(t' \equiv z)$. Consequently, the map $\mathrm{rep}(y,u \mid -)$ transforms

$$(z_0) \quad \vdash t(\zeta) \equiv z \rightarrow BBJ_\zeta(t(\zeta) \equiv z) \qquad \text{into}$$

$$(z_1) \quad \vdash t'(\zeta') \equiv z \rightarrow BBJ_{\zeta'}(t'(\zeta') \equiv z) .$$

Hence Lemma 6 holds for $t'(\zeta')$ and z if it did hold for $t(\zeta)$ and z , i.e. if z not in ζ. If z' is a variable not in ζ' but occurs in ζ then I start from (z_0), with a variable z neither in ζ nor in ζ', and proceed to (z_1). As z is not in ζ', Lemma 5(c) shows

$$\vdash \mathrm{rep}(z,z' \mid GG_{\zeta'}(t' \equiv z)) \equiv GG_{\zeta'}(t' \equiv z'))$$

whence $\mathrm{rep}(z,z' \mid -)$ transforms (z_1) into

$$(z_2) \quad \vdash t'(\zeta') \equiv z' \rightarrow BBJ_{\zeta'}(t'(\zeta') \equiv z') .$$

It now remains to show the

LEMMA 8 Every PR-term t is good .

The proof is by induction on the generation of PR-terms. It follows from what was shown above that the initial PR-terms are good. Assume now that $t(\xi)$ is good and that χ is a sequence of good terms; I want to show that $t(\chi)$ is good. Let z be the variable (not occurring in $t(\chi)$) for which I wish to prove Lemma 6. If $t'(\xi')$ arises from $t(\xi)$ by renaming the variables ξ then also $t'(\xi')$ is good, and there still holds $t(\chi) = t'(\chi)$; I now choose a renaming such that the variables in ξ' are different from z and do not occur in the terms of χ. This being done, I write again $t(\xi)$ for $t'(\xi')$. There now holds

(g_0) $\vdash t(\xi) \equiv z \to BB(t(\xi) \equiv z)$ since $t(\xi)$ is good

(g_1) $\vdash \chi_i \equiv \xi_i \to BB(\chi_i \equiv \xi_i)$ for $i<k$ since χ_i is good .

The equality axioms imply

$$\vdash \bigwedge <\chi_i \equiv \xi_i \mid i<k> \to (t(\chi) \equiv z \to t(\xi) \equiv z) \quad \text{whence}$$

(g_2) $\vdash \bigwedge <\chi_i \equiv \xi_i \mid i<k> \to (t(\chi) \equiv z \to BB(t(\xi) \equiv z))$

by (g_0). Analogously, the equality axioms imply

$$\vdash \bigwedge <\chi_i \equiv \xi_i \mid i<k> \to (t(\xi) \equiv z \to t(\chi) \equiv z) \quad \text{whence}$$

$$\vdash \bigwedge <BB(\chi_i \equiv \xi_i) \mid i<k> \to (BB(t(\xi) \equiv z) \to BB(t(\chi) \equiv z))$$

by Lemma 7(d) when the conjunction is resolved into iterated implications bracketed to the right. Thus by (g_1)

$$\vdash \bigwedge <\chi_i \equiv \xi_i \mid i<k> \to (BB(t(\xi) \equiv z) \to BB(t(\chi) \equiv z)) \quad \text{whence}$$

$$\vdash \bigwedge <\chi_i \equiv \xi_i \mid i<k> \to (t(\chi) \equiv z \to BB(t(\chi) \equiv z)) \quad \text{by } (g_2) .$$

This I rewrite as

$$\vdash (\chi_0 \equiv \xi_0 \to (\chi_1 \equiv \xi_1 \to \dots \to (t(\chi) \equiv z \to BB(t(\chi) \equiv z)) \dots)) \quad \text{whence}$$

$$\vdash \exists \xi_0 (\chi_0 \equiv \xi_0) \to (\chi_1 \equiv \xi_1 \to \dots \to (t(\chi) \equiv z \to BB(t(\chi) \equiv z)) \dots)$$

since the assumptions made on ξ secure that ξ_0 does not occur behind the first implication sign. Thus $\vdash \exists \xi_0 (\chi_0 \equiv \xi_0)$ implies

$$\vdash (\chi_1 \equiv \xi_1 \to \dots \to (t(\chi) \equiv z \to BB(t(\chi) \equiv z)) \dots) ,$$

and after (k-1) analogous steps I arrive at $\vdash t(\chi) \equiv z \to BB(t(\chi) \equiv z)$.

There remains the case of a term $f(\xi,x)$ defined by primitive recursion from good terms $g(\xi)$ and $r(\xi,x,y)$:

$$(f_0) \qquad f(\xi,\kappa_0) \equiv g(\xi) \,,$$

$$(f_1) \qquad f(\xi, s(x)) \equiv r(\xi, x, f(\xi, x)) \,,$$

where x is not in ξ; in the following I shall often write sx in place of $s(x)$. So I have to prove the formula

$$B(\xi,x,z) : f(\xi,x) \equiv z \rightarrow BB(f(\xi,x) \equiv z)$$

for z not in ξ and different from x, and by logic this provably equivalent to

$$A(\xi,x) : \forall z\, f(\xi,x) \equiv z \rightarrow BB(f(\xi,x) \equiv z) \,.$$

This formula I shall prove by PA–induction, though for $A(\xi,\kappa_0)$ I return to $B(\xi,\kappa_0,z)$:

$\vdash f(\xi,\kappa_0) \equiv z \rightarrow g(\xi) \equiv z$	by (f_0),
$\vdash g(\xi) \equiv z \rightarrow BB(g(\xi) \equiv z)$	as $g(\xi)$ is good
$\vdash f(\xi,\kappa_0) \equiv z \rightarrow BB(g(\xi) \equiv z)$	propositionally
$\vdash BB(f(\xi,\kappa_0) \equiv g(\xi))$ from (f_0)	by Lemma 7(a)
$\vdash BB(g(\xi) \equiv z) \rightarrow BB(f(\xi,\kappa_0) \equiv z)$	by Lemma 7(f) ,

and now lines 3 and 5 imply $\vdash B(\xi,\kappa_0,z)$.

Before turning to the induction scheme

$$A(\xi,x) \rightarrow A(\xi, sx)$$

I choose a new variable y and observe that Lemma 5(c) gives both

$$\vdash \operatorname{rep}(z,y \mid GG_{\zeta}{}'(f(\xi,x) \equiv z)) \equiv GG_{\zeta}{}'(f(\xi,x) \equiv y))$$

$$\vdash \operatorname{rep}(y,z \mid GG_{\zeta}{}'(f(\xi,x) \equiv y)) \equiv GG_{\zeta}{}'(f(\xi,x) \equiv z)) \,.$$

Hence $\operatorname{rep}(z,y \mid BB(f(\xi,x) \equiv z)) = BB(f(\xi,x) \equiv y)$ and analogously for z,y replaced by y,z . Consequently, $BB(f(\xi,x) \equiv z)$ and $BB(f(\xi,x) \equiv y)$ are provably equivalent whence $A(\xi,x)$ becomes provably equivalent to

$$\forall z\, f(\xi,x) \equiv z \rightarrow BB(f(\xi,x) \equiv y)$$

and to

$$A_y(\xi,x) : \forall y\, f(\xi,x) \equiv y \rightarrow BB(f(\xi,x) \equiv y) \,.$$

So the induction schema for $A(\xi,x)$ will be proven if I can show

$$(f_2) \qquad \vdash A_y(\xi,x) \rightarrow A(\xi, sx) \,.$$

In doing so, I shall use the abbreviations

$V(x)$ for $f(\xi,x) \equiv z$, $\qquad V_y(x)$ for $f(\xi,x) \equiv y$,

$C(x)$ for $BB(V(x))$, $\qquad C_y(x)$ for $BB(V_y(x))$,

D for $BB(r(\xi,x,y) \equiv z)$,

and write *pp* as abbreviation for 'propositionally'. I then reason as follows:

a. $\vdash f(\xi,x) \equiv y \to (f(\xi, s(x)) \equiv z \to r(\xi,x,y) \equiv z)$ by (f_1)

b. $\vdash f(\xi,x) \equiv y \to (r(\xi,x,y) \equiv z \to f(\xi, s(x)) \equiv z)$ by (f_1)

c. $\vdash r(\xi,x,y) \equiv z \to BB(r(\xi,x,y) \equiv z)$ as $r(\xi,x,y)$ is good

d. $\vdash V_y(x) \to (V(sx) \to D)$ by a, c

e. $\vdash C_y(x) \to (D \to C(sx))$ by b, Lemma 7(d)

f. $\vdash C_y(x) \to ((V_y(x) \to (V(sx) \to D)) \to (V_y(x) \to (V(sx) \to C(sx))))$ *pp*

g. $\vdash C_y(x) \to (V_y(x) \to (V(sx) \to C(sx)))$ by d,f

h. $\vdash V_y(x) \to (C_y(x) \to B(\xi, sx, z))$ by g and the definition of B

i. $\vdash (V_y(x) \to C_y(x)) \to (V_y(x) \to B(\xi,sx,z)))$ *pp* from h

j. $\vdash A_y(\xi,x) \to (V_y(x) \to C_y(x))$ quantifier axiom

k. $\vdash A_y(\xi,x) \to (V_y(x) \to B(\xi, sx, z))$ from i, j

l. $\vdash V_y(x) \to (A_y(\xi,x) \to B(\xi, sx, z))$ *pp* from k

m. $\vdash \exists y\, V_y(x) \to (A_y(\xi,x) \to B(\xi, sx, z))$ quantifier rule

n. $\vdash \exists y\, f(\xi,x) \equiv y$ equality logic

o. $\vdash A_y(\xi,x) \to B(\xi, sx, z)$ from m, n

p. $\vdash A_y(\xi,x) \to \forall z(B(\xi, sx, z))$ quantifier rule

and this is (f_2).

This completes the proof of Lemma 8, hence of Lemma 6, hence of Theorem 4A, hence of Theorem 4, and therefore the proof of Theorem 1.

Epilogue

The name 'mathematical logic' specifies the substantive 'logic' by the adjective 'mathematical'. In usual understanding this will be read as *logic presented with mathematical methods*. Yet there is a second meaning, namely that of *logic as used in mathematics*.

The mathematician, at first sight, may claim that he uses no other kind of logic than does his colleague in the law faculty. Yet as an illustration, let me recall his use of logic in the lemma that every sequence α of real numbers contains a monotonic subsequence, as taught in the beginner's course of analysis. The mathematician's logic says that

either	for every n : there is an i>n : for every j>i : $\alpha(i) \geq \alpha(j)$
or	there is an n : for every i>n : there is a j>i : $\alpha(i) < \alpha(j)$.

In the first case, he defines a sequence β by $\beta(0) = 0$ and $\beta(m+1)$ the smallest i above $\beta(m)$; then the sequence $\alpha \cdot \beta$ descends monotonically. In the second case he defines β by $\beta(0) = n+1$ and $\beta(m+1)$ the smallest j above $\beta(m)$; then the sequence $\alpha \cdot \beta$ ascends monotonically. The alternative between the first statement and its classically formed negation clearly presumes perfect information about the infinitely many numbers $\alpha(0)$, $\alpha(1)$, $\alpha(2)$, ... If this presumption were not made, there should remain only the non-classical provabilities described in Chapter 2.9.6-7 . - The lawyer, on the other hand, argues about members of a fixed, finite domain - the inhabitants of the earth at most - where it is not completely unreasonable to assume perfect information . And his quantifiers may be replaced by finite conjunctions or disjunctions.

So it is the logic used in mathematics for which the use of quantifiers becomes essential. It was the analysis of mathematical arguments which lead FREGE 79, p.21, to the insight that, when speaking about infinite domains, the task to deduce $M \Longrightarrow \forall x w$

(1) is carried out *schematically* (or uniformly) by the use of *variables*, and

(2) requires necessarily the consideration of the *language* used to argue *about* mathematical contents.

I have explained (1) already in connection with the eigenvariable condition (EV) during the discussion of the rules

$$(I\forall)\ \frac{M \Longrightarrow v(y)}{M \Longrightarrow \forall x\ w}\ (EV) \qquad\qquad (E\exists)\ \frac{v(y),\ M \Longrightarrow u}{\exists x\ w,\ M \Longrightarrow u}\ (EV)$$

in Book 2, p.203. As for (2), FREGE l.c. wrote explicitly that y is a *latin letter* (which does not occur in M nor in w except at the place indicated);

Frege did not use the word "variable" and had "a" printed where here stands "y". Thus for Frege y is *not* an object of the mathematics dealt with in the statements under consideration, but an object of the *language* in which such statements are *expressed.*

In the special case of arithmetic, the same observations (1) and (2) must be made for deductions $M \Longrightarrow \forall x w$ with help of the induction rule (IRA) which, rewritten for a sequent calculus, becomes

$$(\mathrm{IRA}_s) \quad \frac{M \Longrightarrow w(\kappa_0) \qquad M \Longrightarrow w(y) \to w(s(y))}{M \Longrightarrow \forall x w(x)}$$

with y satisfying the eigenvariable condition that it does not occur free in the conclusion. Again, the hypothesis $M \Longrightarrow w(y) \to w(s(y))$ contains y as a free variable, and this in general will mean that it has been derived by *free variable arguments* set up for y – for instance those used in school to conclude from $\Sigma\langle i^2 \mid i \leq y\rangle = (^1/_6)\cdot y\cdot(y+1)\cdot(2y+1)$ upon $\Sigma\langle i^2 \mid i \leq y+1\rangle = (^1/_6)\cdot(y+1)\cdot(y+2)\cdot(2(y+1)+1)$.

In this way then, instances of critical quantifier rules or of induction rely on *schematical* proofs of their hypotheses. This may, at first sight, be viewed as a strenghtening of the premisses of these rules, hence as only a *weak form* of the rules themselves – though it is not at all clear what would be supposed to take the place of this condition on uniformly asserted hypotheses. Of course, we always may conceive of beings, angels say or demons, able to wield more powerful tools such as ω–rules in order to make accessible to them *the full truth.*

Still, we are aware that in the semantical treatment of truth and satisfaction no such uniformities are mentioned: a valuation φ into an infinite structure is defined to satisfy $\forall x w$ if each of the variants φ' of φ (i.e. $\varphi' =_x \varphi$) satisfies w without any montion of uniformity, though I cannot possibly perform a test of each and every single one of these infinitely many variants. Yet the attentive reader may be tempted to recall that Theorem 2.10.1, relying on the semantical studies of Book 1, implied the *completeness* result: the sequent $C \Longrightarrow \forall x w$ is (classically) derivable if, and only if, $\forall x w$ is a cs–consequence of C, i.e. if every valuation, in every structure, satisfies $\forall x w$ if it satisfies C. Thus schematical, finitary provability already suffices to *always* secure infinitary semantical consequence. Again, at first sight this may appear as astonishing, but closer inspection of the completeness result, as carried out in Chapter 1.1, makes it clear that its (mathematical) proof itself made use of perfect information about the infinite set of deductions. The completeness result recaptures semantical verifications as finitary schematical deductions, but only at the price of indirect and infinitary arguments in its very own proof (taking place at the same level as the arguments about verifications themselves, but one level above that of the formal deductions).

Recapitulating: quantifier and induction rules with eigenvariables appear as places where the *seam* connecting mathematics and language becomes openly visible. It is here that my freely conceived imaginations are confronted with the sober facts of language; I am forced to acknowledge that my argumentation depends on linguistic facilities because I must present finite proofs also for concepts involving the infinite.

As for the schematical character of mathematical induction, namely the necessity to prove the induction hypothesis schematically, it is absent if the terminology about sets is used to describe the natural numbers as the *smallest set* N containing 0 and *closed* under the successor map s , where s is injective and 0 not a value under s ; the induction principle then reads

for every M : if $0 \epsilon M$ and $s(M) \subseteq M$ then $N \subseteq M$.

In its application, the verification of $s(M) \subseteq M$ is likely to use a schematical argument again, but the set formulation disregards the means by which such proof is carried out.

It is this, perhaps, one of the reasons behind the success of DEDEKIND's set language which *objectifies* linguistic notions (from Kronecker's indeterminates to properties expressed by formulas), replacing them with sets as mathematical objects. Linguistic relationships about provabilities then acquire the appearance of mathematical truths about set theoretical objects, enabling the mathematician to disregard the connection between the concepts he speaks about and the language he employs to do so.

HISTORICAL SUPPLEMENT. While it is not necessary for the analysis leading to (1) and (2) to have available a symbolic language such as FREGE's, it certainly was conductive to develop these insights. The radical novelty of FREGE's ideas, as well as his use of a 2-dimensional notation for proof figures, left them unnoticed for two decades; it required PEANO 's 91 linear notation to have RUSSELL 03 make them slowly gain recognition. In the meantime, MITCHELL 83 and PEIRCE 85 had found it suitable to introduce quantifying operators (writing them as Π and Σ) which they treated as generalized conjunctions and disjunctions without seeing the need for *rules* governing their use in proofs. In RUSSELL 08 and WHITEHEAD-RUSSELL 10 the quantifier rules are stated and used, yet the continuous interweaving of syntactical symbolism and its semantical background may leave the reader in doubt whether the deductions presented are painstakingly written protocols of mental activities, or are formal transformations of strings of characters. That deductions are viewed as sequences of character strings, generated under the rules from axioms, became explicit with HILBERT 22. It was this aspect which later lead to speak of the possibility to verify *mechanically* whether a sequence of strings be a deduction.

That induction can be expressed with help of the notion of the *smallest* set closed under the successor map, as mentioned above, was carried out by

DEDEKIND 88 ; yet it is fully present already in FREGE 79 (cf. his footnote on p.64) and FREGE 84 , p.92-96, only that Frege never speaks about sets but about extensions of concepts. [However, the hard work filling the pages of Dedekind and Frege concerns something quite different, namely the definition of the order relation, and the proof of its properties, from the weak assumptions made in the definition; FREGE called the technique he developed for this purpose his *Reihenlehre*, while DEDEKIND named his very similar approach that of a *Kettenlehre*.] The article of PEANO 89 gave rise to the name of *Peano's axioms*, although Peano did nothing but rewrite Dedekind's definition in a symbolic manner - and stated the recursive definitions of the arithmetic operations as axioms without mentioning that DEDEKIND 88 had given the justification for such definitions in his theorem on recursion (shown here in Chapter 1.0). But also the name of *Peano arithmetic*, accepted in this Book, is historically midleading in that Peano stayed with Dedekind's set theoretical induction principle stated above and did not use the induction schema (ISA), introduced e.g. in HILBERT 22 as Axiom 9 .

References

R. Dedekind: Was sind und was sollen die Zahlen. Braunschweig 1888

G. Frege: Begriffsschrift, eine der arithmetischen nachgebildete Formelsprache des reinen Denkens. Halle 1879

G. Frege: Die Grundlagen der Arithmetik, eine logisch-mathematische Untersuchung über den Begriff der Zahl. Breslau 1884

D. Hilbert: Neubegründung der Mathematik. Erste Mitteilung. Abh. aus dem Math.Semin.der Hamburg Univ. 1 (1922) 157–177

O.H. Mitchell: On a new algebra of logic. In: Studies in Logic: By Members of the Johns Hopkins University. Boston 1883 , pp. 22–106

G. Peano: Arithmetices principia, nova methodo exposita. Torino 1889

G. Peano: Notations de logique mathématique; introduction au formulaire de mathématique. Torino 1894

C.S.Peirce: On the algebra of Logic; a Contribution to the Philosophy of Notation. American.J.of Math. 7 (1885) 180–202

B.A. Russell: Principles of Mathematics. I . Cambridge 1903

B.A. Russell: Mathematical Logic as based on the Theory of Types. American.J.of Math. 30 (1908) 222–262 .

A.N.Whitehead, B.A.Russell: Principia Mathematica, vol. I , Cambridge 1910

Index of concepts and names

Index of symbolic notations

(arithmetizations)

Milton Keynes UK
Ingram Content Group UK Ltd.
UKHW020022071024
449327UK00032B/2884

9 780367 398576